ENVIRONMENTAL SCIENCE, ENGINEERING AND TECHNOLOGY

# SOURCES AND REDUCTION OF GREENHOUSE GAS EMISSIONS

# ENVIRONMENTAL SCIENCE, ENGINEERING AND TECHNOLOGY

Additional books in this series can be found on Nova's website under the Series tab.

Additional E-books in this series can be found on Nova's website under the E-books tab.

ENVIRONMENTAL SCIENCE, ENGINEERING AND TECHNOLOGY

# SOURCES AND REDUCTION OF GREENHOUSE GAS EMISSIONS

STEFFEN D. SALDANA
EDITOR

**Nova Science Publishers, Inc.**
*New York*

Copyright © 2010 by Nova Science Publishers, Inc.

**All rights reserved.** No part of this book may be reproduced, stored in a retrieval system or transmitted in any form or by any means: electronic, electrostatic, magnetic, tape, mechanical photocopying, recording or otherwise without the written permission of the Publisher.

For permission to use material from this book please contact us:
Telephone 631-231-7269; Fax 631-231-8175
Web Site: http://www.novapublishers.com

## NOTICE TO THE READER

The Publisher has taken reasonable care in the preparation of this book, but makes no expressed or implied warranty of any kind and assumes no responsibility for any errors or omissions. No liability is assumed for incidental or consequential damages in connection with or arising out of information contained in this book. The Publisher shall not be liable for any special, consequential, or exemplary damages resulting, in whole or in part, from the readers' use of, or reliance upon, this material. Any parts of this book based on government reports are so indicated and copyright is claimed for those parts to the extent applicable to compilations of such works.

Independent verification should be sought for any data, advice or recommendations contained in this book. In addition, no responsibility is assumed by the publisher for any injury and/or damage to persons or property arising from any methods, products, instructions, ideas or otherwise contained in this publication.

This publication is designed to provide accurate and authoritative information with regard to the subject matter covered herein. It is sold with the clear understanding that the Publisher is not engaged in rendering legal or any other professional services. If legal or any other expert assistance is required, the services of a competent person should be sought. FROM A DECLARATION OF PARTICIPANTS JOINTLY ADOPTED BY A COMMITTEE OF THE AMERICAN BAR ASSOCIATION AND A COMMITTEE OF PUBLISHERS.

Additional color graphics may be available in the e-book version of this book.

LIBRARY OF CONGRESS CATALOGING-IN-PUBLICATION DATA
Sources and reduction of greenhouse gas emissions / editor, Steffen D. Saldana.
p. cm.
Includes index.
ISBN 978-1-61668-856-1 (hardcover)
1. Greenhouse gas mitigation--Government policy. 2. Greenhouse gas mitigation--Economic aspects. I. Saldana, Steffen D.
TD885.5.G73S678 2010
363.738'74--dc22
2010014114

*Published by Nova Science Publishers, Inc.* ✦ *New York*

# CONTENTS

| | | |
|---|---|---|
| **Preface** | | **vii** |
| **Chapter 1** | Methane Capture: Options for Greenhouse Gas Emission Reduction<br>*Kelsi Bracmort, Jonathan L. Ramseur, James E. McCarthy,*<br>*Peter Folger and Donald J. Marples* | **1** |
| **Chapter 2** | Nitrous Oxide from Agricultural Sources: Potential Role<br>in Greenhouse Gas Emission Reduction and Ozone Recovery<br>*Kelsi Bracmort* | **21** |
| **Chapter 3** | An Overview of Greenhouse Gas (GHG) Control Policies<br>in Various Countries<br>*Jane A. Leggett, Richard K. Lattanzio, Carl Ek and Larry Parker* | **31** |
| **Chapter 4** | Anaerobic Digestion: Greenhouse Gas Emission Reduction<br>and Energy Generation<br>*Kelsi Bracmort* | **75** |
| **Chapter 5** | Farm-Based Anaerobic Digesters as an Energy and Odor Control<br>Technology: Background and Policy Issues<br>*William F. Lazarus* | **89** |
| **Chapter 6** | The Costs of Reducing Greenhouse-Gas Emissions<br>*Congressional Budget Office* | **119** |
| **Chapter 7** | Emissions of Greenhouse Gases in the United States 2008<br>*United States Energy Information Administration* | **137** |
| **Chapter 8** | The Use of Agricultural Offsets to Reduce Greenhouse Gases<br>*Joseph Kile* | **233** |
| **Chapter Sources** | | **247** |
| **Index** | | **249** |

# PREFACE

Human activities around the world are producing increasingly large quantities of greenhouse gases, particularly carbon dioxide ($CO_2$), resulting from the consumption of fossil fuels and deforestation. Most experts expect that the accumulation of such gases in the atmosphere will result in a variety of environmental changes over time, including a gradual warming of the global climate, extensive changes in regional weather patterns, and significant shifts in the chemistry of the oceans. Although the magnitude and consequences of such developments are highly uncertain, researchers generally conclude that a continued increase in atmospheric concentrations of greenhouse gases would have serious and costly effects. This book explores a comprehensive response to this problem.

Chapter 1 - Research on climate change has identified a wide array of sources that emit greenhouse gases (GHGs). Among the six gases that have generally been the primary focus of concern, methane is the second-most abundant, accounting for approximately 8% of total U.S. GHG emissions in 2007. Methane is emitted from a number of sources. The most significant are agriculture (both animal digestive systems and manure management); landfills; oil and gas production, refining, and distribution; and coal mining.

As Congress considers legislation to address climate change by capping or reducing GHG emissions, methane capture projects offer an array of possible reduction opportunities, many of which utilize proven technologies. Methane capture projects (e.g., landfill gas projects, anaerobic digestion systems) restrict the release of methane into the atmosphere. The methane captured can be used for energy or flared. Methane capture challenges differ depending on the source. Most methane capture technologies face obstacles to implementation, including marginal economics in many cases, restricted pipeline access, and various legal issues.

Some of the leading methane capture options under discussion include market-based emission control programs, carbon offsets, emission performance standards, and maintaining existing programs and incentives. At present, methane capture technologies are supported by tax incentives in some cases, by research and demonstration programs in others, by regulation in the case of the largest landfills, and by voluntary programs. Congress could decide to address methane capture in a number of different ways, including (1) determining the role of methane capture in climate change legislation; (2) determining whether methane capture should be addressed on an industry-by-industry basis; and (3) determining if current methane capture initiatives will be further advanced with legislative action regardless of other facets of the climate change policy debate. What role methane capture would play in prospective legislation to control GHGs—whether methane sources would be included among those covered by a cap-and-trade system, for example, whether they would be a source of emission offsets from sources not covered by cap-and-trade, or whether their emissions might be subject to regulation—is among the issues that Congress faces.

A few government programs have supported the capture of methane to mitigate climate change. The Methane-to-Markets Partnership, administered by the Environmental Protection Agency (EPA), is an international initiative to reduce global methane emissions. EPA also oversees a variety of voluntary programs related to the Methane-to-Markets initiative (e.g., Coalbed Methane Outreach Program, Natural Gas STAR Program, Landfill Methane Outreach Program, AgSTAR Program).

This chapter discusses legislative alternatives for addressing methane capture, sources of methane, opportunities and challenges for methane capture, and current federal programs that support methane recovery.

Chapter 2 - Gases other than carbon dioxide accounted for nearly 15% of U.S. greenhouse gas emissions in 2007, yet there has been minimal discussion of these other greenhouse gases in climate and energy legislative initiatives. Reducing emissions from non-carbon dioxide greenhouse gases, such as nitrous oxide ($N_2O$), could deliver short-term climate change mitigation results as part of a comprehensive policy approach to combat climate change.

Nitrous oxide is 298 times more potent than carbon dioxide in its ability to affect climate change; and moreover, results of a recent scientific study indicate that nitrous oxide is currently the leading ozone-depleting substance. Thus, legislation to restrict nitrous oxide emissions could contribute to both climate change protection and ozone recovery.

The primary human source of nitrous oxide is agricultural soil management, which accounted for two-thirds of the $N_2O$ emissions reported in 2007 (approximately 208 million metric tons $CO_2$ equivalent). One proposed strategy to lower $N_2O$ emissions is more efficient application of synthetic fertilizers. However, further analysis is needed to determine the economic feasibility of this approach as well as techniques to measure and monitor the adoption rate and impact of $N_2O$ emission reduction practices for agricultural soil management.

As Congress considers legislation that would limit greenhouse gas emissions (both H.R. 2454 and S. 1733 would require that greenhouse gas emissions be reduced by 83% in 2050), among the issues being discussed is how to address emissions of non-$CO_2$ greenhouse gases. Whether such emissions should be subject to direct regulation, what role EPA should play using its existing Clean Air Act authority, whether the sources of $N_2O$ should be included among the covered entities of a cap-and-trade system, whether $N_2O$ reductions should be considered offsets to be purchased by the covered entities of a cap-and-trade system, and what role USDA should play in any $N_2O$ reduction scheme are among the issues being discussed. How these issues are resolved will have important implications for agriculture, which has taken a keen interest in climate change legislation.

Chapter 3 - As Congress considers legislation to address climate change, and follows negotiations toward a new international agreement to reduce greenhouse gas (GHG) emissions, the question of the comparability of actions across countries frequently arises. Concerns are raised about what the appropriate sharing of efforts should be among countries, as well as the potential trade implications if countries undertake different levels of GHG reductions and, therefore, incur varying cost impacts on trade-sensitive sectors. This chapter summarizes the GHG control policies in effect or under consideration in the European Union

(EU) and various other large countries, and offers a brief set of initial observations. It gives particular emphasis to how particular trade-sensitive sectors may be treated in the context of each national program.

All countries examined have in place, or are developing, some enforceable policies that serve to reduce GHG emissions. Most are at some stage of making their programs more stringent. The wealthiest countries have all taken on GHG limitation or reduction targets under the Kyoto Protocol. Some of the emerging economies have voluntarily stated GHG targets, though none have yet accepted legally binding obligations in an international agreement. The forms of targets, and their stringencies, vary widely across countries.

The scope of specific GHGs and economic sectors covered by national (or sub-national) reduction measures is generally, but not completely, similar. All have policies that affect carbon dioxide emissions; most have some measures that cover the additional five gases covered under the Kyoto Protocol (methane, nitrous oxide, sulfur hexafluoride, perfluorocarbons, and hydrofluorocarbons).

The programs and measures used vary across countries. Even when some measures have similar names (e.g., voluntary programs and voluntary action plans), the measures may differ in important ways that may influence their effectiveness and impacts on trade competiveness. Within sectors of a country, emission rates and control requirements may vary widely. A country may have some facilities with emission rates (or energy intensities) comparable to the best globally, even if the country's sectoral average as a whole has, for example, a significantly higher energy intensity than the global average.

This chapter presents an overview of GHG control policies within individual countries. It does not present a rigorous assessment of the comparability of GHG control policies across countries or within specific sectors. The criteria for assessing comparability internationally are not widely agreed, and could encompass a range of considerations, not all quantitatively measurable.

This chapter summarizes the greenhouse gas (GHG) control policies in effect or under consideration in a number of large countries, and offers a brief set of initial observations. This overview allows preliminary comparison across countries. Because of congressional interest in the comparability of countries' actions, and in the potential trade ramifications of differential policies, these country fact sheets give emphasis to how particular trade-sensitive sectors may be treated in the context of each national program. Where specific industries are not listed in a country's fact sheet, no further information was found.

The European Union's policies are presented first, followed by any additional rules or policies under consideration in several of the largest EU Member States (i.e., France, Germany, the United Kingdom). A number of additional large-emitting countries follow in alphabetical order. Finally, the Appendix provides a comparison of early 2009 vehicle efficiency standards across countries, which may be a useful reference for a sector that emits a large portion of global GHG emissions.

Chapter 4 - Anaerobic digestion technology may help to address two congressional concerns that have some measure of interdependence: development of clean energy sources and reduction of greenhouse gas emissions. Anaerobic digestion technology breaks down a feedstock—usually manure from livestock operations—to produce a variety of outputs including methane. An anaerobic digestion system may reduce greenhouse gas emissions because it captures the methane from manure that might otherwise be released into the atmosphere as a potent greenhouse gas. The technology may contribute to the development of clean energy because the captured methane can be used as an energy source to produce heat or generate electricity.

Anaerobic digestion technology has been implemented sparingly, with 125 anaerobic digestion systems operating nationwide. Some barriers to adoption include high capital costs, questions about reliability, and varying payment rates for the electricity generated by anaerobic digestion systems. Two sources of federal financial assistance that may make the technology more attractive are the Section 9007 Rural Energy for America Program of the Food, Conservation, and Energy Act of 2008 (2008 farm bill, P.L. 110-246), and the Renewable Electricity Production Tax Credit (26 U.S.C. §45).

Congress could decide to encourage development and use of the technology by (1) identifying the primary technology benefit, so as to determine whether it should be pursued in the framework of greenhouse gas emission reduction or clean energy development; (2) determining if the captured methane will count as a carbon offset; and (3) considering additional financing options for the technology.

This chapter provides information on anaerobic digestion systems, technology adoption, challenges to widespread implementation, and policy interventions that could affect adoption of the technology.

Chapter 5 - This chapter summarizes the existing literature and analytical perspectives on farm-based digesters, highlights major efforts in the United States and Europe to expand digester usage, and discusses key policy issues affecting digester economics. The study was largely a review of the "gray literature" on digesters, and it serves as a snapshot overview of the industry. Digesters are fairly capital-intensive when viewed primarily as an energy source. On a strictly market basis, current U.S. average electricity prices do not appear to provide sufficient economic justification for digesters to move beyond a fairly limited niche. Digesters make the most sense today where the odor and nutrient management benefits are important, or where the electricity or heat has a higher-than-average value. Digester biogas is mainly methane, which is destroyed when flared or used for electricity. This methane destruction is beneficial in terms of climate change. The associated carbon credits may become a more significant farm revenue source in the future.

Chapter 6 - Human activities around the world are producing increasingly large quantities of greenhouse gases, particularly carbon dioxide ($CO_2$) resulting from the consumption of fossil fuels and deforestation. Most experts expect that the accumulation of such gases in the atmosphere will result in a variety of environmental changes over time, including a gradual warming of the global climate, extensive changes in regional weather patterns, and significant shifts in the chemistry of the oceans. Although the magnitude and consequences of such developments are highly uncertain, researchers generally conclude that a continued increase in atmospheric concentrations of greenhouse gases would have serious and costly effects.

A comprehensive response to that problem would include a collection of strategies: research to better understand the scientific processes at work and to develop technologies to address them; measures to help the economy and society adapt to the projected warming and other expected changes; and efforts to reduce emissions, averting at least some of the potential damage to the environment and attendant economic losses. Those strategies would all present technological challenges and entail economic costs.

Reducing emissions would impose a burden on the economy because it would require lessening the use of fossil fuels and altering patterns of land use. This issue brief discusses the economic costs of reducing greenhouse-gas emissions in the United States, describing the main determinants of costs, how analysts estimate those costs, and the magnitude of estimated costs. The brief also illustrates the uncertainty surrounding such estimates using studies of a recent legislative proposal, H.R. 2454, the American Clean Energy and Security Act of 2009.

Chapter 7 - This chapter, the seventeenth annual report, presents the Energy Information Administration's latest estimates of emissions for carbon dioxide, methane, nitrous oxide, and other greenhouse gases. Documentation for these estimates is available online at www.eia.doe.gov/oiaf/ggrpt.

Chapter 8 - This chapter is edited and excerpted testimony by Joseph Kile before the Subcommittee on Conservation, Credit, Energy, and Research, Committee on Agriculture on December 3, 2009.

In: Sources and Reduction of Greenhouse Gas Emissions
Editor: Steffen D. Saldana

ISBN: 978-1-61668-856-1
© 2010 Nova Science Publishers, Inc.

*Chapter 1*

# METHANE CAPTURE: OPTIONS FOR GREENHOUSE GAS EMISSION REDUCTION

## *Kelsi Bracmort, Jonathan L. Ramseur, James E. McCarthy, Peter Folger and Donald J. Marples*

### ABSTRACT

Research on climate change has identified a wide array of sources that emit greenhouse gases (GHGs). Among the six gases that have generally been the primary focus of concern, methane is the second-most abundant, accounting for approximately 8% of total U.S. GHG emissions in 2007. Methane is emitted from a number of sources. The most significant are agriculture (both animal digestive systems and manure management); landfills; oil and gas production, refining, and distribution; and coal mining.

As Congress considers legislation to address climate change by capping or reducing GHG emissions, methane capture projects offer an array of possible reduction opportunities, many of which utilize proven technologies. Methane capture projects (e.g., landfill gas projects, anaerobic digestion systems) restrict the release of methane into the atmosphere. The methane captured can be used for energy or flared. Methane capture challenges differ depending on the source. Most methane capture technologies face obstacles to implementation, including marginal economics in many cases, restricted pipeline access, and various legal issues.

Some of the leading methane capture options under discussion include market-based emission control programs, carbon offsets, emission performance standards, and maintaining existing programs and incentives. At present, methane capture technologies are supported by tax incentives in some cases, by research and demonstration programs in others, by regulation in the case of the largest landfills, and by voluntary programs. Congress could decide to address methane capture in a number of different ways, including (1) determining the role of methane capture in climate change legislation; (2) determining whether methane capture should be addressed on an industry-by-industry basis; and (3) determining if current methane capture initiatives will be further advanced with legislative action regardless of other facets of the climate change policy debate. What role methane capture would play in prospective legislation to control GHGs—whether methane sources would be included among those covered by a cap-and-trade system, for example, whether they would be a source of emission offsets from sources not covered by cap-and-trade, or whether their emissions might be subject to regulation—is among the issues that Congress faces.

A few government programs have supported the capture of methane to mitigate climate change. The Methane-to-Markets Partnership, administered by the Environmental Protection Agency (EPA), is an international initiative to reduce global methane emissions. EPA also oversees a variety of voluntary programs related to the Methane-to-Markets initiative (e.g., Coalbed Methane Outreach Program, Natural Gas STAR Program, Landfill Methane Outreach Program, AgSTAR Program).

This chapter discusses legislative alternatives for addressing methane capture, sources of methane, opportunities and challenges for methane capture, and current federal programs that support methane recovery.


# INTRODUCTION

In the climate change policy debate, methane capture projects have garnered attention for their ability to mitigate greenhouse gas emissions. Methane capture projects prevent the release of methane, a potent greenhouse gas, into the atmosphere. The captured methane is generally flared or used for energy purposes.[1] The U.S. Environmental Protection Agency (EPA) has identified four sources of methane with the greatest potential for capture in the near term: landfills, coal mines, agriculture, and oil and gas systems. The amount of methane captured from each will depend on legislative developments, economics, technology, and outreach.

Methane ($CH_4$) constituted approximately 8% of U.S. greenhouse gas emissions in 2007.[2] Anthropogenic (human-related) sources of methane in the United States include enteric fermentation,[3] landfills, natural gas systems, coal mines, and manure management. Efforts to reduce emissions of methane—the second-most important greenhouse gas after carbon dioxide ($CO_2$)—could play a significant role in climate change mitigation.

This chapter will discuss the policy options for addressing methane capture (and their implications), legislative proposals for methane capture, domestic and international sources of methane, opportunities and challenges for methane capture, and federal programs that support methane capture.


# POLICY OPTIONS FOR ADDRESSING METHANE CAPTURE

Congress may employ multiple strategies to encourage or require methane capture as part of climate change legislation: market-based approaches, such as a cap-and-trade program or emissions fees; carbon offsets or credits as a complementary design element of a market-based approach; emission performance standards; and/or maintaining existing programs and incentives.[4] Policymakers may consider using different strategies for different methane emission sources. These strategies and related issues are discussed below.

## Market-Based Emission Control Programs

One option for policymakers is to include methane emission sources as covered entities in a market-based greenhouse gas (GHG) emission control program. Market-based mechanisms that limit GHG emissions can be divided into two types: those that focus on quantity control (e.g., a cap-and-trade program) and those that focus on price control (e.g., emissions fees, often called a carbon tax). Although each approach has its own set of advantages and disadvantages,[5] both would place a price on methane emissions from covered sources. To the extent that they are able, covered entities (those subject to the cap or fee) would likely pass the emissions price through to consumers. For example, if solid waste landfills were subject to a cap or fee based on methane emissions, the landfill operators would likely raise the price of waste disposal to account for the new cost of emissions.

**Table 1. Selected Sources of U.S. Methane Emissions and Potential Number of Entities Subject to Emission Control Program**

| Methane Emission Source | Percentage of U.S. GHG Emissions (2006 data) | Potential Applications | |
|---|---|---|---|
| | | Entity | Number |
| $CH_4$ from livestock (enteric fermentation) | 1.8 | Cattle operations[a] | 967,440 |
| $CH_4$ from landfills | 1.8 | Landfills[b] | 1,800 |
| $CH_4$ from natural gas systems | 1.5 | Natural gas processors | 530 |
| $CH_4$ from coal mines | 0.8 | Active coal mines[c] | 1,374 |
| $CH_4$ from manure management | 0.6 | Cattle operations; Swine operations[d] | 967,440 65,640 |

Source: CRS analysis of data from USDA and EPA.

a. U.S. Department of Agriculture, *Farms, Land in Farms, and Livestock Operations: 2007 Summary* (2008).

b. EPA, *Inventory of U.S. Greenhouse Gas Emissions and Sinks: 1990-2006* (April 2008), citing BioCycle, *15th Annual BioCycle Nationwide Survey: The State of Garbage in America* (2006).

c. Methane from underground mines, which accounts for about 61% of coal mine methane, is removed through ventilation systems for safety reasons. These emissions would be easier to monitor under an emission control program than aboveground coal mine methane emissions. Number of active coal mines from Energy Information Administration (EIA), "Coal Production and Number of Mines by State and Mine Type," at http://www.eia.gov.

d. U.S. Department of Agriculture, *Farms, Land in Farms, and Livestock Operations: 2007 Summary* (2008). Other animals—chickens, horses, and sheep—contribute approximately 10% of the total emissions from manure (EPA, *Inventory of U.S. Greenhouse Gas Emissions and Sinks: 1990-2006* (April 2008), table 6-6).

Recent cap-and-trade and carbon tax proposals[6] have generally not applied to methane emissions from the primary sources of such emissions. A main argument for excluding some of these groups concerns the administrative costs of covering them under an emissions program. As Table 1 indicates, the number of methane emission sources is relatively large compared to their total contribution to U.S. GHG emissions. This is particularly the case for methane emissions from the agriculture sector.

Although an even larger number of sources (e.g., industries, automobiles, buildings) generate $CO_2$ emissions, the vast majority of $CO_2$ emissions can be addressed by subjecting a relatively small number of entities to an emissions cap. This opportunity exists for $CO_2$ emissions, because policymakers could apply the emissions cap *upstream* of the actual emissions, typically where the emission inputs are produced or enter the U.S. economy.[7] Under this approach, policymakers could address $CO_2$ emissions from fossil fuel combustion and non-energy uses—in aggregate 82% of U.S. GHG emissions—by covering fewer than 2,500 entities.[8] For most methane sources, particularly in the agriculture sector, an analogous opportunity does not exist.

In addition, some of the source categories identified in Table 1 may be more amenable to emissions coverage than others. For example, roughly 25% of the methane emissions from natural gas systems comes from field production,[9] which may be impractical to monitor and measure accurately.[10] The remaining 75% primarily involves accidental releases sometimes referred to as fugitive emissions.[11] Landfill methane may offer fewer challenges in terms of measurement, but the largest landfills are already reducing methane emissions pursuant to landfill gas reduction requirements established by the Clean Air Act (42 U.S.C. 7401 *et seq.*).[12]

## Carbon Offsets

Policymakers could encourage methane capture activities by allowing methane abatement as an eligible offset project or as an emission (or tax) credit in a GHG emission control program, such as a cap-and–trade system or carbon tax. A carbon offset is a measurable reduction, avoidance, or sequestration of GHG emissions from an emission source not covered by a cap-and-trade system. Most of the recent cap-and-trade proposals have allowed offsets (under varying conditions) as a compliance alternative.[13]

Offsets would likely make an emissions program more cost-effective by (1) providing an incentive for non-regulated sources to generate emission reductions and (2) expanding emission compliance opportunities for regulated entities. The main concern with offset projects is whether or not they represent real emission reductions. For offsets to be real, a ton of $CO_2$-equivalent emissions reduced from an offset project should equate to a ton emitted from a capped source, such as a smokestack or exhaust pipe, and would not have occurred without the regulatory incentive. This objective presents challenges because some offset projects are difficult to measure.

However, some methane capture projects, such as those from landfills or coal mines, are generally considered to be of higher quality (more credible) than other offset types. These projects are relatively easy to measure and verify, and in many cases would likely not occur if not for the financing provided by an offset market. Therefore, the challenge of proving "additionality" is easier to overcome.[14]

The advantage some methane capture projects have over other GHG mitigation activities may spur policymakers to control these methane releases directly (via some of the options discussed), instead of encouraging abatement through an offset market. Moreover, allowing certain activities as offsets, while imposing emission controls or caps on others, may raise issues of fairness. For example, why should specific GHG emission sources, such as electricity generators, be capped while other sources, such as landfill or animal feedlot methane, have the potential to generate financial gain for owners and/or operators through the offset market?

## Emission Performance Standards

Another option for policymakers is to require emission performance standards for particular methane emission sources. This approach has historically represented the core of U.S. federal air pollution policy. New legislation would not be required to pursue the standards approach. The ability to limit methane emissions already exists under various Clean Air Act authorities that Congress has enacted, a point underlined by the Supreme Court in an April 2007 decision, *Massachusetts v. EPA*. Although the current EPA Administrator has stated a preference for controlling GHG emissions through new legislation, the agency has begun to take actions that could lead to GHG emission performance standards from particular sources.[15]

Pursuant to Clean Air Act authority, EPA would achieve emission reductions by setting emission performance standards on each source of pollution, or requiring that sources use a particular type of technology, such as the "best available control technology." Although emission performance standards have proven to be effective through decades of experience, source-by-source regulation often cannot achieve, by itself, a desired emission reduction target at the least collective cost. Moreover, performance standards can be difficult to adjust as circumstances (e.g., technologies) change. On the other hand, they may be less expensive where measurement, administrative, or transaction costs are high relative to emission control costs. This approach may be a practical option for certain specific sources of methane emissions.

## Maintain Existing Programs/Incentives

As discussed later in this chapter, the federal government currently supports several programs that stimulate methane capture. In addition to these initiatives, which are generally voluntary in nature, since 1996 the Clean Air Act has imposed air emission standards on large solid waste landfills. However, as discussed below, the vast majority of landfills are not covered under the Clean Air Act, and there is room to increase the amount of methane captured from solid waste landfills. Moreover, the primary objective of these standards is to reduce the hazardous air pollutants and non-methane organic compounds contained in landfill gas, not to reduce methane emissions for climate-related reasons. Regardless, as mentioned above, the existing Clean Air Act authorities could be used to address a wider universe of methane sources, for the express purpose of controlling GHG emissions.

Because methane can be used as an energy source, the existing marketplace provides some incentive to capture methane for this purpose. If a GHG emission control program were enacted, such a program would increase this incentive by raising the price of traditional high-carbon energy sources (e.g., coal) relative to captured methane. The strength of the incentive would depend on the stringency of the enacted emission control program.

# Table 2. Selected Legislation Proposed in the 111th Congress Relevant to Methane

| Bill (Short Title) | General Purpose | Comments |
|---|---|---|
| **A. Bills to Capture Methane** | | |
| H.R. 1158 (Biogas Production Incentive Act of 2009) | To promote biogas produ-ction, and for other purposes. | The biogas must contain at least 52% *methane*. |
| H.R. 3202 (Water Protection and Reinvestment Act of 2009) | To establish a Water Prote-ction and Reinvestment Fund to support investments in clean water and drinking water infrastructure, and for other purposes. | Considers the installation of small renewable energy generators for methane capture as an eligible activity for climate change adaptation and mitigation grants. |
| H.R. 1342 (Landfill Greenhouse Gas Reduction Act) | To amend the Solid Waste Disposal Act to provide for the reduction of greenhouse gases, and for other purposes. | Allows for the collection of a fee on solid waste received by a solid waste landfill and requires the local government to use the revenues generated by such fees for entities to undertake approved GHG reduction projects within its jurisdiction (e.g landfill gas recovery projects). |
| H.R. 3534 (Consolidated Land, Energy, and Aquatic Resources Act of 2009) | To provide greater efficienc-ies, transparency, returns, and accountability in the administration of Federal mineral and energy resources by consolidating administra-tion of various Federal energy minerals management and leasing programs into one entity to be known as the Office of Federal Energy and Minerals Leasing of the Department of the Interior, and for other purposes. | Any coal lease issued on lands for which the United States owns bo-th the coal and gas resources shall include a requirement that the les-see recover the coal mine methane associated with the leased coal re-sources to the maximum feasible extent, taking into account the economics of both the mining and methane capture operations. |
| H.R. 2454 (American Clean Energy and Security Act of 2009) | The four titles of the legisl-ation cover clean energy, energy efficiency, reducing global warming pollution, transitioning to a clean energy economy, and adaption to climate change. | Coal mine *methane* used to gener-ate electricity at or near the mine mouth is considered a qualifying energy resource (e.g., source of usable energy). The carbon dioxide-equivalent of 1 ton of *methane* is 25 metric tons. Sec. 732, "Establishment of an Offsets Program," may include reenhouse gas reductions achieved |
| | | through the destruction of *methane* and its conversion to carbon dioxide. Under Title VIII, "Additional Greenhouse Gas Standards," the Administrator shall include in the inventory each source category that is responsible for at least 10% of the uncapped *methane* emissions in 2005. The inventory required by this section shall not include sources of enteric fermentation.[a] |

| B. Other Bills Concerning Methane | | |
|---|---|---|
| H.R. 1426 | To amend the Clean Air Act to prohibit the issuance of permits under Title V of that act for certain emissions from agricultural production. | No permit shall be issued under a permit program under this title for any carbon dioxide, nitrogen ox-ide, water vapor, or *methane* emi-ssions resulting from biological processes associated with lives-tock production. |
| H.R. 2996 (Departm-ent of the Interior, Environment, and Related Agencies Appropriations Act, 2010) | Making appropriations for the Department of the Interior, environment, and related agencies for the fiscal year ending September 30, 2010, and for other purposes. | Section 420 of the Senate comm.-ittee-reported bill prohibited funds in the bill and other acts from being used to promulgate or implement any regulation require-ing the issuance of permits under Title V of the Clean Air Act for carbon dioxide, nitrous oxide, water vapor, or *methane* emiss-ions resulting from biological processes associated with livestock production. |
| H.R. 3505 (American Energy Production and Price Reduction Act) | To increase the supply of American made energy, reduce energy costs to the American taxpayer, provide a long-term energy framework to reduce dependence on foreign oil, tap into Amer-ican sources of energy, and reduce the size of the Federal deficit. | Definition of Air Pollutant- Section 302(g) of the Clean Air Act (42 U.S.C. 7602(g)) is amen-ded by adding the following at the end thereof: `The term `air pollut-ant' shall not include carbon dioxide, water vapor, *methane*, nitrous oxide, hydrofluorocar-bons, perfluorocarbons, or sulfur hexafluoride. |
| S. 719 (Surface Estate Owner Notification Act) | To direct the Secretary of the Interior to notify surface estate owners in cases in which the leasing of federal | Lease may mean a lease that provides for development of oil and gas resources (including coalbed *methane*) owned by the United States. |
| | minerals underlying the land are to be used for oil and gas development. | |
| S. 1462 (American Clean Energy Leadership Act of 2009) | To promote clean energy technology development, enhanced energy efficiency, improved energy security, and energy innovation and workforce development, and for other purposes. | Contains amendments to the Methane Hydrate Research and Development Act of 2000 (P.L. 106-193). |

Source: Prepared by CRS.
a. Inventory refers to the annual tracking of greenhouse gas emissions and removals from various sources.

# LEGISLATIVE PROPOSALS CONCERNING METHANE CAPTURE

Members of the 111th Congress have introduced more than 40 bills related to methane emissions. One group of bills would specify methane as a greenhouse gas, promote biogas production, support landfill gas recovery projects, and address or promote methane capture.[16] Another set of bills not related to methane capture would, among other provisions, for example, prohibit permit issuance under the Clean Air Act for methane emissions from biological processes associated with livestock operations, or expand methane hydrate research. [17] Table 2 provides a summary of selected legislation pertaining to methane capture and methane in general, and shows the range of objectives addressed.

H.R. 2454, which passed the House on June 26, 2009,[18] contains numerous energy provisions, including a GHG emission cap-and-trade system. If enacted, the cap-and-trade program may allow some methane capture activities to generate offsets. However, some methane sources may be subject to emission performance standards. One enacted piece of legislation (P.L. 111-5, the American Recovery and Reinvestment Act of 2009) extended and expanded existing incentives for open-loop biomass and landfill gas electricity production and created a new incentive for the same activities.

# METHANE: A PRIMER

Methane—a colorless, odorless gas with the molecular formula $CH_4$—is produced by "methanogenic" bacteria that decompose organic matter in the absence of oxygen. Sometimes referred to as "marsh gas," methane is flammable, can cause suffocation, and can be explosive in low concentrations in air. It is the primary component (70%-90%) of natural gas fuel. Roughly 24% of total U.S. energy consumed in 2008 was natural gas. Consumption is spread across a wide array of economic sectors, with electric power generation and industrial consumption accounting for 28% of total consumption; residential use, 21%; and commercial use, 13%.[19]

## Global Warming Potential

Global warming potential (GWP) is an estimate of how much a greenhouse gas affects climate change over a quantity of time relative to $CO_2$, which has a GWP value of 1. Methane is a potent greenhouse gas with a global warming potential of 25.[20] Over a 100-year timeframe, methane is 25 times more effective than $CO_2$ at trapping heat in the atmosphere. In other words, it takes 25 tons of $CO_2$ to equal the effect of 1 ton of CH4. Methane has a relatively short atmospheric lifetime (approximately 12 years) when compared to the atmospheric lifetime of carbon dioxide; thus efforts to capture methane from anthropogenic sources provide more near-term climate change abatement than capturing or reducing comparable amounts of $CO_2$, but less multi-decadal abatement.

Once methane or other greenhouse gases are converted, using GWP or other methods, they can be expressed in a common unit of measurement: carbon dioxide-equivalent ($CO_2$-eq. or $CO_2$e). $CO_2$e both takes into account the potency of each gas and expresses the quantity of the gas. Carbon dioxide-equivalent has been adopted as a principal unit of measurement to aggregate or make comparisons across greenhouse gases. $CO_2$e expresses the tons of a greenhouse gas in the equivalent effect of tons of $CO_2$ on climate change (more specifically, on "radiative forcing").[21] Once all gases are converted to $CO_2$e, they can be compared or added together.

# Sources of Methane

## *Domestic*

The top three anthropogenic sources of the roughly 585 million metric tons $CO_2e$ of methane emitted in 2007 were enteric fermentation, landfills, and natural gas systems.[22] These three sources combined were responsible for about 64% of total U.S. methane emissions (see Figure 1). There are also natural sources of methane emissions, such as wetlands, and releases of natural gas from geologic formations. Natural sources of methane are generally assumed to account for 30% of an annual methane emissions inventory that includes natural and anthropogenic sources.[23]

## *International*

Methane accounted for nearly 17% of global greenhouse gas emissions in 2005.[24] Asia is reported as having emitted the most methane on a regional basis. China, India, the United States, the European Union, and Brazil are the top five methane-emitting countries (see Table 3). The agriculture sector is the leading source of methane emissions for the world (see Appendix).[25]

One analysis of global average atmospheric concentrations for methane indicates that, while growth leveled off for approximately a 15-year period beginning in the early 1990s, methane concentrations may have begun to increase again in 2007, possibly due to warmer temperatures in the Arctic and increased precipitation in the tropics.[26] Global methane emissions from natural sources are estimated at approximately 225 million metric tons of methane per year.[27]

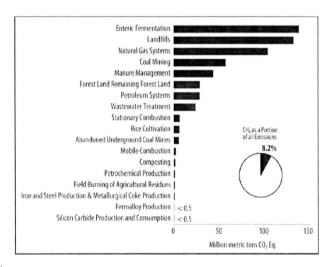

Source: U.S. EPA.
Note: The "forest land remaining forest land" category contains forest land that stays forest land based on IPCC guidance for defining inventory categories. Methane emissions from the category "forest land remaining forest land" are attributed to wildfires and prescribed fires on managed forest land.

Figure 1. 2007 U.S. Sources of Anthropogenic Methane Emissions
Source: Climate Analysis Indicators Tool (CAIT) Version 6.0 (Washington, DC: World Resources Institute, 2009).
Notes: Excludes land use change.

**Table 3. Top Five Methane-Emitting Countries in 2005**

| Country | Million MT (Tg) $CO_2e$ | % of World Total |
|---|---|---|
| China | 853 | 13 |
| India | 548 | 9 |
| United States | 521 | 8 |
| European Union | 449 | 7 |
| Brazil | 389 | 6 |

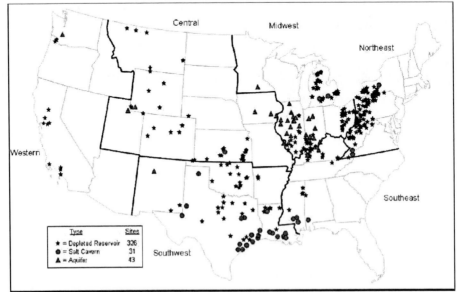

Source: Energy Information Administration, Office of Oil & Gas, Natural Gas Division, Gas Transportation Information System, December 2008.
Notes: There are no natural gas storage facilities in Alaska or Hawaii.

Figure 2. U.S. Underground Natural Gas Storage Facilities, Close of 2007

## Methane Use and Storage

Methane may be captured in its pure form or as a component of biogas, depending on the source.[28] The methane captured can be "flared" (combusted without commercial purpose) or used to generate heat or electricity. Flaring the gas destroys the methane and yields carbon dioxide ($CO_2$) and water.[29] The release of carbon dioxide as a result of flaring is less risky in terms of climate forcing than releasing the methane or biogas as is into the atmosphere.

Captured methane is stored chiefly underground as a constituent of natural gas. Underground storage options include depleted gas or oil fields, aquifers, or salt cavern formations[30] (see Figure 2). A less common option is the storage of natural gas in liquid form. Liquefied natural gas (LNG) is roughly one six-hundredth the volume of gaseous natural gas, allowing for transport by ship to areas that are inaccessible via a natural gas pipeline.[31]

# OPPORTUNITIES AND CHALLENGES FOR METHANE CAPTURE

Capturing methane from various sectors of the U.S. economy requires different strategies because some strategies may be more economically feasible for specific emission sources or locations. Policy laid out in forthcoming climate change proposals may further provide technical and economic incentives to overcome barriers—past and present—to methane capture. The following section summarizes opportunities and challenges for methane capture from the top four sources of methane: agriculture, landfills, oil and natural gas systems, and coalbed methane (see Table 4).

## Agriculture

Methane emissions from the U.S. agriculture sector are mostly attributable to enteric fermentation and manure management, the largest and fifth-largest sources of methane emissions in 2007, respectively.[32] Ruminant animals (e.g., cattle, sheep) are the major emitters of methane via enteric fermentation, a non-point source of methane emissions. The amount of methane emitted from enteric fermentation depends on the feed quality and amount of feed ingested by the animal. Options to reduce methane emissions from enteric fermentation include improved animal productivity and feed management.[33]

Some manure management systems (e.g., storage of liquid or slurry manure in a waste storage structure) are a point-source of methane emissions. Methane released from the anaerobic decomposition of manure depends mainly on the storage temperature, storage time, and manure composition. Methane emissions from some manure management systems may be captured with an anaerobic digestion system (AD system) that flares the gas or uses it for energy purposes.[34] Barriers to methane capture from manure management include limited technology and information exchange between agricultural producers and the technology transfer community, high up-front capital costs for AD systems, unsatisfactory technology reliability, and low rates paid by some utilities for the electricity generated.

## Landfill Gas

Landfills were the second-largest U.S. source of methane emissions in 2007.[35] Landfill gas—a mixture of roughly 50% methane and 50% carbon dioxide, but including small amounts of other gases—is released into the atmosphere if not captured. The amount of gas produced at any given landfill depends on the amount of organic material in the waste, the landfill's design, the climate at the site of the landfill, and the operating practices used by the site's operator. In general, large amounts of organic waste and high levels of moisture in a landfill lead to greater gas production.

Landfill gas is captured at the nation's largest landfills.[36] A 1996 Clean Air Act regulation known as the "Landfill Gas Rule" established New Source Performance Standards and Guidelines that require landfills with a 2.5 million metric ton design capacity that accepted waste after November 8, 1987, to capture and burn the gas. The gas can either be flared or used for energy production— often it is used as fuel for electricity generation. As mentioned above, flaring is less damaging to the atmosphere than release of the methane.

**Table 4. U.S. Methane Emissions by Source (million metric tons $CO_2e$)**

| Source | 2000 | 2005 | 2006 | 2007 |
|---|---|---|---|---|
| Agriculture—Enteric Fermentation | 134.4 | 136.0 | 138.2 | 139.0 |
| Landfills | 122.3 | 127.8 | 130.4 | 132.9 |
| Natural Gas Systems | 130.8 | 106.3 | 104.8 | 104.7 |
| Coal Mining | 60.5 | 57.1 | 58.4 | 57.6 |
| Agriculture—Manure Management | 34.5 | 41.8 | 41.9 | 44.0 |

Source: U.S. EPA, *U.S. Emissions Inventory 2009: Inventory of U.S. Greenhouse Gas Emissions and Sinks: 1990-2007*. See http://epa.gov/climatechange/emissions/usinventoryreport.html.

In promulgating the 1996 rule, EPA said that the 2.5 million metric ton minimum "corresponds to cities greater than 100,000 people." The agency also stated that the regulations "will only affect less than 5 percent of all landfills" but would reduce emissions of methane by 37% at new landfills, and by 39% at existing facilities.

In fact, partly as a result of tax incentives and voluntary programs, landfill gas capture projects are in operation at approximately 480 landfill sites as of December 2008.[37] This represents roughly 27% of the 1,754 municipal solid waste landfills reported in operation in 2007.[38]

Whatever success existing regulations, tax incentives, and voluntary programs may be having, a significant amount of methane continues to be emitted even at landfills subject to the Landfill Gas Rule. In addition, there are few methane capture projects at smaller landfills and at landfills that ceased operation before November 1987 (those not covered under the Clean Air Act). The latter group, numbering in the tens of thousands of sites, poses a particular challenge. Often, there is no responsible party who might implement a methane collection system if the site's original owner is no longer in business. At other sites (e.g., sites owned by local governments), there may be no continuing stream of revenue to support installation and operation of the necessary equipment, since the landfill has closed. Further barriers to additional landfill gas capture may include high capital costs for equipment, low rates paid for the gas captured and/or electricity generated, permitting requirements, and liability concerns.[39]

## Oil and Natural Gas

Natural gas systems were the third-largest U.S. source of methane emissions in 2007. Methane can be released from natural gas systems during normal operations, maintenance, and unexpected system disorder. An array of technologies and suggested strategies to reduce methane emissions from various stages of natural gas system production is available.[40]

Additionally, methane is emitted during oil production, transportation, and refining. Options to reduce methane emissions from the oil sector include flaring, direct use, and reinjection of methane into oil fields. Offshore oil operations (oil platforms) tend to use captured methane directly because flaring is economically unattractive. Onshore oil operations usually inject the captured methane into a pipeline. Captured methane can also be injected into an oil production field to enhance future oil recovery. One analysis estimated the reduction efficiency (which is the percentage reduction achieved with adoption of a mitigation option) for flaring, direct use, and reinjection of methane to be 98%, 90%, and 95%, respectively.[41] The equipment used for abatement has a technical lifetime of 15 years.[42] Barriers to methane capture from oil and natural gas systems include federal and state economic regulations, financial constraints, abatement technology cost, and abatement technology availability.

## Coalbed Methane

The coal mining sector was the fourth-largest source of U.S. methane emissions in 2007.[43] Most methane emissions from coal mining occur during the mining process in underground mining operations. The amount of methane released depends chiefly on the coal mine type (e.g., underground mine, surface mine, abandoned mine) and the mining operation type. Two techniques are available to capture methane emissions from coal mines: degasification (including enhanced degasification) and ventilation air methane systems.

A degasification system facilitates the removal of methane gas from a mine by ventilation and/or by drainage. Methane is captured through a series of vertical wells, horizontal boreholes, or gob wells drilled into the mine before or after mining operations.[44] A sizeable portion of the methane captured from degasification systems can be injected into a pipeline directly for energy purposes. Enhanced degasification uses the same approach as degasification systems, but has the capacity to extract lower-quality methane that must be cleaned and upgraded to meet "pipeline quality" gas criteria. Ventilation air methane (VAM) systems flush air into underground mines to keep methane concentration levels at or below 1%. VAM systems are necessary to provide safe working environments for miners because methane can be explosive in low concentrations in air. Methane captured from degasification systems has a higher methane concentration (30%-90%) than methane captured from ventilation air systems.

Methane captured from coal mines using the methods described above can be used to generate electricity on-site or for sale to utility companies. Of the estimated 9,294 coal mines (active underground, active surface, and abandoned underground) in the United States, about 580 are currently active underground coal mines, of which 50 have methane capture projects.[45] Barriers to methane capture from coal mines include legal issues, economic circumstances (e.g., high capital costs for equipment, low electricity prices), restricted pipeline capacity for transporting coalbed methane from the mines to natural gas markets, and difficulties with technology development. A primary barrier to methane recovery from coal mines is uncertainty regarding coalbed methane ownership, which exists in part because coalbed methane is located in the same stratum as the coal reserves, making a clear distinction for ownership difficult.[46] Older leases may not clearly specify whether the owner of the coal rights is also the owner of the coalbed methane. Ownership may lie with the owner(s) of the coal rights, owner(s) of the oil and gas rights, or surface owner(s). Ownership may also be an issue for federal lands in the West because developers of federally owned coalbed methane must apply for a gas lease to implement a coal mine methane project via competitive leasing procedures open to all.

## Concerns Applicable to All Sources

Two impediments to methane capture cross-cut the top four anthropogenic sources of methane emissions: pipeline capacity, and the price offered by the electric power industry for electricity generated by captured methane. In addition to capacity, another issue is pipeline access for those wanting to purchase captured methane but not immediately adjacent to the methane capture source. In addition to price, other electricity industry issues of concern are competitiveness and the sale of excess power generated from captured methane.

## FEDERAL SUPPORT FOR METHANE CAPTURE

Periodic reports to Congress from the executive branch, as well as hearing testimony, have conveyed the significance of methane capture since the early 1990s.[47] Congress and the executive branch have supported methane capture projects through voluntary programs, energy management programs, and research and development programs. This section highlights existing efforts.

## Methane-to-Markets Partnership

The Methane-to-Markets Partnership is an international initiative for methane capture and reuse from four sources: oil and gas, coal mines, landfills, and agriculture.[48] The partnership is administered by the U.S. Environmental Protection Agency (EPA), which supports the voluntary efforts of the 29 country partners. National governments, research institutions, and the private sector have collaborated since 2004 to develop cost-effective, near-term methane capture projects globally. The partnership receives its legal authority from the Clean Air Act, Section 103 (42 U.S.C. § 7403), and the National Environmental Policy Act (NEPA, 42 U.S.C. §§ 4321-4347). Approximately $4.5 million was appropriated to the partnership for FY2009. Supplemental funding for the partnership is received from the U.S. Department of State. Other U.S. government partners—the Department of Energy, the Department of Agriculture, the Agency for International Development, and the Trade and Development Agency—have the discretion to provide funds to support the partnership. Financial support from government partners varies in amount and by fiscal year.

## Voluntary Methane Programs

EPA facilitates a number of voluntary programs related to the Methane-to-Markets initiative that seek to reduce domestic methane emissions from different sectors. Many of these programs receive broad legislative authority from the Clean Air Act, Section 103 (42

U.S.C. § 7403). EPA provides some technical assistance and educational material. The AgSTAR Program supports biogas capture and use at livestock operations managing liquid and slurry manures.[49] The Coalbed Methane Outreach Program (CMOP) works with the coalbed methane industry to reduce coal mine methane emissions via methane capture and reuse.[50] The Natural Gas STAR Program specializes in promoting the reduction of methane emissions from the oil production and natural gas sector.[51] The Landfill Methane Outreach Program (LMOP) encourages landfill gas energy projects.[52] The Natural Gas STAR Program, Coalbed Methane Outreach Program, and the Landfill Methane Outreach Program combined reduced methane emissions by approximately 64 million metric tons of $CO_2e$ in 2007 out of the roughly 324 million metric tons of $CO_2e$ of the methane emissions reported for the landfills, natural gas systems, petroleum systems, and coal mining categories.[53]

## Federal Energy Management Program

The Department of Energy's (DOE's) Federal Energy Management Program (FEMP) addresses energy management at federal facilities and DOE, as well as fleet and transportation management.[54] One component of the program is converting landfill gas to energy for use at federal facilities. DOE has implemented three landfill gas recovery projects. FEMP receives its legislative authority from the Energy Independence and Security Act of 2007 ( P.L. 110-140) and was appropriated $22 million for FY2009.

## Tax Incentives

Several federal tax incentives subsidize methane capture from landfill and agriculture sources. These tax incentives are broadly broken down into three categories: (1) incentives to produce electricity from captured methane gas; (2) incentives to build facilities that produce electricity from captured methane gas; and (3) incentives to produce alternative fuels using captured methane gas.

Two federal tax incentives subsidize the production of electricity from methane. The production tax credit is allowed for the production of electricity from qualified energy resources at qualified facilities, including open-loop biomass and municipal solid waste facilities.[55] In general, open-loop biomass and municipal solid waste facilities placed in service after August 8, 2005, and before December 31, 2013, may claim a tax credit equal to 1 cent per kilowatt-hour of electricity generated during the first 10 years of production.[56] In addition, a one-time investment tax credit equal to 30% of eligible investment costs is available, in lieu of the production tax credit, for open-loop biomass and municipal solid waste facilities placed in service after December 31, 2008.[57]

Three tax-preferred bond finance options exist to help finance methane capture facilities used to produce electricity. Qualified Energy Conservation Bonds (QECBs), Clean Renewable Energy Bonds (CREBs), and New Clean Renewable Energy Bonds (New CREBs) are a type of bond instrument, tax credit bonds, that offers the holder a federal tax credit instead of interest.[58] The rate of credit for CREBs is intended to be set such that the bonds need not be sold at a discount (for a price less than the face value) or with interest costs to the issuer, while the credit rate for QECBs and New CREBs is set for a credit rate of 70%. All three bond options are available to finance qualified energy production projects, including open-loop biomass facilities and landfill gas facilities. QECBs, CREBs, and New CREBs are all subject to national limits, $2.4 billion, $1.2 billion, and $2.4 billion, respectively. CREBs and New CREBs are allocated by the Secretary of the Treasury to eligible projects in inverse to their size, while QECBs are allocated to the states based upon their share of total U.S. population. Issuing authority for QECBs is without expiration, while CREB and New CREB authority expires at the end of 2009.

In addition, two tax incentives are available where methane gas is used to as a production input for alternative fuels. Facilities with binding construction contracts in place before December 31, 2010, and placed in service before January 1, 2014, are eligible to expense one-half of the cost of qualified property in the facilities first year of service.[59] The remaining 50% of the cost is depreciated under an accelerated five-year depreciation period. Further, compressed or liquefied gas and liquid fuel derived from biomass is eligible for the 50 cent per gallon alternative fuel tax credit for fuel produced through December 31, 2009.[60]

## DOE Methane Hydrate Research and Development

Methane is not captured from naturally occurring gas hydrates because it is bound in the gas and not released. However, recent attention has been directed toward the extraction of methane from gas hydrates as a potential source of energy.[61] The objective of the DOE methane hydrate research and development program is to develop knowledge and technology to allow commercial production of methane from gas hydrates by 2015. The DOE program completed a Gulf of Mexico offshore expedition in May 2009 aimed at validating techniques for locating and assessing commercially viable gas hydrate deposits.[62] The program is planning a two-year Alaska production test beginning in the summer of 2009 to provide critical information about methane flow rates and sediment stability during gas hydrate dissociation. Both projects have international and industry partners. Since the enactment of the Methane Hydrate Research and Development Act of 2000 (P.L. 106-193), DOE has spent $87.3 million through FY2008, or approximately 78% of the $112.5 million authorized by law. The Omnibus Appropriations Act, 2009 (P.L. 111- 8), provided $20 million in FY2009 for natural gas technologies R&D, to include no less than $15 million for gas hydrates R&D. The Obama Administration has requested $25 million for the program in FY2009, or 62.5% of the $40 million authorized by the Energy Policy Act of 2005 (P.L. 109-58). The gas hydrate R&D program is authorized through FY2010 under current law.

## APPENDIX. WORLD METHANE EMISSIONS BY SECTOR IN 2005

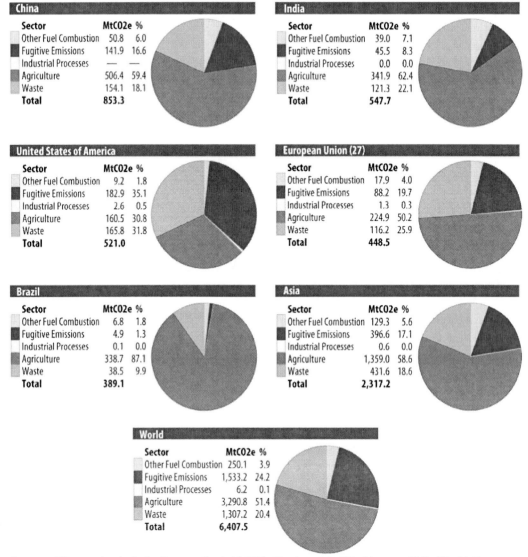

Source: Climate Analysis Indicators Tool (CAIT) Version 6.0 (Washington, DC: World Resources Institute, 2009).

## End Notes

[1] Flaring is the combustion of the gas without commercial purposes. Flaring produces carbon dioxide which is a less potent greenhouse gas than methane.

[2] Environmental Protection Agency, *2009 U.S. Greenhouse Gas Inventory Report*, April 2009, http://www.epa.gov/climatechange/emissions/usinventoryreport.html.

[3] Enteric fermentation is the production and release of methane via eructation (burping) and flatulence as ruminant animals digest their feed.

[4] The climate-changing impact of multiple greenhouse gases is commonly measured and compared using their global warming potential as expressed in units of carbon dioxide equivalent. Therefore, many concepts and

actions are preceded with the word carbon which may actually account for an assortment of greenhouse gases in both quantity and quality (e.g., carbon tax, carbon offset).

[5] See CRS Report R40242, *Carbon Tax and Greenhouse Gas Control: Options and Considerations for Congress*, by Jonathan L. Ramseur and Larry Parker.

[6] See CRS Report R40556, *Market-Based Greenhouse Gas Control: Selected Proposals in the 111th Congress*, by Jonathan L. Ramseur, Larry Parker, and Brent D. Yacobucci.

[7] An upstream approach would apply the cap to fossil fuels when they enter the U.S. economy, either at the mine, wellhead, or another practical "chokepoint" in the production chain. Imported fuels would be addressed at their point of entry into the United States.

[8] For more on these issues, see CRS Report R40242, *Carbon Tax and Greenhouse Gas Control: Options and Considerations for Congress*, by Jonathan L. Ramseur and Larry Parker.

[9] As described by EPA, "wells are used to withdraw raw gas from underground formations. Emissions arise from the wells themselves, gathering pipelines, and well-site gas treatment facilities such as dehydrators and separators. Fugitive emissions and emissions from pneumatic devices account for the majority of CH4 emissions. Flaring emissions account for the majority of the non-combustion $CO_2$ emissions." EPA, *Inventory of U.S. Greenhouse Gas Emissions and Sinks: 1990-2006* (April 2008).

[10] See, e.g., Gilbert Metcalf and David Weisbach, *The Design of a Carbon Tax* (June 2008), Tufts University and the University of Chicago.

[11] EPA, *Inventory of U.S. Greenhouse Gas Emissions and Sinks: 1990-2006* (April 2008).

[12] Landfill gas New Source Performance Standards (NSPS) went into effect in 1996. See U.S. Environmental Protection Agency, "Standards of Performance for New Stationary Sources and Guidelines for Control of Existing Sources: Municipal Solid Waste Landfills," 61 *Federal Register* 9914, March 12, 1996.

[13] For more information pertaining to carbon offsets, see CRS Report RL34436, *The Role of Offsets in a Greenhouse Gas Emissions Cap-and-Trade Program: Potential Benefits and Concerns*, by Jonathan L. Ramseur.

[14] Additionality refers to whether the offset project represents an activity that is beyond what would have occurred under a business-as-usual scenario. In other words, would the emission reductions or sequestration have happened anyway? Additionality is generally considered to be the most significant factor that determines the integrity of the offset.

[15] For more information on these developments, see CRS Report R40585, *Climate Change: Potential Regulation of Stationary Greenhouse Gas Sources Under the Clean Air Act*, by Larry Parker and James E. McCarthy; Environmental Protection Agency, "EPA Finds Greenhouse Gases Pose Threat to Public Health, Welfare / Proposed Finding Comes in Response to 2007 Supreme Court Ruling ," press release, April 17, 2009, http://yosemite.epa.gov/opa/admpress.nsf/0/ 0EF7DF675805295D8525759B00566924.

[16] Biogas consists of 60%-70% methane, 30%-40% carbon dioxide, and trace amounts of other gases.

[17] Methane hydrates—a mixture of water and natural gas—are a potentially huge global energy resource.

[18] See CRS Report R40643, *Greenhouse Gas Legislation: Summary and Analysis of H.R. 2454 as Passed by the House of Representatives* , coordinated by Mark Holt and Gene Whitney.

[19] For more information on market conditions for natural gas, see CRS Report R40487, *Natural Gas Markets: An Overview of 2008*, by Robert Pirog; and the Energy Information Administration report *Natural Gas Year-In-Review 2008*, April 2009, http://www.eia.doe.gov/pub/oil ngyir2008.html#consumption.

[20] The Intergovernmental Panel on Climate Change (IPCC) assigns methane a carbon dioxide equivalent, or global warming potential, of 25. Intergovernmental Panel on Climate Change, Climate Change 2007: The Physical Science Basis (2007), p. 212.

[21] "Radiative forcing" is defined as the change in the difference between incoming and outgoing radiation at the top of the troposphere. $CO_2e$ is not exactly equivalent to radiative forcing, but it is similar and easier to understand for policy purposes than the main alternative, watts per square meter ($W/m^2$).

[22] 1 teragram = 1 million metric tons. A Tg $CO_2e$ (teragram of carbon dioxide equivalent) is a principal unit of measurement across greenhouse gases. See footnote 3 for the definition of enteric fermentation.

[23] Kathleen Hogan, *Current and Future Methane Emissions from Natural Sources*, United States Environmental Protection Agency, EPA 430-R-93-01 1, Washington , DC, August 1993.

[24] World Resources Institute, *Climate Analysis Indicators Tool (CAIT) Version 6.0.*, Washington, DC, 2009. Data quality for the global methane emission estimates reported varies due to uncertainty and possible inconsistency depending on reporting agencies adherence to data collection and interpretation for standardized definitions and measurements for each sector and territory.

[25] The World Resources Institute includes methane emissions from the following activities for the agriculture sector: enteric fermentation from livestock, livestock manure management, rice cultivation, and other agricultural sources. The sole exception, according to CAIT data compiled for 2005, is the United States, where the leading sources of methane—the fugitive emission sector and the waste sector (e.g., landfills, wastewater treatment)—surpass the agriculture sector slightly. However, 2007 data from EPA shown in Figure 1 depicts the agriculture sector as the largest U.S. methane emission source. The World Resources Institute

includes methane emissions from the following activities for the fugitive emission sector: oil and natural gas systems, and coal mining.

[26] For more information, see CRS Report RL34266, *Climate Change: Science Highlights*, by Jane A. Leggett, and E.J. Dlugokencky, L. Bruhwiler, and J.W.C. White, et al., "Observational Constraints on Recent Increases in the Atmospheric CH4 Burden," *Geophysical Research Letters*, August 18, 2009.

[27] EPA prepared analysis using data from the Intergovernmental Panel on Climate Change, Climate Change 2007: The Physical Science Basis Contribution of Working Group I to the Fourth Assessment Report of the Intergovernmental Panel on Climate Change, 2007. http://epa.gov/methane/sources.html#natural

[28] Biogas consists of 60%-70% methane, 30%-40% carbon dioxide, and trace amounts of other gases.

[29] Stoichiometric equation for biogas combustion: $CH_4 + 2O_2 \rightarrow CO_2 + 2H_2O$.

[30] For more information on underground storage options, see http://www.naturalgas.org/naturalgas/storage.asp.

[31] Energy Information Administration, Department of Energy, *The Global Liquefied Natural Gas Market: Status & Outlook*, DOE/EIA-0637, December 2003.

[32] In other parts of the world methane emissions from rice cultivation are a major concern because rice is grown on flooded fields that produce anaerobic conditions to release methane. U.S. methane emissions from rice cultivation are minimal because the United States is not a major producer of rice.

[33] L. E. Chase, "Methane Emissions from Dairy Cattle," Mitigating Air Emissions from Animal Feeding Operations Conference, IA, May 2008, http://www.ag.iastate.edu/wastemgmt/Mitigation_Conference_proceedings/CD_proceedings/Animal_Housing_Diet/Chase-Methane_Emissions.pdf.

[34] For more information on anaerobic digestion systems, see CRS Report R40667, *Anaerobic Digestion: Greenhouse Gas Emission Reduction and Energy Generation*, by Kelsi Bracmort.

[35] Environmental Protection Agency, *2009 U.S. Greenhouse Gas Inventory Report*, April 2009, http://www.epa.gov/ climatechange/emissions/usinventoryreport.html.

[36] A common landfill gas capture system consist of an arrangement of vertical wells and horizontal collectors usually installed after a landfill cell has been capped. Without a gas collection system, the landfill gas would escape into the atmosphere.

[37] Environmental Protection Agency, *Landfill Methane Outreach Program* , Energy Projects and Candidate Landfills, http://www.epa.gov/lmop/proj/index.htm.

[38] Environmental Protection Agency, *Municipal Solid Waste in the United States: 2007 Facts and Figures*, http://www.epa.gov/epawaste/nonhaz/municipal/pubs/msw07-rpt.pdf. The most recent report contains 2007 data.

[39] Lenders may hesitate to provide funding for landfill gas capture projects due to unease about possibly having to remediate a landfill under CERCLA (Comprehensive Environmental Response, Compensation, and Liability Act; 42 USC 9607).

[40] Environmental Protection Agency, *Natural Gas STAR Program: Cost-Effective Opportunities to Recover Methane*, http://www.epa.gov/gasstar/basic-information/index.html#sources.

[41] Environmental Protection Agency, *Global Mitigation of Non-CO_2 Greenhouse Gases*, EPA 430-R-06-005, June 2006.

[42] Technical lifetime is the length of time the equipment is expected to perform as intended.

[43] Environmental Protection Agency, *2009 U.S. Greenhouse Gas Inventory Report*, April 2009, http://www.epa.gov/ climatechange/emissions/usinventoryreport.html.

[44] A gob well allows for the extraction of methane from the gob area of a mine.

[45] Environmental Protection Agency, *Methane to Markets Partnership Country Specific Strategy for the United States*, October 2008, http://www.methanetomarkets.org/resources. Most of these coal mines have been abandoned (8,000).

[46] Environmental Protection Agency, *Coalbed Methane Extra: Coal Mine Methane Ownership Issues*, EPA-430-N-00- 004, 2007, http://www.epa.gov/cmop/docs/fall_2007.pdf.

[47] Environmental Protection Agency, *Options for Reducing Methane Emissions Internationally, Volume 1: Technological Options for Reducing Methane Emissions*, EPA 430-R-93-006, July 1993; Environmental Protection Agency, *Opportunities to Reduce Methane Emissions in the United States Report to Congress*, EPA 430-R-93-012, October 1993. Nineteen hearings pertaining to methane have been held since 1990 including U.S. Congress, Senate Committee on Environment and Public Works, hearing on S. 1772, *Gas Petroleum Refiner Improvement and Community Empowerment Act*, 109th Cong., 1st sess., October 18, 2005, S.Hrg. 109-1001; and U.S. Congress, House Committee on Science, *Energy Research, Development, Demonstration, and Commercial Application Act of 2006*, 109th Cong., 2nd sess., July 28, 2006, H.Rept. 109-611.

[48] For more information on the Methane-to-Markets Partnership, visit http://www.methanetomarkets.org.

[49] For more information on the AgSTAR Program, see http://www.epa.gov/agstar.

[50] For more information on the Coalbed Methane Outreach Program, see http://www.epa.gov/cmop/index.html.

[51] For more information on the Natural Gas STAR Program, see http://www.epa.gov/gasstar/index.html.

[52] For more information on the Landfill Methane Outreach program, see http://www.epa.gov/lmop.

[53] AgSTAR Program reduced methane emissions—not available. However, the program has provided technical assistance for 135 operational anaerobic digestion systems, which generated approximately 332,100 megawatt-

hours (MWh) of energy in 2008. Environmental Protection Agency, *ENERGY STAR and Other Climate Protection Partnerships 2007 Annual Report*, October 2008, http://www.epa.gov/appdstar/pdf/2007AnnualReportFinal.pdf; U.S. EPA, U.S. Emissions Inventory 2009: Inventory of U.S. Greenhouse Gas Emissions and Sinks: 1990-2007.

[54] For more information on the Federal Energy Management Program, see http://www1.eere.energy

[55] Internal Revenue Code (I.R.C.) Section 45. Municipal solid waste covers two types of power facilities: trash combustion facilities that burn trash directly to generate power, and landfill gas facilities that first produce methane, which is then burned to generate electricity. Anaerobic digestion systems are an example of an open-loop biomass system.

[56] The credit rate is adjusted each year for inflation. In addition, facilities placed in service prior to August 8, 2005, may claim the credit for the first five years of production. Further, the date in service break-point for certain open-loop biomass facilities is October 22, 2004.

[57] I.R.C. Section 48.

[58] Qualified Energy Conservation Bonds (QECBs), Clean Renewable Energy Bonds (CREBs) and New Clean Renewable Energy Bonds (New CREBs) are defined in I.R.C. Sections 54D, 54C and 54, respectively. See CRS Report R40523, *Tax Credit Bonds: Overview and Analysis*, by Steven Maguire, for more information on tax credit bonds.

[59] I.R.C. Section 179C.

[60] I.R.C. Section 45K.

[61] Methane hydrates—a mixture of water and natural gas—are a potentially huge global energy resource. For more information on the DOE methane hydrate R&D program, see CRS Report RS22990, *Gas Hydrates: Resource and Hazard*, by Peter Folger; and http://www.netl.doe.gov/technologies/oil-

[62] On May 6, 2009, DOE announced it had completed a 21-day drilling expedition in the Gulf of Mexico in collaboration with the USGS and the Minerals Management Service. A preliminary announcement of the expedition results is available at http://www.netl.doe.gov/technologies/oil GOMJIP_Leg2Announcement.html.

In: Sources and Reduction of Greenhouse Gas Emissions       ISBN: 978-1-61668-856-1
Editor: Steffen D. Saldana                                   © 2010 Nova Science Publishers, Inc.

*Chapter 2*

# NITROUS OXIDE FROM AGRICULTURAL SOURCES: POTENTIAL ROLE IN GREENHOUSE GAS EMISSION REDUCTION AND OZONE RECOVERY

## *Kelsi Bracmort*

### ABSTRACT

Gases other than carbon dioxide accounted for nearly 15% of U.S. greenhouse gas emissions in 2007, yet there has been minimal discussion of these other greenhouse gases in climate and energy legislative initiatives. Reducing emissions from non-carbon dioxide greenhouse gases, such as nitrous oxide ($N_2O$), could deliver short-term climate change mitigation results as part of a comprehensive policy approach to combat climate change.

Nitrous oxide is 298 times more potent than carbon dioxide in its ability to affect climate change; and moreover, results of a recent scientific study indicate that nitrous oxide is currently the leading ozone-depleting substance. Thus, legislation to restrict nitrous oxide emissions could contribute to both climate change protection and ozone recovery.

The primary human source of nitrous oxide is agricultural soil management, which accounted for two-thirds of the $N_2O$ emissions reported in 2007 (approximately 208 million metric tons $CO_2$ equivalent). One proposed strategy to lower $N_2O$ emissions is more efficient application of synthetic fertilizers. However, further analysis is needed to determine the economic feasibility of this approach as well as techniques to measure and monitor the adoption rate and impact of $N_2O$ emission reduction practices for agricultural soil management.

As Congress considers legislation that would limit greenhouse gas emissions (both H.R. 2454 and S. 1733 would require that greenhouse gas emissions be reduced by 83% in 2050), among the issues being discussed is how to address emissions of non-$CO_2$ greenhouse gases. Whether such emissions should be subject to direct regulation, what role EPA should play using its existing Clean Air Act authority, whether the sources of $N_2O$ should be included among the covered entities of a cap-and-trade system, whether $N_2O$ reductions should be considered offsets to be purchased by the covered entities of a cap-and-trade system, and what role $N_2O$ reduction scheme are among the issues being discussed. How these issues are resolved will have important implications for agriculture, which has taken a keen interest in climate change legislation.

## INTRODUCTION

Policymakers are dedicating considerable attention to climate change mitigation, primarily discussing options for carbon dioxide ($CO_2$) emission reduction.[1] Less frequently addressed in proposed legislation is emission reduction for non-$CO_2$ greenhouse gases, such as nitrous oxide ($N_2O$). However, $N_2O$ reduction efforts have the potential to mitigate climate change. Moreover, $N_2O$ emission sources may be regulated under the existing Clean Air Act as a class I or class II ozone-depleting substance at the discretion of the Environmental Protection Agency (EPA) Administrator. No new legislation needs to be passed to regulate $N_2O$ for climate protection and ozone recovery.

The five non-$CO_2$ greenhouse gases regularly monitored but not entirely regulated by EPA (methane, nitrous oxide, hydroflourocarbons, perflourocarbons, and sulfur hexaflouride) accounted for nearly 15% of U.S. greenhouse gas (GHG) emissions in 2007, as measured by total tons of $CO_2$ equivalent.[2] Nitrous oxide—the third most abundant greenhouse gas—was responsible for roughly 4% of total U.S. GHG emissions in 2007 by weight. Although they comprise a smaller portion of GHG emissions, non-$CO_2$ greenhouse gases, including $N_2O$, are more potent than $CO_2$. The gases identified above are 25 to 22,800 times more effective than an equivalent weight of $CO_2$ at trapping heat in the atmosphere, with $N_2O$ being 298 times more potent by weight.[3]

In addition to being one cause of climate change, $N_2O$ is an ozone-depleting substance (ODS).[4] Indeed, scientific analysis suggests that $N_2O$ is now the leading ODS, as other substances have been reduced significantly owing to regulations enacted in the late 1980s, in the Montreal Protocol on Substances that Deplete the Ozone Layer.[5] $N_2O$ emission reduction could thus play a compelling role in recovery of the ozone layer as well as in climate change remediation.

The agriculture sector is the primary anthropogenic source of nitrous oxide.[6] The bulk of U.S. $N_2O$ emissions stem from fertilizing agricultural soils for crop production. Strategies or technologies designated for $N_2O$ emission reduction are limited.[7] This is partly due to the dispersed nature of $N_2O$ emission sources.

In the agriculture sector, the majority of $N_2O$ is released as a consequence of specific nitrogen cycle processes (nitrification and denitrification) when large amounts of synthetic nitrogen fertilizers are used for crop production. More efficient application of synthetic fertilizers (e.g., precision agriculture, nitrogen inhibitors, nitrogen sensors, controlled-release fertilizer products) is one way to reduce excess amounts of nitrogen available for bacterial processing and eventual release to the atmosphere as $N_2O$. High costs and difficulty in measuring these products' efficacy, among other deterrents, have hampered widespread adoption of practices to reduce $N_2O$ emissions.

This chapter focuses on the contributions of $N_2O$ to climate change and ozone depletion. Policy options for $N_2O$ emission reduction, sources of $N_2O$, and federal support to lower $N_2O$ emissions are discussed.

# NITROUS OXIDE: A PRIMER

Nitrous oxide ($N_2O$), familiar to some as "laughing gas," contributes to climate change and ozone depletion. Once released, $N_2O$ lingers in the atmosphere for decades (its atmospheric lifetime is approximately 114 years) and is 298 times more effective at trapping heat in the atmosphere over a 100-year time frame than carbon dioxide ($CO_2$).[8] $N_2O$ emission quantity estimates have remained fairly constant over the last few years, hovering around 312 million metric tons carbon dioxide equivalent ($CO_2e$). See Table 1.

## Sources of $N_2O$ Emissions

Nitrous oxide is emitted from anthropogenic (manmade) and natural sources. Oceans and natural vegetation are the major natural sources of $N_2O$. Agricultural soil management (e.g. fertilization, application of manure to soils, drainage and cultivation of organic soils) is responsible for two- thirds of anthropogenic U.S. $N_2O$ emissions.[9] In 2007, $N_2O$ emissions from agricultural soil management totaled more than 200 million metric tons of $CO_2e$.[10] Other anthropogenic sources of $N_2O$ are combustion by mobile sources (cars, trucks, etc.), nitric acid production, and manure management.[11]

Figure 1 depicts the origination and passage of nitrogen (N) that leads to $N_2O$ emissions from agricultural soil management. The amount of $N_2O$ emitted from cropland soils largely depends on the amount of nitrogen applied to a crop, weather, and soil conditions. Corn and soybean crops emit the largest amounts of $N_2O$, respectively, due to vast planting areas, plentiful synthetic nitrogen fertilizer applications, and, in the case of soybeans, high nitrogen fixation rates (Figure 2).[12]

**Table 1. U.S. Greenhouse Gas Emissions (million metric tons $CO_2e$)**

| Gas / Source | 2005 | 2006 | 2007 | Avg. Contribution[a] |
|---|---|---|---|---|
| Carbon dioxide ($CO_2$) | 6,090.8 | 6,014.9 | 6,103.4 | 85% |
| Methane ($CH_4$) | 561.7 | 582.0 | 585.3 | 8% |
| *Nitrous oxide ($N_2O$)* | *315.9* | *312.1* | *311.9* | *4%* |
| Hydroflourocarbons (HFCs) | 116.1 | 119.1 | 125.5 | 1.7% |
| Perflouruocarbons (PFCs) | 6.2 | 6.0 | 7.5 | <1% |
| Sulfur hexaflouride ($SF_6$) | 17.9 | 17.0 | 16.5 | <1% |
| Total | 7,108.6 | 7,051.1 | 7,150.1 | |

Source: Environmental Protection Agency, *2009 U.S. Greenhouse Gas Inventory Report*.
a. Average contribution to total U.S. greenhouse gas inventory based on data provided for 2005 to 2007.

## The Nitrogen Cycle

Comprehension of the nitrogen cycle (Figure 3) is beneficial when crafting policy to reduce $N_2O$ emissions from anthropogenic sources. Nitrogen, an essential element required by organisms to grow, is found throughout the atmosphere in various forms. The nitrogen cycle portrays the routes in which nitrogen moves through the soil and atmosphere in both organic and inorganic form. Certain processes within the nitrogen cycle convert the nitrogen into a form that can be taken up by plants. Four of the major processes are:

- nitrogen fixation—conversion of nitrogen gas ($N_2$) to a plant-available form;
- nitrogen mineralization—conversion of organic nitrogen to ammonia (NH3);
- nitrification—conversion of ammonia ($NH_3$) to nitrate ($NO_3$-) via oxidation (that is, by being combined with oxygen); and
- denitrification—conversion of nitrates back to nitrogen gas.

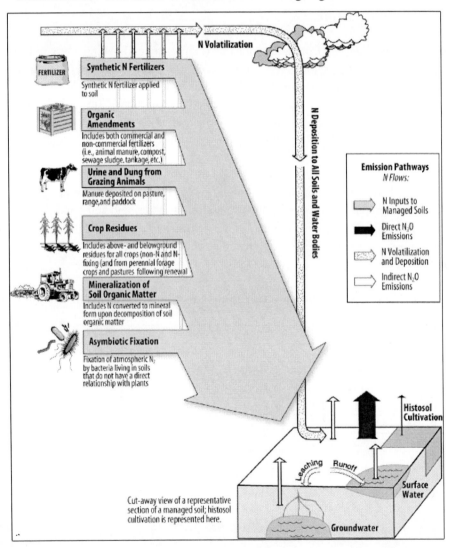

Source: U.S. Environmental Protection Agency, *2009 U.S. Greenhouse Gas Inventory Report*, EPA 430-R-09-004, Chapter 6, April 2009. Adapted by CRS.

Figure 1. Sources and Pathways of Nitrogen (N) Resulting in $N_2O$ Emissions from Agricultural Soil Management

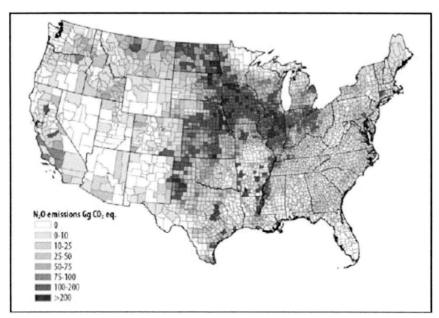

Source: USDA, *U.S. Agriculture and Forestry Greenhouse Gas Inventory: 1990-2005.* Adapted by CRS.
a. 1 Gigagram (Gg) is equivalent to 1,000 metric tons.
b. Major crops are defined as corn, soybean, wheat, hay, sorghum, and cotton.

Figure 2. County-Level $N_2O$ Emissions from Major Cropped Soils in 2005

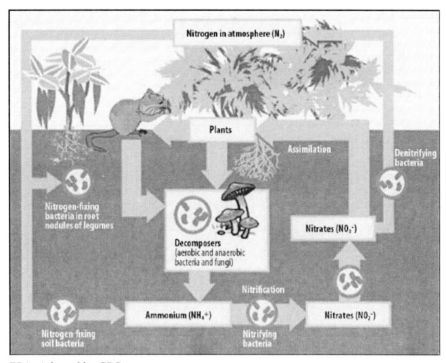

Source: EPA. Adapted by CRS

Figure 3. The Nitrogen Cycle

Nitrous oxide is a byproduct of nitrification and denitrification. Both processes occur naturally. Excess application of nitrogen fertilizer can lead to increased nitrification, which can cause nitrate to leach into groundwater or surface runoff (in turn, this causes eutrophication, which can damage aquatic environments).

## OPPORTUNITIES AND CHALLENGES FOR NITROUS OXIDE EMISSION REDUCTION

$N_2O$ emission mitigation options are available for agricultural soil management and nitric acid production. Nitric acid is a chemical compound used to make synthetic fertilizers. $N_2O$ abatement options for nitric acid production include a high-temperature catalytic reduction method, a low- temperature catalytic reduction method, and nonselective catalytic reduction.[13] The estimated reduction efficiencies (the percentage reduction achieved with adoption of a mitigation option) are 90%, 95%, and 85%, respectively.

Agricultural soil management mitigation options recommended by researchers and technology transfer specialists to discourage excess application of nitrogen fertilizers and soil disturbance (Table 2) are not generally being practiced. Fertilizer and soil best management practices aim to provide the crop with the nutrient and soil conditions necessary for crop production, and prevent nutrient and soil loss from the crop field (e.g., erosion, leaching).[14] Some may consider less money spent towards fertilizer use an economic incentive for agricultural producers.[15] Others may be concerned to ensure that crop yields meet expected feed, fiber, and fuel mandates (e.g., for corn ethanol), which may be difficult to attain with less fertilizer use.[16] Monitoring reduced nitrogen fertilization applications on a large scale for climate change mitigation purposes may be difficult; it is not clear how such a program could be managed at a national level.[17] Enforcement options could include voluntary verification, third-party verifiers, or government intervention.

### Table 2. Select $N_2O$ Mitigation Alternatives for Agricultural Soil Management

| Mitigation Alternative | Description |
| --- | --- |
| Split fertilization | Application of same amount of nitrogen fertilizer as in baseline but divided into three smaller increments during crop uptake period to better match nitrogen application with crop demand and reduce nitrogen availability for leaching, nitrification, denitrification, and volatilization. |
| Simple fertilization reduction (10%, 20%, or 30%) | Reduction of nitrogen-based fertilizer from one-time baseline application of 10%, 20%, or 30%. |
| Nitrification inhibitor | Reduces conversion of ammonium to $NO_3$-, which slows the immediate availability of nitrate (nitrate is water soluble). The inhibition of nitrification reduces nitrogen loss and increases overall plant uptake. |
| No-till | Conversion from conventional tillage to no-till, where soils are disturbed less and more crop residue is retained. |

Source: EPA, *Global Mitigation of Non-CO₂ Greenhouse Gases*, http://www.epa.gov/climatechange/ economics downloads/GM_SectionV_Agriculture.pdf.

Reporting $N_2O$ emissions from agricultural soil management was not included in the Final Mandatory Reporting of Greenhouse Gases Rule issued by EPA on September 22, 2009.[18] EPA's reasoning behind this decision was that no low-cost or simple direct $N_2O$ measurement methods exist. Additionally, EPA released a proposed rule requiring new or modified facilities that could trigger Prevention of Significant Deterioration (PSD) permitting requirements to apply for a revision to their operating permits to incorporate the best available control technologies and energy efficiency measures to minimize GHG emissions.[19]

## FEDERAL SUPPORT FOR NITROUS OXIDE EMISSION REDUCTION

USDA provides some financial and technical assistance for nutrient management through its conservation programs.[20] Moreover, USDA's Agricultural Research Service (ARS) is studying the relationship between agricultural management practices and nitrous oxide emissions.[21]

In addition to the agriculture sector, work is being done in the transportation sector to reduce $N_2O$ emissions. Mobile combustion was responsible for nearly 10% of $N_2O$ emissions reported in 2007.[22] One $N_2O$ emission reduction initiative, proposed by EPA and the Department of Transportation, is to cap tailpipe $N_2O$ emissions at 0.010 grams per mile as part of a wider effort to reduce greenhouse gas emissions and improve fuel economy in tandem, via a $CO_2$ emission standard for light-duty vehicles.[23] EPA has allocated financial resources to quantify $N_2O$ emissions for the greenhouse gas inventory (e.g., DAYCENT model).[24]

## POLICY OPTIONS FOR NITROUS OXIDE EMISSION REDUCTION

Congress has begun to investigate the reduction of non-$CO_2$ greenhouse gas emissions, including $N_2O$ emissions, as one strategy to mitigate climate change. Some contend that $N_2O$ emissions reduction could serve as a short-term response in the larger, long-term scheme of mitigation and adaptation efforts.[25] It may be viewed as a short-term response because $N_2O$ emissions make up a small amount of the GHG inventory compared to $CO_2$ emissions. Any substantial approach to mitigate climate change is likely at some point to have to address sources that emit $CO_2$.

Congress could approach $N_2O$ emissions reduction as part of a comprehensive GHG emission strategy offering economically attractive abatement alternatives to discourage actions leading to climate change. For example, a cap or fee on $N_2O$ emissions could spur innovative methods for agricultural producers to limit excess synthetic fertilizer application. Congress could also examine the tools necessary to identify $N_2O$ emission abatement options, assess their cost, and determine their economic impact for full incorporation into climate change legislation.

Besides mitigating climate change, reducing $N_2O$ emissions could lead to ozone recovery. Congress could explore the co-benefits that may arise from restricting $N_2O$ emissions for climate change purposes. $N_2O$ is not regulated as an ODS under the Clean Air Act, Title VI, Stratospheric Ozone Protection (as guided by the Montreal Protocol). As emissions of other ODSs (e.g., chlorofluorocarbon- 11, halon- 1211) have declined due to regulation, $N_2O$ has emerged as the dominant ODS.[26] The first-ever published ozone depletion potential (ODP) value assigned to $N_2O$, 0.0 17, is less than the ODP value of 1.0 for the reference gas chlorofluorocarbon 11 (CFC11). While some may not see a cause for alarm based on the ODP value alone, the quantity of $N_2O$ emissions and its potency as a GHG can lead to serious harm (see Table 1).

The ODP value for $N_2O$ does not allow for its mandatory inclusion as a class I substance for regulation under the Clean Air Act.[27] However, $N_2O$ could be listed as a class II substance at the direction of the EPA Administrator or regulated under Section 615 of the act.[28] Class I substances have an ODP of 0.2 or more and are more harmful to stratospheric ozone molecules than Class II substances, which have an ODP of less than 0.2.

With or without ODP substance listing, Congress may find it useful to incorporate the ozone depletion impacts of $N_2O$ into its climate change policy proposals both to reduce greenhouse gas emissions and to further ozone recovery achievements. Classifying $N_2O$ emission reduction as an eligible offset type, including $N_2O$ as a covered entity within a cap-and-trade program, or directing EPA to use existing authority under the Clean Air Act to regulate $N_2O$ are other available options to reduce $N_2O$ emissions for ozone or climate protection. Any option chosen to reduce $N_2O$ emissions will more than likely require an improvement of $N_2O$ estimation, measurement, and reporting methods and possible financial incentives.

Congress could apply lessons learned from previous international agreements that are intended to abolish harmful compounds. The outcomes of the Montreal Protocol, put into action in the late 1980s, may prove useful to Congress in understanding the long-term implications of certain climate change policy options, specifically cap-and-trade. A number of gases were phased out under the Protocol, which allowed for each country to establish a regulatory framework to monitor and reduce ODSs. Certain ozone-depleting substances, such as $N_2O$, were not included in the Protocol partly because their threat was not perceived as urgent at the time. However, one unintended consequence of the success of the Protocol reducing targeted ODSs is that $N_2O$ has emerged as the leading ODS.

## End Notes

[1] For more information on $CO_2$ emission reduction techniques, see CRS Report RL33801, *Carbon Capture and Sequestration (CCS)*, by Peter Folger. For more information on legislative proposals to address climate change and regulation of greenhouse gases under the Clean Air Act, see CRS Report R40556, *Market-Based Greenhouse Gas Control: Selected Proposals in the 111$^{th}$ Congress*, by Jonathan L. Ramseur, Larry Parker, and Brent D. Yacobucci; and CRS Report R40585, *Climate Change: Potential Regulation of Stationary Greenhouse Gas Sources Under the Clean Air Act*, by Larry Parker and James E. McCarthy.

[2] U.S. Environmental Protection Agency, *2009 U.S. Greenhouse Gas Inventory Report*, EPA 430-R-09-004, April 2009, http://epa.gov/climatechange/emissions/usinventoryreport.html.

[3] The potency of a greenhouse gas is described by its global warming potential (GWP), an estimate of how much a greenhouse gas affects climate change over a quantity of time relative to $CO_2$, which has a GWP value of 1. Intergovernmental Panel on Climate Change, *Climate Change 2007: The Physical Science Basis* (2007), p. 212.

[4] An ozone-depleting substance is a compound that contributes to stratospheric ozone depletion by releasing chlorine or bromine atoms into the atmosphere when broken down, leading to the destruction of ozone, a substance necessary to prevent harmful UVB rays from reaching Earth.

[5] The Montreal Protocol is an international treaty crafted to protect the stratospheric ozone layer by gradually eliminating a number of ozone-depleting substances.

[6] Also in the agriculture sector, animal digestive systems and manure management account for a large portion of U.S. methane emissions. The Intergovernmental Panel on Climate Change (IPCC) assigns nitrous oxide and methane a global warming potential of 298 and 25, respectively.

[7] Strategies and technologies for $N_2O$ emission reduction are limited in comparison to resources expended for methane capture. Methane capture technologies, as well as financial and technical support, for point sources have been available for decades. For more information on methane capture, see CRS Report R40813, *Methane Capture: Options for Greenhouse Gas Emission Reduction*, by Kelsi Bracmort et al.

[8] S. Solomon, D. Qin, and M. Manning et al., *Climate Change 2007: The Physical Science Basis. Contribution of Working Group I to the Fourth*, IPCC, IPCC WG1 AR4 Report, New York, NY, 2007, http://ipcc-wg1.ucar.edu/wg1/ wg1-report.html. The IPCC assigned $N_2O$ a global warming potential of 298 over a 100-year time horizon.

[9] Agricultural soil management includes practices that add to, or create an environment conducive to the release of, mineral nitrogen (N).

[10] U.S. Environmental Protection Agency, *2009 U.S. Greenhouse Gas Inventory Report*, EPA 430-R-09-004, April 2009, http://epa.gov/climatechange/emissions/usinventoryreport.html.

[11] Mobile (fuel) combustion leads to $N_2O$ being emitted as a byproduct. $N_2O$ is released as a byproduct during the oxidation of ammonia for production of nitric acid, a primary component of synthetic fertilizers and some explosives. $N_2O$ emissions are generally released in large amounts from dry manure handling systems (e.g., pasture, solid storage)..

[12] U.S. Department of Agriculture, *U.S. Agriculture and Forestry Greenhouse Gas Inventory: 1990-2005* , Technical Bulletin No. 1921, 2008, http://www.usda.gov/oce/global_change/AFGGInventory1990_2005.htm. Nitrogen fixation is the conversion of nitrogen gas to ammonia.

[13] U.S. Environmental Protection Agency, *Global Mitigation of Non-CO₂ Greenhouse Gases*, EPA-430-06-005, 2006. Catalytic reduction methods use a catalyst to reduce nitrous oxides in exhaust gas at varying temperatures.

[14] C. S. Snyder, *Fertilizer Nitrogen BMPs to Limit Losses That Contribute to Global Warming* , International Plant Nutrition Institute, Ref. # 08057, June 2008, http://www.ipni.net/ipniweb/portal.nsf/0/ 6D54ABC2C92D9 AFA8525749B0074FF59.

[15] According to the Government Accountability Office, natural gas is the highest-priced factor when producing nitrogen fertilizer. Thus, natural gas prices impact nitrogen fertilizer costs. U.S. General Accounting Office, *Natural Gas: Domestic Nitrogen Fertilizer Production Depends on Natural Gas Availability and Prices*, GAO-03-1148, September 2003, http://www.gao.gov/new.items/d031148.pdf.

[16] For example, the Renewable Fuel Standard (RFS) is a provision established by the Energy Policy Act of 2005 requiring gasoline to contain a minimum amount of fuel produced from renewable biomass (including corn). For more information on the RFS, see CRS Report R40 155, *Selected Issues Related to an Expansion of the Renewable Fuel Standard (RFS)*, by Brent D. Yacobucci and Randy Schnepf.

[17] For more information on monitoring carbon in agriculture, see CRS Report RS22964, *Measuring and Monitoring Carbon in the Agricultural and Forestry Sectors*, by Ross W. Gorte and Renée Johnson.

[18] For more information on the agricultural implications of the Mandatory Reporting of Greenhouse Gases Rule, see CRS Report RL32948, *Air Quality Issues and Animal Agriculture: A Primer*, by Claudia Copeland.

[19] For more information on the proposed PSD rule, see CRS Report R40585, *Climate Change: Potential Regulation of Stationary Greenhouse Gas Sources Under the Clean Air Act*, by Larry Parker and James E. McCarthy; and EPA, *Proposed Rule: Prevention of Significant Deterioration and Title V Greenhouse Gas Tailoring Rule*, 2009, http://www.epa.gov/NSR/fs20090930action.html.

[20] For more information on agricultural conservation programs, see CRS Report R40763, *Agricultural Conservation: A Guide to Programs* , by Megan Stubbs. For more information on technical assistance for nutrient management, see USDA, Natural Resources Conservation Service, *Conservation Practice Standard— Nutrient Management Code 590*, August 2006, ftp://ftp-fc.sc.egov.usda.gov/NHQ/practice-standards

[21] For more information on the efforts underway at ARS, visit the Air Quality of Agricultural Systems Research Unit website at http://www.ars.usda.gov/main/site_main.htm?modecode=36-25-15-15, or the Air Quality National Program website at http://www.ars.usda.gov/research/programs/programs.htm?NP_CODE=203.

[22] U.S. Environmental Protection Agency, *2009 U.S. Greenhouse Gas Inventory Report*, EPA 430-R-09-004, April 2009, http://epa.gov/climatechange/emissions/usinventoryreport.html.

[23] U.S. Environmental Protection Agency, *EPA and NHTSA Propose Historic National Program to Reduce Greenhouse Gases and Improve Fuel Economy for Cars and Trucks* , EPA-420-F-09-047, September 2009, http://www.epa.gov/otaq/climate/regulations

[24] EPA uses the DAYCENT ecosystem model for the U.S. greenhouse gas inventory "to estimate direct $N_2O$ emissions from mineral cropland soils that are managed for production of major crops—specifically corn, soybeans, wheat, alfalfa hay, other hay, sorghum, and cotton."

[25] Shilpa Rao and Keywan Riahi, "The Role of Non-CO₂ Greenhouse Gases in Climate Change Mitigation: Long-Term Scenarios for the 21st Century," *Energy Journal*, vol. 27 (2006), pp. 1-26; Mario Molina, Durwood Zaelke, and K. Madhava Sarma et al., "Reducing Abrupt Climate Change Risk Using the Montreal Protocol

and Other Regulatory Actions to Complement Cuts in $CO_2$ Emissions," *Proceedings of the National Academy of Sciences of the United States of America*, October 12, 2009.

[26] A. R. Ravishankara, John S. Daniel, and Robert W. Portmann, "Nitrous Oxide ($N_2O$): The Dominant Ozone-Depleting Substance Emitted in the 21st Century," *Science Express*, August 27, 2009.

[27] 42 U.S.C. § 7671a. The EPA Administrator may add to the list of class I substances any substances that the Administrator determines as having an ozone depletion potential of 0.2 or greater.

[28] 42 U.S.C. § 767 1n. The EPA Administrator has the authority to promptly promulgate regulations respecting the control of an ODS by submitting notice of the proposal and promulgation of such regulation to the Congress.

In: Sources and Reduction of Greenhouse Gas Emissions    ISBN: 978-1-61668-856-1
Editor: Steffen D. Saldana    © 2010 Nova Science Publishers, Inc.

*Chapter 3*

# AN OVERVIEW OF GREENHOUSE GAS (GHG) CONTROL POLICIES IN VARIOUS COUNTRIES

## *Jane A. Leggett, Richard K. Lattanzio, Carl Ek and Larry Parker*

### ABSTRACT

As Congress considers legislation to address climate change, and follows negotiations toward a new international agreement to reduce greenhouse gas (GHG) emissions, the question of the comparability of actions across countries frequently arises. Concerns are raised about what the appropriate sharing of efforts should be among countries, as well as the potential trade implications if countries undertake different levels of GHG reductions and, therefore, incur varying cost impacts on trade-sensitive sectors. This chapter summarizes the GHG control policies in effect or under consideration in the European Union (EU) and various other large countries, and offers a brief set of initial observations. It gives particular emphasis to how particular trade-sensitive sectors may be treated in the context of each national program.

All countries examined have in place, or are developing, some enforceable policies that serve to reduce GHG emissions. Most are at some stage of making their programs more stringent. The wealthiest countries have all taken on GHG limitation or reduction targets under the Kyoto Protocol. Some of the emerging economies have voluntarily stated GHG targets, though none have yet accepted legally binding obligations in an international agreement. The forms of targets, and their stringencies, vary widely across countries.

The scope of specific GHGs and economic sectors covered by national (or sub-national) reduction measures is generally, but not completely, similar. All have policies that affect carbon dioxide emissions; most have some measures that cover the additional five gases covered under the Kyoto Protocol (methane, nitrous oxide, sulfur hexafluoride, perfluorocarbons, and hydrofluorocarbons).

The programs and measures used vary across countries. Even when some measures have similar names (e.g., voluntary programs and voluntary action plans), the measures may differ in important ways that may influence their effectiveness and impacts on trade competiveness.

Within sectors of a country, emission rates and control requirements may vary widely. A country may have some facilities with emission rates (or energy intensities) comparable to the best globally, even if the country's sectoral average as a whole has, for example, a significantly higher energy intensity than the global average.

This chapter presents an overview of GHG control policies within individual countries. It does not present a rigorous assessment of the comparability of GHG control policies across countries or within specific sectors. The criteria for assessing comparability internationally are not widely agreed, and could encompass a range of considerations, not all quantitatively measurable.

This chapter summarizes the greenhouse gas (GHG) control policies in effect or under consideration in a number of large countries, and offers a brief set of initial observations. This overview allows preliminary comparison across countries. Because of congressional interest in the comparability of countries' actions, and in the potential trade ramifications of differential policies, these country fact sheets give emphasis to how particular trade-sensitive sectors may be treated in the context of each national program. Where specific industries are not listed in a country's fact sheet, no further information was found.

The European Union's policies are presented first, followed by any additional rules or policies under consideration in several of the largest EU Member States (i.e., France, Germany, the United Kingdom). A number of additional large-emitting countries follow in alphabetical order. Finally, the Appendix provides a comparison of early 2009 vehicle efficiency standards across countries, which may be a useful reference for a sector that emits a large portion of global GHG emissions.

## SYNTHESIS OBSERVATIONS

- All countries examined have in place, or are developing, some enforceable policies that serve to reduce greenhouse gas (GHG) emissions. Most are at some stage of making their programs more stringent.
- The scope of specific GHGs and economic sectors covered by national (or sub-national) reduction measures is generally, but not completely, similar. All have policies that affect carbon dioxide ($CO_2$) emissions; most have some measures that cover the additional five gases covered under the Kyoto Protocol, including methane ($CH_4$), nitrous oxide ($N_2O$), sulfur hexafluoride ($SF_6$), perfluorocarbons (PFC), and hydrofluorocarbons(HFC).
- The programs and measures used vary across countries. Even when some measures have similar names (e.g., voluntary programs and voluntary action plans), the measures may differ in important ways that may influence their effectiveness and impacts on trade competiveness. For example, many countries support "voluntary programs" or "voluntary action plans." Some of these voluntary efforts may provide technical assistance with few requirements from participants; other programs may include formal emission reduction targets, reporting, and governmental pressure to achieve targets.
- Within economic sectors of a country, emission rates and control requirements may vary widely. A country may have some facilities with emission rates (or energy

intensities) comparable to the best globally, even if the country's sector as a whole has, for example, an energy intensity significantly higher than the global average for that sector. Such discrepancies often occur in emerging economies wherein an older, less-efficient industrial sector is being replaced by new infrastructure.

- Most of the programs include provisions to assist or exempt trade-sensitive sectors, but the definition of what is trade-sensitive, and the approaches to assisting or protecting the sectors, vary widely. "Trade-sensitivity" is a continuing phenomenon. Companies become more or less competitive on an international market according to a host of factors, including productivity, market demand, resource costs, labor costs, exchange rates, and the like. The addition of a carbon control regime to this competitive dynamic has raised concerns that, in the absence of similar policies among competing nations, trade-exposed industries that must control their emissions, or face increased costs passed-through by suppliers, may be less competitive and may lose global market share to competitors in countries lacking comparable carbon policies.[1] These concerns have led many countries to consider specific provisions for exposed sectors.
- Assessing the comparability of GHG control policies across countries and in specific sectors could be difficult, and the results could be subject to debate. How well alternative policy directions and methods could stand up under possible challenges against border adjustments under the World Trade Organization (WTO) may merit further investigation. However, consideration of specific methods to assess comparability, and their implications, is beyond the scope of this chapter.

# EUROPEAN UNION[2]

## 1. Overall GHG Emission Target, If Any, and Timing

Under the Kyoto Protocol, the European Union (EU) agreed to reduce GHG emissions of its 15 Member states in 1997 (EU-15) in aggregate by 8% below 1990 levels during the first commitment period of 2008-2012. (There is no collective target for the EU-27, the current 27 Member states of the EU.) In 2007 and 2008, EU-15 GHG emissions were approximately 5% and 6%, respectively, below 1990 levels. In November 2009, The European Commission projected that the EU-15 will surpass its obligation to reduce GHG emissions under the Kyoto Protocol.[3] The EU-15 will have reduced their domestic GHG emissions to about 7% below 1990 levels during 2008-2012. Plans by EU-15 Member states to acquire international credits through the Kyoto Protocol's three market-based mechanisms would provide another 2.2% GHG reduction, while acquisitions by operators in the EU Emission Trading Systems may provide an additional 1.4% GHG reduction, and enhancement of carbon removals by sinks may offer another 1.0%. With additional policies and measures, the Commission projects that the EU-15 may be around 13% below 1990 levels in 2008-2015.

For the post-Kyoto period (beyond 2012), the European Council adopted on April 23, 2009 the "20-20-20" Policy—a climate and energy package to require by 2020:

- a 20% reduction in GHG emissions from 1990 levels,

- a 20% share of renewable energy in the European Union's final consumption figures (including a 10% share in each Member State's transport sector), and
- a 20% reduction in energy consumption.[4]

The legislation also committed to scale up the GHG emission reduction target to 30% if other developed countries make comparable efforts under a new international agreement. The purpose is to limit the global temperature rise to no more than 2°Celsius above preindustrial levels.

## 2. Principal Policy Instrument(s)

a. Expansion of current European Union Emissions Trading System (EU ETS).[5]
b. Effort-sharing relationships among Member States to reduce emissions in sectors not covered by the EU ETS. It will be left to Member States to define and implement policies in such sectors, although a number of EU-wide measures in areas such as efficiency standards, passenger car emission standards, and a landfill directive for waste disposal will contribute. The European Community infringement procedures and mechanisms for corrective action under the effort- sharing decision are to be put in place to monitor progress.[6]
c. Regulations stipulating mandatory national targets for the overall shares of energy from renewable sources in gross final consumption of energy, taking into account differing starting points for each Member.[7] It will be left to Member States to determine renewable share allocation among sectors.

At the national level, several EU Member states also impose carbon emission fees to some degree. Carbon fees exist in Denmark, Finland, and Sweden; they have been proposed to begin on Jan. 1, 2010 in France. Spain and Ireland reportedly have also signaled that they may consider domestic carbon fees in addition to EU and other national policies.[8] In addition, on October 5, 2009, an EU Taxation Commissioner revealed that in early 2010 the European Commission plans to propose an expansion of existing energy taxation in order to charge $CO_2$ emission fees as well.[9] The new carbon tax would cover sectors not under the EU ETS (see below), such as agriculture, households, and transport. The proposal explicitly is intended to help the EU achieve compliance with its law to reduce GHG emissions to 20% below 1990 levels by 2020. All taxation proposals, to pass into law, require unanimous agreement of the 27 EU Member states, which may be difficult to achieve, and the assent of the European Parliament.

## 3. Covered Gases and Sectors

The only greenhouse gas covered under the original 2003 EU ETS was $CO_2$. The expanded EU ETS to take effect in 2013 will add $N_2O$ emissions from nitric, adipic, and glyoxalic acid production, and PFC emissions from the aluminum sector. Gases not stipulated in the EU ETS, but defined as "greenhouse gases" in Annex II of DIRECTIVE 2003/87/EC include $CH_4$, HFC, and $SF_6$. These gases will be controlled under guidelines for sectors not covered by the EU ETS.

Sectors originally covered in the 2003 EU ETS were: power and combustion installations (exceeding 20 megawatts (MW)); petroleum refineries; coke ovens; metal ore production installations; iron and steel production installations (exceeding 2.5 tons of product per hour); factories for cement (exceeding 50 tons per day), glass (exceeding 20 tons per day); ceramics including tiles, bricks, stoneware, porcelain (exceeding 75 tons per day); and production of pulp, paper and board (exceeding 20 tons per day). The expanded EU ETS will increase the scope of covered sectors beginning in 2013 to include primary and secondary aluminum production facilities; ferrous, ferro-alloy, and non-ferrous metal production facilities; mineral wool and gypsum plants; ammonia, petro-chemical and chemical plants including carbon black organics, nitric acid, adipic acid, glyoxal, organic chemicals (exceeding 100 tons per day), hydrogen (exceeding 25 tons per day), soda ash, and sodium bicarbonate. Additionally, certain categories of aviation will be incorporated into the ETS involving commercial flights departing or arriving in a territory of a Member State.[10] In the EU ETS, Member states decide a National Allocation Plan (NAP), subject to review by the EU, to give emission allowances to individual plants. In the first pilot trading period, some Member states allocated more emission allowances than needed to companies, so that revisions to the scheme in Phase III, beginning in 2013, have been adopted to avoid over-allocation, including increasing rates of auctioning allowances.

Sectors not covered by the EU ETS but covered by adopted legislation include transport, housing, agriculture, and waste (see the following discussion).

## 4. Allocation of GHG Reductions to Various Sectors

The European Union's programs call for a 21% reduction in EU ETS sector emissions compared to 2005 and a 10% reduction in non-EU ETS sector emissions compared to 2005. This is expected to achieve an overall reduction of 14% compared with 2005, which is equivalent to a reduction of 20% compared with 1990. The EU ETS covers electricity generation and the main energy-intensive industries—power stations, refineries, iron and steel, cement and lime, paper, food and drink, glass, ceramics, engineering, and vehicles. Initially, countries allocate allowances to covered sectors, but limited auctioning of permits is planned for the future (e.g., maximum 10% of allowances are auctioned in Phase II).

Phase III ETS: Emissions from sectors covered in the EU ETS will be cut 21% from 2005 levels by 2020. A single EU-wide cap on emissions will be set for EU ETS covered sectors. Allowances will be allocated on the basis of rules harmonized across Member states. The tentative annual cap figure will begin at 1,974 million tons $CO_2$ in 2013 and decrease annually. The total number of allowances (one allowance equals permission to emit one ton) in 2013 will begin at the average total quantity issued for the 2008-20 12 period and will decrease annually at a rate of 1.74%. Free allocation of emission allowances will be progressively replaced by auctioning allowances by 2020. Auctioning will begin in 2013 at 20% and gradually rise to 70% in 2020 and to 100% in 2027. Power producers must acquire all allowances at auction in order to prevent windfall profits (following experience under the pilot trading period). Member States that are highly dependent on fossil fuels and/or States insufficiently connected to the grid (these include Bulgaria, Cyprus, Czech Republic, Estonia, Hungary, Latvia, Lithuania, Malta, Poland and Romania) are allowed to apply for a derogation procedure of reduced auctioning rates for power production of 30% in 2013, gradually rising to 100% in 2020, as long as producers invest in clean technologies to the market value of the permits. Furthermore, less affluent states (the 10 above plus Greece and Portugal) will receive an increased amount of emission permits to auction amounting to 12% more than their actual share to assist in revenue generation. Each Member state will be allowed to determine use of revenue with a suggested investment of 50% toward clean technologies and pollution abatement.

Non-ETS: Sectors not covered by the EU ETS are transport, housing, agriculture and waste. The 2009 Directive proposes to cut emission in these sectors by 10% EU-wide from 2005 levels by 2020. Targets will be mandated according to each Member states' relative wealth (based on GDP per capita and economic growth prospects) with figures ranging from -20% to +20%. Targets are binding on Member states and are enforceable through the usual EU infringement procedure.[11] If a country exceeds its annual objective, it must implement corrective measures, and will be penalized via a deduction from the following year's $CO_2$ allowance. Several flexibility measures are available including the possibility of trading emission cuts across countries; carrying forward ("banking") extra emission reductions; and using a limited amount of credit from developing countries (through an offsets mechanism similar to the Kyoto Protocol's Clean Development Mechanism).

The transportation sector has legally-binding standards for $CO_2$ emissions from new passenger cars to apply as of 2012 in order the meet the 20% emission reduction by 2020.[12] Reductions are required to achieve 120 grams carbon dioxide per kilometer ($CO_2$/km) for 65% of fleet in 2012, 75% in 2013, 80% in 2014 and 100% starting in 2015. A target of 95 grams $CO_2$/km is set for 2020. Enforcement is set through financial penalties against the car manufacturers depending on how far their fleet exceeds the targets.[13]

A renewable energy mandate sets mandatory national targets for each Member state in accordance with each country's different starting points. The purpose of mandates is to provide certainties for investment. Each country will report to the European Council by June 2010 regarding how each Member has allocated the renewable target among transport, electricity, heating and cooling sectors. A 10% target for renewable energy in the transportation sector is set at the same level for all countries.

## 5. Any Regulations or Exemptions Specific to Trade-sensitive Sectors

The climate and energy package in the 2009 Directive provides that the risk of "carbon leakage"[14] may be reduced by allotting free carbon allowances to businesses exposed to "significant risk of carbon leakage" (SRCL) by the cost of compliance with the EU ETS. (The European Commission must adopt a list of sectors deemed exposed to a significant risk of carbon leakage no later than December 31, 2009. A draft list was proposed in September 2009, discussed below.) However, any free allowances will not be decided until 2011. The list may be revised before 2014, based on reanalysis of trade figures, and identification of countries that make firm commitments to reduce their GHG emissions.

If international negotiations on climate change in Copenhagen do not lead to a comprehensive international agreement, several criteria permit an EU ETS-covered industrial sector to allege SRCL:

- if the industry can demonstrate that purchasing permits increases its costs (more than 5% of gross value added) and faces international competition (non-EU trade intensity above 10%), or
- if the industry can demonstrate that purchasing permits significantly increases its costs (more than 30% of gross value added), or
- if the industry faces international competition (non-EU trade intensity above 3 0%), then it can qualify for the free allocation of allowances.

Free allocation of permits typically will not be at 100% of needs for SRCL facilities, however. Free allowances will be adjusted according to Community-wide ex-ante benchmarks so as to ensure incentives for GHG reduction. The benchmarks will be set at the average performance of the 10% most GHG emissions-efficient installations in a sector in 2007-2008. Only the most efficient businesses in a sector, therefore, have a chance to receive all of their allowances free. If a business emits more than this benchmark allocation, it will need to acquire allowances up to its actual emissions.

As of September 2009, EU analysis assessed the industries and productions potentially exposed to carbon leakage risks. Assuming that 100% of allowances were auctioned (which will not occur initially), the analysis concluded that 146 sectors (out of 258) and five additional product categories meet the EU's criteria for being exposed to SRCL.[15] Outside of these sectors, 13 subsectors and products may be exposed to risk: food processing industries; industrial gases; nonmetallic mineral products; glass fibers (filament glass fibers); and, colors and similar preparations for ceramics/glass etc.[16] The EU analysis estimates that the listed sectors now constitute about 75% of GHG emissions covered by the EU ETS.

An alternative approach to issues of competitiveness in trade sensitive sectors put forward by the European Commission is the integration of importers into the EU ETS. Under an integrated emission trading regime, foreign producers would purchase emission certificates for their imports according to the emissions produced. In a speech in London on January 21, 2008, the President of the European Commission, Jose Manuel Barroso, said: "I think we should also be ready to [ ... ] require importers to obtain allowances alongside European competitors, as long as such a system is compatible with WTO requirements...." Beyond these measures, French President Nicolas Sarkozy, with possible interest from German Chancellor Angela Merkel, has indicated interest in potentially charging carbon levies against imports from countries that do not meet stringent environmental standards. (See fact sheet on France.)

(See also the Appendix, comparing EU efficiency standards for motor vehicles with those of other countries.)

# FRANCE[17]

(Policies and statements if substantially different from the European Union)

## 1. Overall GHG Emission Target, If Any, and Timing

Under the Kyoto Protocol, France's share of the EU target is not to exceed the 1990 level during the period 2008-2012.

France has a stated long-term national GHG emissions target of 75% below the 1990 level by 2050. A law is planned to reduce energy consumption of existing buildings by 38% by 2020.

## 2. Principal Policy Instrument(s): (See "EU ETS").

October 2007, French President Nicolas Sarkozy called for a plan to institute a national "carbon tax" on global-warming pollutants. Policy considerations ranged from a freeze on the building of new highways and airports, to a vast plan to shift freight traffic from road to rail, to a commitment to slash pesticide use by half within 10 years by Europe's biggest farm producer. Tramway and TGV high-speed train networks are to be extended, and drivers encouraged to buy cleaner cars through bonuses and penalties. The carbon tax apparently is set to move to Parliamentary action in 2009 (see below).

## 3. Covered Gases and Sectors: (See "European Union").

## 4. Allocation of GHG Reductions to Various Sectors

About half of French industry's GHG emissions are covered by the EU ETS, including large emission sources in the power generation, iron, steel, glass, cement, pottery and brick sectors.

In September 2009, President Nicolas Sarkozy stated that the proposed carbon tax would begin in January 2010. Because Sarkozy's party holds a majority in its parliament, expectations are that the new carbon levy will be enacted into law. Initially set at 17 Euros (US$25)[18] per ton of emitted $CO_2$, the tax on the use of oil, natural gas and coal would nudge up the cost of a liter of gasoline by US$0.06 (US$0.23 a gallon). It would apply to households as well as enterprises, but not to the heavy industries and power companies in France that are covered by the EU's emissions trading scheme (see the EU ETS under "European Union). Revenues from the new tax would be returned to taxpayers through cuts in income tax and other taxes. France's Le Monde newspaper says the tax will cover 70% of the country's carbon emissions (e.g., from vehicles) and bring in about 4.3 billion Euros (US$6.4 billion) of revenue annually. Sweden, Denmark, Finland, Norway and Switzerland already impose similar taxes, although Sweden's is levied at a much higher emission fee (108 Euros/ton of CO2, or US$161/ton).

## 5. Any Regulations or Exemptions Specific to Trade-sensitive Sectors

French President Nicolas Sarkozy has promoted a European levy on carbon-intensive imports from countries outside the Kyoto Protocol. The United States could be subject to such proposed fees should it not adopt legally enforceable GHG controls domestically. The Economist has said, "That leads some to suspect that his ultimate objective is to create a pretext for protectionism."[19]

In addition, President Sarkozy, along with German Chancellor Angela Merkel, has called for the United Nations to support "appropriate adjustment measures" to be levied against countries that do not join or implement an international agreement being negotiated for agreement in Copenhagen in December 2009.[20]

**Motor Vehicles:** A law is planned to cut GHG emissions from transport by 20% by 2020; it would include a goal of 7% bio-fuels by 2010 and EU emissions limit for new cars— 130g/km— to be phased in from 2012.

# GERMANY[21]

(Policies and statements if substantially different from the European Commission)

## 1. Overall GHG Emission Target, If Any, and Timing

Under the Kyoto Protocol, Germany's share of the EU target is to reduce GHG emissions to 21% below 1990 levels during the period 2008-20 12. (Germany was able to take on such a deep target because of its reunification with East Germany, taking on East Germany's high emissions baseline and reducing emissions by closing and improving many inefficient installations.)

The German government approved a new package of climate change measures in June 2008 that are a legal transposition of the EU's Integrated Climate Change and Energy Programme.[22] The German measures aim at a $CO_2$ emission reduction of 40% by 2020 compared to 1990 levels. The legislative package focuses on the transport and construction sectors.

## 2. Principal Policy Instrument(s): (See "EU ETS")

The Integrated Climate Change and Energy Programme: In 2007, the German government, working from the general guidelines of European policy decisions, implemented a concrete program of measures at the national level. Through 29 measures, the program addresses a wide range of matters, including combined heat and power generation, the expansion of renewable energies in the power sector, carbon capture and sequestration (CCS) technologies, smart metering, clean power station technologies, the introduction of modern energy management systems, support programs for climate protection and energy efficiency (apart from buildings), energy efficient products, provisions on the feed-in of biogas to natural gas grids, an energy savings ordinance, a modernization program to reduce $CO_2$ emissions from buildings, energy efficient modernization of social infrastructure, the Renewable Energies Heat Act program for the energy efficient modernization of federal buildings, a carbon dioxide strategy for passenger cars, the expansion of the bio-fuels market, reform of vehicle tax on the basis of carbon dioxide, energy labeling of passenger cars, the reduction of emissions of fluorinated greenhouse gases, procurement of energy efficient products and services, energy research and innovation, increased electric mobility, international projects on climate protection and energy efficiency, reporting on energy and climate policy by German embassies and consulates, and a transatlantic climate and technology initiative. In June 2008, the program was enacted with a package of measures to double electricity generated by combined heat and power technology (CHP) to 25%. The share of renewable electricity will also be increased to 20%, especially through subsidizing off-shore wind farm development. At the same time the package has set a target of producing half of Germany's electricity from renewable energy sources or super-efficient plants by 2020. The package aims for an 11% reduction in electricity consumption by 2020.

Loans for energy efficiency and $CO_2$ reduction measures in the domestic sector have been available as an economic recovery measure.

## 3. Covered Gases and Sectors: (See "European Union").

## 4. Allocation of GHG Reductions to Various Sectors: (See "European Union")

## 5. Any Regulations or Exemptions Specific to Trade-sensitive Sectors

Germany wants to give companies in globally-traded sectors bigger EU allowance quotas in the EU ETS to soften the cost impact of Europe's climate change policy.

Germany has been a vocal opponent of auctioning emissions allowances, although the EU has decided to move forward with limited auctioning. As examples of Germany's past stance, in January 2008, Environment Minister Sigmar Gabriel critiqued the European Commission's plan to commence auctioning emissions permits that are currently distributed for free, stating that "The European Union cannot ignore the question of how to preserve the international competitiveness of industries that consume lots of energy," such as cement, steel and chemicals, all key sectors of the Germany economy.[23] Sectors "which have reached their average for reductions of carbon dioxide emissions must be able to obtain free emission rights to be able to remain in Europe," claiming that many European industries could be forced to relocate elsewhere in order to maintain competitive prices in international markets. German

Economy Minister Michael Glos has also criticized the plan to auction emission rights.[24] Gabriel also condemned the weakness of the commission's project in terms of developing renewable energies, which he said threatened national support for such energies. Gabriel nonetheless reiterated German opposition to EU plans to reduce new car emissions to 120 grams of $CO_2$/km by 2012 without distinguishing by the class of vehicle (German car makers produce many powerful automobiles which emit high levels of $CO_2$).

# UNITED KINGDOM[25]

(Policies and statements if substantially different from the European Commission)

## 1. Overall GHG Emission Target, If Any, and Timing

Under the Kyoto Protocol, the United Kingdom's (UK) share of the EU target is to reduce GHG emissions to 12.5% below 1990 levels during the period 2008-2012.

Climate Change Act of 2008 introduced a legally binding long-term target to cut emissions by at least 80% by 2050 and at least 34% by 2020 compared to 1990 levels.[26] Major provisions of the act include the setting of legally binding targets, the establishment of a carbon budgeting system, and the creation of a Committee on Climate Change. The carbon budgeting system establishes caps on GHG emissions over five-year periods, with three budget periods being set at a time, charting progress to 2050. The act also requires that the government amend the act to include emissions from shipping and aviation by December 31, 2012. The act states that a reduction of power sector emissions by 40% should be achievable by 2020.

Goal to reduce $CO_2$ emissions from new houses to zero by 2016.

## 2. Principal Policy Instrument(s): (See "EU ETS")

The Carbon Budgeting System is outlined in the 2008 Climate Change Act. In it, the Secretary of State is authorized to set an amount for the net UK carbon account (the "carbon budget") for successive periods of five years each ("budgetary periods"), beginning with the period 2008- 2012.

The Carbon Reduction Commitment (CRC)[27] applies to non-energy intensive sectors not covered by the EU ETS. It will apply a mandatory emissions cap and trading program to cut carbon emissions from large commercial and public sector organizations (including supermarkets, hotel chains, government departments, large local authority buildings using more than 6,000 megawatt hours (MWh) of electricity through mandatory half hourly meters) by 1.1 million tons of carbon per year by 2020. Allowances in the CRC system would be sold by auction. The revenue raised from the sale of Carbon Reduction Commitment allowances are to be recycled back into the scheme through bonuses and penalties meant to stimulate organizations to reduce their levels of emissions. Any bonus or penalty administered to an organization are to be based on their ranked position on performance in three metrics (gross emissions, growth, and early compliance actions).[28]

The Carbon Emissions Reduction Target (CERT) came into effect on April 1, 2008, and will run until 2011 as an obligation on energy suppliers to achieve targets for promoting reductions in carbon emissions in the household sector. As reported by the Energy Savings Trust, an independent UK-based non-governmental organization, "it was originally estimated that CERT would stimulate approximately £2.8 billion (US$4.7 billion)[29] of investment by energy suppliers in carbon reduction measures. In September 2008, the Government announced that the level of funding available from the energy suppliers would be increased by £560 million" (US$893 million). The investment would increase the program's lifetime carbon savings to 185 million tons (Mt) $CO_2$ (31 Mt $CO_2$ more than under the original CERT target of 154 Mt $CO_2$).[30]

The Renewable Energy Strategy: The Department of Energy and Climate Change (DECC) details how the UK plans to hit its target of getting 15% of energy (electricity, heat and transport) from renewable sources by 2020. In order to achieve the target, 30% of electricity must come from renewable energy sources, including nuclear power (a five-fold increase from today's rate of ~5%), 12% of heat must be generated by renewables, and 10% of transport energy must be from renewables. The main instrument to achieve these targets for renewable (and nuclear) electricity generation are "Non-Fossil Fuel Obligations" (NFFO), begun in 1989, now Renewables Obligations," requiring operators of the distribution grid to purchase quotas of renewable and nuclear electricity. The prices are subsidized by a Climate Change Levy.[31]

The Climate Change Levy was established in the UK under the Finance Act 2000 (2000 c: 17): a tax on most fuels, including natural gas, electricity (including nuclear) and solid fuels, but not on vehicle or household users, nor renewable energy or cogeneration.[32] Revenues are used to help fund employment insurance, and to fund the Carbon Trust.[33] In addition, energy-intensive businesses qualify for a levy reduced by 80% if they signed voluntary Climate Change Agreements to improve energy efficiency or reduce GHG emissions. Although the Climate Change Levy initially was a fixed rate, the 2006 UK budget tied the rates to account for inflation beginning in 2007.

## 3. Covered Gases and Sectors: (See "European Union").

## 4. Allocation of GHG Reductions to Various Sectors

The EU ETS covers electricity generation and the main energy intensive industries—power stations, refineries, iron and steel, cement and lime, paper, food and drink, glass, ceramics, and engineering and vehicles. Overall, these account for around 50% of UK $CO_2$ emissions. Non- energy intensive, large-scale, commercial and public sectors are covered by the CRC policy (amounting to 25% of the business sector). Household emissions are covered by the CERT policy.[34]

## 5. Any Regulations or Exemptions Specific to Trade-sensitive Sectors

The UK's "Low Carbon Industrial Strategy" states a vision that the nation "must create the conditions for the UK to be—and be recognised as—the leading location in the world for growing an innovative low carbon business and developing new low carbon products and services."[35] The UK strategy appears oriented toward supporting identified opportunities in "green" businesses and technologies, aiding them through:

- a Low Carbon Investment Fund, (with financing of £405 million—US$674 million);
- a business-led Technology Strategy Board;
- an Energy Technologies Institute (ETI), serving as a private/public partnership to invest in development of low carbon energy technologies;
- R&D tax credits;
- a Carbon Trust to support development and deployment of new and emerging low carbon technologies; and
- a UK "innovation infrastructure," including intellectual property systems and procedures, standards, and a National Measurement System.

# AUSTRALIA[36]

## 1. Overall GHG Emission Target, If Any, and Timing

Under the Kyoto Protocol, Australia accepted a target to limit its net GHG emission increase to 8% above 1990 levels. It has also proposed that, under a new international agreement, it would take on a target to reduce its GHG emissions by 25% below 2000 levels by 2020 if "the world agrees to an ambitious global deal to stabilise levels of $CO_2$ equivalent in the atmosphere at 450 parts per million (ppm) or lower."[37]

## 2. Principal Policy Instrument(s)

The Australian government proposed a Carbon Pollution Reduction Scheme (CPRS) to be phased in beginning July 1, 2011. A one-year period would occur from 2011-12, during which carbon emission permits would be sold at a fixed price Aus$10 per ton of carbon (US$9.20);[38] these may not be banked for use in later periods. The full cap-and-trade system would be in effect by 2012, by which time all covered businesses must purchase carbon permits at market prices. The Senate did not pass this proposal on its first reading in August 2009; the proposal was reintroduced on October 22 for consideration in the week of November 16 in Australia's Senate, although Climate Change Minister Penny Wong has suggested that passage may be difficult in 2009.[39]

The Australian program also includes a Renewable Energy Target, and investment in carbon capture and storage. Up to 5 percentage points of its offered 25% target for 2020 could be met by purchase of international emission reduction credits using CPRS revenue, though no earlier than 2015. Eligible businesses also may receive government funding for energy efficiency investments, available from a Aus$200 million (US$184 million) portion of a Climate Change Action Fund.

While the Australian Senate did not pass the carbon reduction proposal in August, it passed a Renewable Energy Target (RET) into law[40] that establishes a system of tradable Renewable Energy Certificates (RECs). It requires that 20% of electricity come from renewable resources by 2020 (projected to require 45 gigawatt hours (GWh)). Currently, about 8% of Australia's electricity is generated with renewables. Among other provisions, the law provides Solar Credits, allowing receipt of a multiple of 2-5 of RECs for qualified installations, that will subsidize the capital costs of small-scale systems, such as household photovoltaic systems. The grants of RECs will depend on the generation of energy, not the installed capacity (which, in some countries, has not stimulated maximizing the use of installed capacity).

## 3. Covered Gases and Sectors

As proposed, the CPRS would initially cover the six GHG of the Kyoto Protocol, and emissions from stationary energy, transport, industrial processes, waste, forestry, and fugitive emissions from oil and gas production.[41] It is expected to cover 75% of Australia's GHG emissions and about 1000 entities (out of 7.6 million registered businesses in Australia).[42] Agriculture eventually may be included.

## 4. Allocation of GHG Reductions to Various Sectors

Permits would be available in 2011 at a fixed price of Aus$10 per ton of carbon-equivalent (US$8.60), after which all covered sources must purchase their permits through auction or the market.

## 5. Any Regulations or Exemptions Specific to Trade-sensitive Sectors

The proposed CPRS includes provisions to assist emissions-intensive, trade-exposed industries (EITE). Eligibility for assistance would be determined by an assessment of all entities conducting a specific activity. First, there would be quantitative and qualitative tests to assess the activity's trade exposure. Second, there would be assessments of greenhouse gas intensity based on the average emissions per million dollars of revenue or emissions per million dollars of value added. The baseline for the emission data would be 2006-07 to 2007-08, while the baseline for revenue/value added data would be 2004-05 to the first half of 2008-09.

The government allocates free permits using an allocation baseline of emissions per unit of output for each EITE activity. This baseline will provide the basis for eligibility at either the 90% or 60% assistance rates. The proposal[43] would set up two initial rates of assistance: (1) 90% allocation of allowances for activity with emissions intensity of at least 2,000 tons of emissions per million dollars revenue or 6,000 tons of emissions per million dollars of value added; (2) 60% allocation of allowances for activity with emissions intensity between 1,000 tons of emissions per million dollars revenue and 1,999 tons of emissions per million dollars revenue or between 3,000 tons and 5,999 tons of emissions per million dollars of value-added. This assistance per unit of production will be reduced by 1.3% annually.

The proposed CPRS would include a five-year Global Recession Buffer as part of an assistance package to EITE. Industries eligible for 60% assistance would receive a "buffer" of 10% free emission permits; industries eligible for 90% assistance would receive a 5% buffer of free emission permits.

Reviews of the EITE scheme would occur every five years, and would consider a list of identified issues, including whether the assisted firms are making progress toward world's best practice efficiencies, and whether "broadly comparable carbon constraints" are imposed in competing economies. Any changes to the system would require five years' advance notice.

The scope of consideration for assistance includes (1) direct emissions covered, (2) related cost increases for electricity and steam use, and (3) related cost increases for upstream emissions from natural gas and its components (e.g., methane and ethane) used as feedstock. The assistance package would include direct emissions and some indirect emissions.

Two amendment bills to the Renewable Energy (Electricity) Act 2000 were passed on August 20, 2009 and received Royal Assent on September 8, 2009. The Renewable Energy Amendments contain provisions to assist electricity-intensive industries and the coal industry. Under these provisions, one or more emissions-intensive trade-exposed activities may be partially exempted from its REC requirements. If resulting Partial Exemption Certificates are taken into account, it would reduce the charge for falling short of RECs that would otherwise be payable.[44] In this law, the definition of "emissions-intensive trade-exposed activity" would be either defined by further regulations, or by regulations under a Carbon Pollution Reduction Scheme Act 2009 if passed. The methods for calculating the amounts of partial exemptions would be defined by regulations.

# BRAZIL[45]

## 1. Overall GHG Emission Target, If Any, and Timing

In November 2009, Dilma Rousseff, chief of staff for Brazilian President Luiz Inácio Lula da Silva, was reported as saying that her country would take a proposal for voluntary GHG emissions reductions of 36-39% by 2020 to the Copenhagen summit.[46] Brazil's emissions would drop to near 1994 levels if the top end of the pledge is met, representing about a 20% cut from the 2.1 million tons emitted in 2005. The emission cuts would be based largely on reducing deforestation rates, and would depend in large part on obtaining "sufficient" financing. President Lula stated in December 2008 that Brazil would slow its rate of deforestation in the state of Amazonas by 70% by 2017, compared to the average rate from 1996 to 2005. In September 2009, the Brazilian government extended this target to an 80% reduction by 2020.[47] Brazil has set a target by 2010 for zero deforestation in its Atlantic Forest.

## 2. Principal Policy Instrument(s)

In December 2008, Brazilian President Luiz Inácio Lula da Silva signed the National Climate Change Plan (PNMC) into effect.[48] Policy measures include:

- Stimulating energy efficiency through best practice, including the implementation of an energy efficiency policy that targets a savings of 106 terawatt hours per year (TWh/y) by 2030; the substitution of renewable charcoal for coal in manufacturing sectors; the replacement of one million old refrigerators per year for 10 years; the deployment of solar power systems for water heating; and the phasing out of the use of fire for the clearing and cutting of sugarcane.
- Retaining a high renewable energy share in the electricity sector, including the increase of the total electricity supply from cogeneration, mainly from sugarcane bagasse, to 11.4% by 2030; the reduction of non-technical losses in electricity distribution at a rate of 1,000 GWh/y over the next 10 years; the addition of 34,460 MW capacity from new hydropower plants over the next 10 years; the increase in electrical supply share from wind and sugarcane bagasse by 7,000 MW by 2010; and the expansion of the national solar photovoltaic industry and its deployment in systems isolated from the grid.
- Increasing the share of bio-fuels in transport matrix, including the attempt to encourage industry to achieve an annual substitution rate of 11% bio-fuels for fossil sources over the next 10 years; and the institution of a 5% bio-fuel to diesel mandate by 2010;
- Reducing deforestation rates and eliminating forest losses, increasing policing against illegal logging and curtailing financing to illegal ranching.
- Continuing the policy measures of prior renewable energy regulations including the 2004 Program of Incentives for Alternative Electricity Sources (PROFINA), coordinated by the Ministry of Mining and Energy and Centrais Elétricas Brasileiras (Eletrobras). The program contains new strategies for the incorporation of renewable resources in Brazil's energy matrix and strengthens the country's policy on diversification and development. On its inception, PROFINA contracted 144 generation stations to benefit 19 states with a combined capacity of 3,300 MW from wind, biomass, and small hydro sources for a potential GHG reduction of 2.8 Mt $CO_2$/year.

Many of Brazil's mitigation strategies involve the reduction of deforestation rates in the Amazon. The current administration has expanded protected areas in the Amazon and implemented new environmental policies. More than 62 natural reserves have been established in the Amazon, bringing the total area of the Brazilian Amazon protected by law to 280,000 square kilometers, the fourth-largest percentage of protected area in relation to territory among all countries. In addition to the aforementioned National Climate Change Plan, Brazil has enacted other laws that address deforestation and sustainable development.

- The Public Forest Management Law encourages sustainable development, places a moratorium on soybean plantings and cattle ranching in the Amazon, and authorizes the creation of a plan to reduce the rate of Amazon deforestation by half. Brazil plans to meet this goal by increasing federal patrols of forested areas, replanting 21,000 square miles of forest, and financing sustainable development projects in areas where the local economy depends on logging.

- The Action Plan for the Prevention and Control of Amazon Deforestation intends to improve the monitoring of the deforestation process, from a regional to a local scale; promotes the presence of public authorities in critical zones; confronts the economic speculation problem involved in public lands; plans the appropriate distribution of public lands according to social and ecological needs; and retains commercial wood exploration while also promoting sustainable forest management.
- The Amazon Fund (a private fund) aims to combat deforestation and to promote sustainable development in the Amazon. In 2008, Norway pledged $1 billion to the fund through 2015, making it the first country to do so, stating that it would donate as much as $130 million in 2009.[49]

The Brazilian government maintains that these efforts have been successful. It has recently been reported that deforestation of the Amazon fell by the largest amount in more than 20 years, dropping 45%, from nearly 5,000 square miles to some 2,700 square miles, in 2008, although there normally is a great deal of year-to-year variability in deforestation rates.[50] A continued emphasis on enforcement coincides with legislation. The enactment of the Prevention of the Use of Illegal Timber in the Building Industry Act, starting January 2009, asks for proof of the legal origin of timber from building companies. As such, the government recovered 1.4 million cubic meters of illegal wood and 700 people were put in prison.[51]

Observers note however that other factors contribute to the rate of deforestation beyond governmental policy measures. Brazilian deforestation is strongly correlated to the economic health of the country. Recent reductions are concurrent with the global economic downturn. Falling commodity prices have stalled the expansion of ranching and agriculture into the Amazon. While these trends have seemed favorable for emission reductions, some commentators still point to what they consider continued deforestation practices by commercial and speculative interests, misguided government policies, inappropriate World Bank projects, and commercial exploitation of forest resources. Others see favorable taxation policies, combined with government subsidized agriculture and colonization programs, as a continued encouragement for the destruction of the Amazon. Still others emphasize the inherent difficulty in measuring, reporting and verifying any GHG emission reductions in the Land Use, Land Use Change and Forestry (LULUCF) sector. Finally, most stress the crucial commitment to local law enforcement policies to sustain any regulatory reform that comes out of the federal government.

## 3. Covered Gases and Sectors

Primarily $CO_2$ in deforestation and other domestic agendas; however, U.N. Clean Development Mechanism projects in Brazil include $CH_4$ and $N_2O$ reductions.

## 4. Allocation of GHG Reductions to Various Sectors

Unlike other developed or developing countries, Brazil holds a unique endowment of natural resources that affects its climate change portfolio in the power generation and transportation fuel sectors. A low contribution of greenhouse gas emissions has been due to both market-driven and governmental decisions to adopt renewable energy sources over the past few decades. The markets for both hydroelectricity and sugarcane products (bagasse for thermal purposes and ethanol for transportation fuel) have expanded 10-fold. During this period there was also an important decrease in wood consumption in the residential and industrial sectors and an increase in charcoal consumption in the industrial sector.

Taken together, however, the sectors of energy, industrial processes, solvents and waste treatment contribute only 25% of total GHG emissions, estimated at approximately 1 billion tons. The rest of Brazilian GHG emissions is tied to the LULUCF sector, and of that total, 90% corresponds to the conversion of forests to other uses, especially agriculture and ranching. For this reason, most of Brazil's mitigation policies have concentrated on the forestry sector.

## 5. Any Regulations or Exemptions Specific to Trade-sensitive Sectors

Not specified.

# CANADA[52]

## 1. Overall GHG Emission Target, If Any

In April 2007, then-Environment Minister John Baird announced that by 2020, Canada would reduce its GHG emissions by 150 million tons, or 20%, from its 2006 level. Beyond this, the government hopes to achieve a 60-70% reduction by 2050.[53] The Kyoto emission reduction targets are scored from 1990 (with a few explicit exceptions); some analysts assert that, since Canada's GHG emissions rose 27% between 1990 and 2004, the government would be able to demonstrate far greater progress if it were able to use 2006 as its base year in the Copenhagen Agreement.[54]

## 2. Principal Policy Instrument(s)

The government's most recent plan for regulating industrial air emissions was announced in March 2008.[55] However, observers note that it remains indefinite. Canada's current Environment Minister, Jim Prentice, is traveling around the country's 10 provinces soliciting ideas on a capand-trade system. There has reportedly been a great deal of pressure on the Minister to develop a plan that will be compatible with whatever may be developed in the United States. For example, the original 2007 Canadian plan called for an "intensity target" rather than a cap. Bilateral discussions over a compatible cap-and-trade system are underway.[56] The effort at cross-border harmonization is likely due to the extensive economic integration between the two countries.

The government aims to complete its policy formulation and present its formal plan before the December 2009 United Nations climate change Conference of the Parties in Copenhagen. Some observers note that the government's ambitions might be delayed or curtailed if a snap election is called; however the prospect of such a vote is believed to be increasingly unlikely.[57]

Recognizing that the transportation sector is responsible for about 27% of GHG emissions, the Canadian government is also set to issue mandatory auto emissions regulations—essentially converting fuel efficiency into CO2 limits—and likely will seek to make its standards compatible with those set by the U.S. Environmental Protection Agency. The Environment Ministry may also issue modified regulations regarding usage of ethanol. These changes would be facilitated by amendments to the Canadian Environmental Protection Act of 1999, which, among other things, can be used to regulate tailpipe emissions and ethanol blending. Regulations have yet to be published; the ministry likely will attempt to match and harmonize its emissions standards on a continental basis.

The federal government can also use its spending power to control pollution. The government has created a climate change "ecoTrust" fund from which the provinces may draw in order to pay for programs to reduce their own GHG emissions. The last two federal budgets have also included significant funding for carbon capture and storage, including a large-scale demonstration facility. This could be one important aspect of the attempt to reduce emissions arising from some provinces' extensive use of coal as an energy source; it also could be used for oil sands.

## 3. Covered Gases and Sectors

Although the details are still being negotiated, Canada's regulations will likely cover the six gases included in the Kyoto Protocol. In reducing GHG emissions in Canada, the government will likely also attempt to co-reduce other pollutants such as sulfur dioxide, particulate matter, and mercury. Specific sectors have yet to be determined.

## 4. Allocation of GHG Reductions to Various Sectors

The government has not yet determined the sectoral allocation of reductions, but it has calculated that 35% of Canada's GHG emissions arise from fossil fuel production, industrial processing and manufacturing; 22% from services, residential, waste and agriculture; 16% from electricity and heat generation; and 27% from transportation.[58]

## 5. Any Regulations or Exemptions Specific to Trade-sensitive Sectors

Canadian government officials maintain that exemptions—if any—and regulations are yet to come, and that Environment Minister Prentice is still attempting to strike agreements with the various provinces.

# CHINA[59]

## 1. Overall GHG Emission Target, If Any, and Timing

The Chinese government has not stated a national target for GHG reductions or carbon reductions. The 11th Five-Year Plan set compulsory energy and pollution targets for 2006-2010 that also slow growth of GHG emissions.

The central government has indicated that it will set carbon-intensity targets in its 12th Five-Year Plan, from 2011-2015, along with ambitious targets for energy intensity, inefficient plant closures, and non-fossil energy development.[60] On September 22, 2009, Chinese President Hu Jintao offered to other world premiers that China "will endeavor to cut carbon dioxide emissions per unit of gross domestic product (GDP) by a notable margin by 2020 from the 2005 level."[61] Although no quantity was revealed, rumors suggest that Chinese leadership may announce a quantitative target at the Copenhagen meeting of the UNFCCC in December 2009.

The Chinese climate change website suggests that Chinese leaders are "mulling" GHG goals of improvement of carbon intensity of 4-5% annually, which could lead to an 85-90% reduction of carbon intensity by 2050 compared to the 2005 rate.[62] (A percentage improvement expressed as carbon intensity would be easier to achieve than the same percentage target expressed as energy intensity, so this rate of annual improvement would be less than the annual energy intensity improvement target in the current five-year plan.)

## 2. Principal Policy Instrument(s)

Edicts specify national, provincial, and plant-specific targets or actions. For example, one national goal is to reduce energy consumption per unit of GDP by 20% from 2006-2010. Each province was given a corresponding target in June 2006, and many local governments were assigned energy conservation targets by the National Development and Reform Commission (NDRC) in July 2006. Some of the key instruments the central government is using to meet its targets for 2010 include:

- reducing or eliminating incentives for energy-intensive exports (e.g., export tax rebates);
- implementing a program of "Large Substitute for Small," closing half of small, inefficient electric power plants by 2010, and banning new small plants;
- removing some subsidies from inefficient or polluting plants;

- setting 2010 energy consumption targets within the Top- 1000 Enterprise Program for each large enterprise (in total representing 33% of national energy use in 2004);
- requiring closure of small and inefficient industrial plants, sometimes with compensatory payments;
- setting electricity dispatch rules to favor low-carbon generation, such as feed-in tariffs for renewably-produced electricity that can reach 25-50% higher than coal-based electricity prices;
- providing large subsidies to help finance some large capital investments in efficient or low-emitting technologies;
- allowing energy prices to rise to international price levels in many cases, and imposing (and reportedly beginning to collect) pollution fees;
- setting new vehicle efficiency standards at the Europe-IV level (tighter than U.S.), and making payments to turn in and destroy older, polluting vehicles (like "cash for clunkers");
- raising investments in inter-city and intra-city rail; and
- tightening building efficiency codes by many municipalities, although enforcement may be spotty.

High-level officials have indicated that the 12th Five-Year Plan will specify carbon-intensity targets, and that several national laws will be amended in the near-term to achieve GHG reductions. Carbon cap-and-trade "pilot" projects will be initiated in "some designated areas and industries."[63] President Hu has summarized additional targets that likely would help to restrain expected growth of GHG: a target to increase non-fossil fuel share of primary energy consumption to 15% by 2020, and to increase forest coverage by 40 million hectares and forest stock volume by 1.3 billion cubic meters by 2020 from 2005 levels. China also requires strict fuel efficiency standards for vehicles.

Some have argued that China's policies may be undermined by incomplete implementation, due to sometimes vague statement of requirements, lack of enforcement resources, poor data, conflicting priorities at the local level, and other factors. Though some argue that reporting and enforcement of the targets and regulations have been irregular, there are indications that the central government is working to improve such weaknesses, and to impose career penalties on officials who do not meet their targets.[64] Others are cautious about the central government's will and ability to gain full implementation of national policies at the provincial and local levels.

## 3. Covered Gases and Sectors

Policies are mostly focused on energy reforms not GHG control, though they also reduce $CO_2$ and methane emissions. Some projects under the Kyoto Protocol's Clean Development Mechanism address many industrial gases (such as hydrofluorocarbons) as well. Sectors addressed include energy, vehicle manufacturing, building, energy-intensive industries, forestry, etc. Agriculture seems engaged only through development of bio-fuels.

# 4. Allocation of GHG Reductions to Various Sectors

Many sectors are covered through various programs. Targets and actions are set by enterprise, not industry-wide.

# 5. Any Regulations or Exemptions Specific to Trade-sensitive Sectors

Many Chinese industry-specific policies seem aimed at eliminating the most energy-intensive and inefficient facilities within a sector. Many of China's exporting firms perform close to or at international energy-intensities. In 2007, China removed or reduced export tax rebates for many types of export products, including for energy-intensive, trade-sensitive industries. These adjustments generally have the effect of reducing incentives to export. Examples of additional programs are provided below.

**Iron and Steel:** The Chinese government has been emphasizing restructuring and improving the overall production efficiency of the iron and steel industry, much of which is likely also to reduce direct and indirect emissions. Closures are mandated in 2006-2010 of 100 million tons of iron production capacity and 55 million tons of steel capacity using inefficient and old technologies.[65] From 2006-2008, 61 million tons of iron and 43 million tons of steel capacity were closed, according to government statistics.[66] Mergers and acquisitions are being encouraged to increase concentration and efficiency in the industry. The adjustment and revitalization plan also envisions shifting the product composition of the sector's production, as well as shifting to integrated capacity.

**Aluminum:** Chinese requirements for energy savings and emissions reductions in its aluminum industry have been estimated to achieve its target of reducing GHG from the industry by 25% by the end of 2010.[67] The central government mandated closures of inefficient aluminum smelting capacity in 2006-2010. China's Ministry of Finance announced it would levy a 15% export tariff on non-alloy aluminum rods and poles, and eliminate the 5% import duty on electrolytic aluminum and many other energy-intensive commodities, in order to "further restrict exports of high energy-consuming and polluting resources products and encourage imports of raw materials," as well as to suppress China's trade surplus.[68]

The Chinese government has removed preferential electricity rates for metal producers, so manufacturers now pay market prices. The (U.S.-based) Aluminum Association also notes, "Additionally, China has invested in alternative energy systems that will begin paying off in 2009, namely solar and hydroelectric power, which will reduce the cost of energy."[69] This is likely also to reduce associated GHG emissions.

**Cement:** China set a target to reduce energy intensity in its cement industry by 20% in the 11th Five-Year Plan (2006-2010), using plant closures and installing state-of-the-art technologies. China's cement production is about 50% of the global total. The central government mandated closures of inefficient cement production capacity in 2006-2010, with closures of about 140 million tons of production capacity achieved from 2006-2008.[70] One program is set to "design an economically-viable, environmentally-friendly alternative fuel and raw materials co-processing program, which will include conducting demonstrations in six Chinese plants, and developing, documenting, and disseminating technical guidelines for co-processing.... [T]ools, training materials, and results from the project will be disseminated to further enhance the capacity building of the entire Chinese cement industry. An integrated national database on energy efficiency and emissions for Chinese cement industry, using worldwide recognized methodologies and tools, will also be established."[71]

**Motor Vehicles:** New vehicle efficiency standards have been set at the Europe-IV level (stricter than US standards). National policy and investment promotes rail rather than road transport.

China has enacted its version of the "Cash for Clunkers" program: from Aug 1, 2009 to June 30, 2010, consumers may receive 3,000-6,000 Yuan (US$440-875)[72] per vehicle to replace "yellow tag" passenger cars, vans, and trucks that exceed emission standards, or are 8-12 years old. Previous changes in vehicle taxes, with higher rates for large cars and lower rates for small ones, resulted in increased small car sales in 2008.

The total trade-in subsidy, mainly targeting light commercial vehicles, is likely to cost the government around 5 billion Yuan.

# INDIA[73]

## 1. Overall GHG Emission Target, If Any, and Timing

The Minister of State for Environment and Forests Jairam Ramesh said in September 2009 that India might domestically set "broad indicative targets" for the 2020-2030 period for domestic policies, but that India would not take on legally binding targets internationally.[74] In a United Nations summit meeting on climate change, he re-announced an intent to domestically legislate voluntary targets for vehicle fuel efficiency in 2011, building codes in 2012, and carbon capture and storage by 2020. The government also pledged that 20% of India's energy would come from renewable resources by 2020, and 15% of India's annual GHG emissions would be taken up by forests by 2030[75] (up from about 11% in 2005[76]). The Indian government has pledged that its emissions per capita would always remain below those of the now-industrialized countries (though expected population increases are substantial). Decisions may be made by the Indian Parliament.

## 2. Principal Policy Instrument(s)

To date, India's national government relies almost exclusively on public information, training of energy auditors, voluntary "declarations" of energy management policies by businesses, and small financial awards as its principal instruments to promote energy efficiency. In concept, Ramesh has said that India might enact a law directing the government to set climate-related, but non-mandatory, targets, with reporting to and review by the Parliament. He has indicated that the new law may be similar to the Fiscal Responsibility and Budget Management law (FRBM), which directs the government to develop targets, and requires reporting to the Parliament, as well as Parliamentary approval. The targets in the FRBM are neither specified nor binding.

Prime Minister Manmohan Singh approved in August 2009 a national energy efficiency plan that would require 714 energy-intensive industrial facilities in nine sectors, accounting for 40% of India's fossil fuel use, to meet energy efficiency targets. The energy efficiency plan is estimated by 2015 to avoid about 5% of India's projected fossil fuel use. The Prime Minister's Office may be contemplating setting up a new National Climate Change Mitigation Authority under the Prime Minister's authority.

Reportedly, the government has initiated greenhouse gas abatement plans in the past several months, including reforestation. An existing voluntary set of efficiency standards is expected to become mandatory by 2010. Stronger standards may be set for energy efficiency for certain appliances and government buildings; an Energy Conservation Building Code (ECBC) for all new government buildings; and monitoring of afforestation. India's Prime Minister Singh announced in late August the intention of introducing an energy efficiency trading system to reduce India's energy consumption by 5% and its $CO_2$ emissions by 100 million tons annually from projected levels by 2015 (about 8% of current emissions).[77] Two funds would be created with about $60 million of funding to provide partial loan guarantees and venture capital. Proposed targets may be set by December 2010.

In 2008, the Prime Minister released a National Action Plan on Climate Change, containing eight "national missions": the National Solar Mission; National Mission for Enhanced Energy Efficiency; Nation Mission on Sustainable Habitat; National Water Mission; National Mission for Sustaining the Himalayan Ecosystem; National Mission for a Green India; National Mission for Sustainable Agriculture; and National Mission on Strategic Knowledge for Climate Change.[78] The most concrete measures aimed at increasing solar energy capacity. In November 2009, the Indian Union Cabinet approved a Jawaharlal Nehru National Solar Mission (NSM) to increase India's solar electric capacity from 5 megawatts (MW) to 20 gigawatts (GW) by 2022 (slipping back two years from the initial target date), at a cost of $19 billion.[79] Some $900 million has been approved for the initial phase, to install 1.1 GW of on-grid and 0.2 GW of off-grid solar capacity by 2012. The NSM will offer financial incentives to investors, including tax breaks, and will boost research. Several existing laws support renewable energy development, according to a report from the Pew Center.

> The Electricity Act (2003) encourages the development of renewable energy by mandating that State Electricity Regulatory Commissions (SERCs) allow connectivity and sale of electricity to any interested person and permit off-grid systems for rural areas. The National Tariff Policy (2006) stipulates that SERCs must purchase a minimum percentage of power from renewable sources, with the specific shares to be determined by each SERC individually. The states of Himachal Pradesh and Tamil Nadu have the highest quotas—20% by 2010 and 10% by 2009, respectively. Under the Rural Electrification Policy (2006) electrification of all villages must be completed by 2012.[80]

India established a program to replace 400 million incandescent light bulbs with efficient compact fluorescents by 2012.

A fund supports the regeneration and sustainable management of forests. The initial capitalization of the fund was proposed to be $2.5 billion, with an annual budget of about $1 billion.[81]

Although India has some pollution control standards in place, enforcement of standards has been low.[82] The current government is planning to establish a new National Environmental Authority,[83] apparently to be modeled after the U.S. EPA.

## 3. Covered Gases and Sectors

Most identified and proposed measures address $CO_2$. The proposed system of "tradable energy efficiency certificates" would apply to 714 energy intensive facilities in the following sectors: fossil fuel-fired electricity generation; fertilizer production; cement; iron and steel; chlor-alkali production; aluminum; rail transport; and textiles.

## 4. Allocation of GHG Reductions to Various Sectors

The Bureau of Energy Efficiency would assign energy efficiency improvement targets to the most energy-intensive industrial plants, based on bench-mark performance "bands." Facilities in the most efficient "band" would have a less stringent improvement target, while those in less efficient "bands" would be required to make greater improvements. Facilities that perform better than the targets would receive energy savings certificates ("ESCerts") that could be sold to companies for compliance with their targets or, potentially, banked to meet future requirements. Facilities that fail to meet targets could be fined.

## 5. Any Regulations or Exemptions Specific to Trade-sensitive Sectors

Reportedly, Indian officials have suggested taxing imports based on the per capita carbon emissions of the exporting country.[84] This could have a large impact on the United States, as its per capita emissions are higher than most countries. (Besides foods and fossil fuels, the United States exports to India a wide variety of products, among which the largest in value are: civilian aircraft and parts, steel and other metal products, synthetic fertilizers, chemicals, electronics and industrial equipment, electronics, and gem diamonds.)[85]

**Motor Vehicles:** In India, high taxes are levied on motor fuels: 52% on gasoline and 32% on diesel in 2007. The Prime Minister's office has directed the Bureau of Energy Efficiency to set fuel efficiency labeling standards for vehicles under the Energy Conservation Act, to become effective by 2011. However, after several years' delay, these standards have not been set. As planned, the standards would require labeling only by 2011, with mandatory performance to be effective later. The Bureau of Energy Efficiency would certify the manufacturers' labels. Reportedly, some representatives of the automobile sector have demanded that the standards be set on the basis of $CO_2$ emissions and legally be put on India's list of "local pollutants."[86]

# JAPAN[87]

## 1. Overall GHG Emission Target, If Any, and Timing

Under the Kyoto Protocol, Japan agreed to reduce its GHG emissions to 6% below 1990 levels in the period 2008-20 12.

In mid-2008, then-Prime Minister Fukuda offered to reduce Japan's GHG by 80% from 2008 levels by 2050, and by 8% below 1990 levels by 2020 (without using international credits). Newly elected Prime Minister Yukio Hatoyama pledged Japan to a GHG target of 25% below 1990 levels by 2020, conditional on all major countries' participation in a new international accord. (The outgoing government's proposed target was equivalent to 8% below 1990 levels. In 2008, Japan's GHG emissions were almost 16% above its Kyoto Protocol target.)

## 2. Principal Policy Instrument(s)

The Japanese Government formulated in 2005 the Kyoto Protocol Target Achievement Plan (KPTAP) to promote measures to cope with global warming. The KPTAP lays out estimated emissions and expected reductions by sector, and for several specific programs, in order for Japan to meet its Kyoto Protocol target. The 2008 review and revision of the plan called for further actions to close the gap between expected emissions and the Kyoto target, including more stringent efficiency standards for equipment, vehicles, and small businesses. The government plan concluded that it would be very difficult to constrain emission reductions associated with the residential and commercial sectors, and therefore relied on expanding the Voluntary Action Plans in the business sector to achieve 80% of the envisaged further GHG reductions.[88] (See section on covered gases and sectors, below.)

Since October 2008, Japan has established an integrated domestic GHG emissions market, comprised of four components: (1) Japan's Voluntary Emission Trading System (J-VETS) capand-trade system, initiated in 2005 for voluntary trading of CO2 emissions from energy and process emissions covering only industries that do NOT have in place a Voluntary Action Program; (2) an Experimental Japanese Emissions Trading System, with emissions targets based on industry-specific Voluntary Action Programs; (3) Domestic Credit Scheme, to allow GHG reduction credits (i.e., "offsets") from small and medium-sized companies; and (4) Kyoto Credits, available through any of the three Kyoto Protocol emissions trading mechanisms.

The new Hatoyama government has indicated it plans to create a mandatory GHG cap-and-trade system, require "feed-in" tariffs as financial incentives for renewable energy generation, and may consider a carbon tax.[89] The Hatoyama campaign, on the other hand, pledged before the election to eliminate highway tolls and a fuel tax of about 25 yen (US$0.28)[90] per liter on gasoline by April 2010, which could raise vehicle GHG emissions by as much as 20%.[91]

The Law Concerning the Promotion of Measures to Cope with Global Warming[92] enacted in 1998, directed the national government to promote GHG emission reductions and to enhance carbon sinks. It also directed local governments and business to take actions to limit emissions. This basic authority also directs the central government to publish Japan's GHG emissions.

The 5,000 largest businesses in Japan have been required to report their energy production and consumption for more than a decade by the Law Concerning the Rational Use of Energy.[93] Consequently, the foundation for calculating the energy-related $CO_2$ emissions from each industrial source is established.

The Act on Promotion of Global Warming Countermeasures and Act on Rational Use of Energy establish authorities to promote energy efficiency in "energy-using" equipment, buildings, factories, and machinery. These and related legislation require efficiency labeling, and allow for low-interest financing, industrial improvement bonds, tax exemptions and other financial incentives to promote efficiency. They also require efficiency measures by industrial facilities and for appliances. The Energy Conservation Center of Japan (ECCJ) is a public-private partnership for research and implementation of energy conservation programs (including Japan's Energy Star program, modeled after the US EPA's), accreditation of energy managers, and information.

## 3. Covered Gases and Sectors

Under Japan's Kyoto Protocol Target Achievement Plan, industry is expected to reduce its GHG emissions to 7% below 1990 levels during the Kyoto first commitment period (2008-2012). The Keidanren Voluntary Action Plan[94] on the Environment (VAP) covers 35 industries, include energy, mining, construction, and at least some manufacturing sectors (e.g. production of vehicles, electronics, steel, cement, etc.)

## 4. Allocation of GHG Reductions to Various Sectors

The Keidanren VAPs include a non-binding target of reducing $CO_2$ emissions in industry and energy-converting sectors "below" their 1990 levels by 2010. In the Keidanren VAPs, different industries' metrics of performance and targets differ. In 2007, about 18 industries tightened their voluntary targets, although some observers have criticized even the more stringent targets as being no more than what was already being accomplished. Others argue that the voluntary targets are costly compared to reductions expected in other countries, such as within the European Union.

## 5. Any Regulations or Exemptions Specific to Trade-sensitive Sectors: (See Figure 1 below)

**Motor Vehicles:** The Japanese government provides tax benefits for "eco-friendly" vehicles and exemptions from taxes for three years for "next-generation" vehicles.[95] Beginning in April 2009, subsidies have been offered to purchasers of eco-friendly vehicles (e.g., for cars: 100,000 yen, or US$1100). These include a "cash-for-clunkers"-type program that offers higher subsidies to owners who scrap vehicles 13 years or older and replace them with eco-friendly vehicles (e.g., for cars: 250,000 yen, or US$2700). The subsidies extend as well to minivans, trucks and buses. One industry official reported that, with the subsidies, "eco-friendly" vehicles accounted for almost half of vehicle sales in Japan. [96]

Japan is reputed to have among the most stringent fuel economy standards for vehicles in the world, at 46.9 miles per gallon by 2015 (see Appendix). These are expected to constrain new passenger vehicle emissions of GHG.

**Keidanren Voluntary Action Plan on the Environment· (Target and Measures of Major Organizations)**

| Name of Organization | Target | Measures to Attain Goals (2008) |
|---|---|---|
| The Federation of Electric Power Companies of Japan (FEPC) | • In the period from FY2008 to FY2012, aim to reduce $CO_2$ emissions intensity (Emission per unit of user end electricity) by an average of approx. 20% or to approx. 0.34kg- $CO_2$ /kWh compared to the FY1990 level. | 1. Promotion of nuclear power generation based on security and confidence-building<br>2. Further improvement of efficiency in thermal power generation and discussion on the management and control of thermal power source<br>3. Diffusion and expansion of renewable energy<br>4. Research and development of technology contributing to energy conservation, $CO_2$ recovery and storage technology. |
| The Japan Iron and Steel Federation (JISF) | • Reduce energy consumption in the production process in FY2010 by 10 % compared to the FY1990 level on the assumption of the crude steel production 100 million-ton level.<br>• As an additional measure, use a million tons of plastic waste in blast furnaces, etc. on the assumption of the establishment of appropriate collection systems and others.<br><br>*Take an average of five years from FY2008 to FY2012 to achieve the above target. | 1. Recovery of waste energy (increase of recovery of by-product gas, steam, CDQ steam etc.)<br>2. Efficiency improvement of facilities (new installation and remodeling of in-house power generators, installation of regenerative burners, introduction of high efficiency oxygen compressor)<br>3. Reduction of process, making the process continuous (introduction of direct rolling etc.)<br>4. Improvement of operation (power saving, compressed air saving, steam saving, fuel saving activities, reduction of ratio of reducing agent)<br>5. Investment on energy conservation such as efficiency improvement at the renewal of facilities |
| Japan Chemical Industry Association (JCIA) | • Aim to reduce energy intensity to 90% of the FY1990 level by FY2010.<br>• Develop the chemical industry's own technologies such as catalytic technology, biotechnology and process technology in harmony with the environment.<br>• Contribute to $CO_2$ emission reduction measures in developing countries as well as transferring energy conservation technology and environmental protection technology which have been developed in the chemical industry. | 1. Efficiency improvement of facilities and equipment (installation of high efficiency facilities, replacement of equipment and materials etc.)<br>2. Rationalization of process (process rationalization, process conversion etc.)<br>3. Recovery of waste energy (recovery of waste heat and cool energy, turning waste fluid/ waste oil/waste gas into fuel etc.)<br>4. Improvement of operation methods (condition change of pressure, temperature, flow etc.)<br>5. Fuel switch and others (fuel switch, product modification etc.) |
| Petroleum Association of Japan (PAJ) | • in the period from FY2008 to FY2012, reduce energy intensity in refineries by an average of 13% compared to the FY1990 level. | 1. Revision of operation management (Improvement of control technology and optimization technology)<br>2. Expansion of mutual utilization of waste heat among facilities (among refining facilities etc.)<br>3. Additional construction of recovering facilities of waste heat and waste energy (waste heat boiler, recovery equipment of furnace exhaust gas heat, etc.)<br>4. Efficiency improvement by appropriate maintenance of facilities<br>5. Adoption of efficient equipment and catalyst<br>6. Participation in "Industrial Complex Renaissance" (sharing of heat energy among neighboring factories in industrial complex) |

# An Overview of Greenhouse Gas (GHG) Control Policies in Various Countries 59

| Name of Organization | Target | Measures to Attain Goals |
|---|---|---|
| Japan Paper Association | • In the period from FY2008 to FY2012, aim to reduce fossil energy intensity per product by an average of 20% and $CO_2$ emissions intensity derived from fossil energy by an average of 16% compared to the FY1990 level.<br>• Strive to promote forestation in Japan and overseas to expand owned or managed forested areas to 0.7 million ha by FY2012. | 1. Introduction of energy conservation equipment (heat recovery equipment, introduction of inverters etc.)<br>2. Introduction of high efficiency facilities (high-temperature high-pressure recovery boilers, high-efficiency cleaning equipment, low-differential pressure cleaner, etc.)<br>3. Revision of manufacturing process (shortening and integration of processes)<br>4. Fuel switch (switch to biomass energy, waste energy) |
| Japan Cement Association (JCA) | • Reduce energy intensity of cement production (Thermal energy for cement production + Thermal energy for private power generation + Purchased electrical energy) in FY2010 by 3.8% compared to the FY1990 level.<br><br>* Take an average of five years from FY2008 to FY2012 to achieve the above target. | 1. Facilities to utilize waste as alternative heat energy source (waste wood, waste plastic etc.)<br>2. Efficiency improvement of facilities (fans, coolers, finishing mills etc.)<br>3. New installation and remodeling of energy conservation equipment (high-efficiency clinker coolers etc.)<br>4. Replacement of facilities (including repair of facilities) |
| Japan Automobile Manufacturers Association, Inc. (JAMA) | • In the period from FY2008 to FY2012, reduce the total $CO_2$ emissions from production plants of 14 member companies by an average of 12.5% compared to the FY1990 level. | 1. Energy supply side measures (introduction of energy conservation facilities, improvement of efficiency of boilers, introduction of cogeneration, introduction of high-efficiency compressors, introduction of wind power generation, high efficiency transformers)<br>2. Energy demand side measures (energy conservation in coating line, introduction of invertors for fans and pumps, energy conservation in lighting and air-conditioners)<br>3. Upgrading energy supply methods and technologies of operation and management (reduction of energy loss during no operation, reduction of air leak etc.)<br>4. Merger, abolition and integration of lines<br>5. Fuel switch |
| Japan Gas Association (JGA) | • In the period from FY2008 to FY2012, reduce $CO_2$ emissions intensity per $1m^3$ of gas in the process of city gas production and supply to an average of $12g\text{-}CO_2/m^3$ from $84g\text{-}CO_2/m^3$ in FY1990 and $CO_2$ emission to an average of 0.54 million ton-$CO_2$ from 1.33 million ton-$CO_2$ in FY1990. | 1. Promotion of switching materials (to make high calorie) to natural gas, etc.<br>2. Further Promotion of energy conservation measures (utilization of LNG cold energy, efficiency improvement of facilities, reduction of heat loss, review of the operation of LNG pumps, speed control of seawater pumps) |
| Scheduled Airlines Association of Japan | • Reduce $CO_2$ emission derived from aviation fuel by 12% per production unit (Available seat kilometers) by FY2010 compared to the FY1990 level. | 1. Replacement for new and fuel efficient airplanes and promotion<br>2. Optimization of routes and time by introducing new air traffic control support system (CNS/ATM), etc.<br>3. Reduction in weight of loaded equipment and goods<br>4. Expansion of engine washing with water |
| Japan Department Stores Association (JDSA) | • Reduce energy intensity in stores (Floor space • •Energy consumption per store hours) by 6% in the target years (from FY2008 to FY2012) compared to the FY1990 level. | 1. Promotion of introduction of ESCO projects<br>2. Setting of top runner standard (further promotion of energy efficiency by comparing with other stores)<br>3. Introduction of energy conservation equipment, rooftop gardening, utilization of natural energy |

Source) Prepared from "Results of the FY2007 Follow-up to the Keidanren Voluntary Action Plan on the Environment (Section on Global Warming Countermeasures, Version Itemized per Business Category), in March 2008" (Website by Keidanren (Japan Federation of Economic Organization))

Note: Table copied from: The Energy Conservation Center, Asia Energy Efficiency and Conservation Collaboration Center, 2008. Available at
http://www.asiaeec-col.eccj.or.jp/eng/e3104keidanren_plan.pdf.

Figure 1. Japanese Regulations or Exemptions Specific to Trade-Sensitive Sectors

# KOREA[97]

## 1. Overall GHG Emission Target, If Any, and Timing

On November 17, 2009, the South Korean cabinet approved a 4% GHG emission reduction target by 2020 as a basis for its current and future climate change efforts. The goal is measured from a 2005 baseline and is equivalent to a 30% reduction from "business-as-usual." The target is the most ambitious of three options recommended by the country's Presidential Committee on Green Growth, which had urged South Korea to voluntarily participate in climate change efforts under a midterm target of either an 8% increase, no change, or a 4% cut. President Lee Myung-bak said in a statement released by his office that the decision was made "to facilitate the country's paradigm shift to low-carbon green growth." He characterized the policy as a "voluntary, independent, and domestic target for unilateral reduction," driven by "environmental technology and renewable energy development."[98]

## 2. Principal Policy Instrument(s)

The November recommendation will empower a governmental committee to prepare industry- specific quotas and implement support measures. Near-term reductions will focus on buildings and transportation to give other industry sectors more time to adjust.

In addition to these recent measures, Korea's policies have involved dialogue with industrial organizations, voluntary plans by participating facilities to save energy and reduce $CO_2$ emissions, and some non-regulatory emissions trading. The government has provided financial incentives and technological assistance. Voluntary agreements cover plants that consume more than 2,000 tons of oil equivalent annually.[99] This process has resulted in some performance benchmarking for industries, collaborative research, and participation in the Kyoto Protocol's Clean Development Mechanism.

South Korea recently said it plans to invest about 2% of its GDP annually in environment-related and renewable energy industries over the next five years, for a total of US$84.5 billion. The government said it would try to boost South Korea's international market share of "green technology" products to 8% by expanding research and development spending and strengthening industries such as those that produce light-emitting diodes, solar batteries and hybrid cars.[100] To meet its pledge of a new, quantitative target, the government has indicated it may use GHGtrading and tax incentives. It has also indicated that financial incentives would increase use of hybrid cars, renewable and nuclear energy, light-emitting diode lighting, and smart grids.[101]

## 3. Covered Gases and Sectors

Sectors included in Korea's "Industrial Organization for UNFCCC Task Force Team" are steel, cement, electricity generation, paper, semi-conductor manufacturing, petrochemicals, oil refining, and automobile manufacturing.

## 4. Allocation of GHG Reductions to Various Sectors

Not yet determined.

## 5. Any Regulations or Exemptions Specific to Trade-sensitive Sectors

**Motor Vehicles:** The automobile manufacturing association reached voluntary agreement with the EU to meet $CO_2$ emission standards of 140grams/km by 2008.[102]

# MEXICO[103]

## 1. Overall GHG Emission Target, If Any, and Timing

Mexico voluntarily plans to cut national GHG emissions by 50 million tons per year beginning in 2012, constituting approximately 8% of Mexico's net GHG emissions in 2008. The government has established a non-binding goal to reduce GHG by 50% by 2050 (to 340 million tons of $CO_2$) below 2000 emissions. The pledge is contingent on availability of international technical and financial support and on successful negotiation of an international agreement consistent with stabilizing $CO_2$-equivalent concentrations at 450 parts per million. Mexico foresees converging by 2050 on global average emissions per capita at or below 2.8 tons of $CO_2$ annually.

## 2. Principal Policy Instrument(s)

In 2007, the Government of Mexico set out a Strategy on Climate Change (NSCC) that identified GHG mitigation opportunities, and vulnerability and adaptation policies. The ensuing Mexico Climate Change Program (MCCP) sets 85 specific goals for mitigating GHG in four emission categories and 12 subcategories. In December 2008, Mexican President Felipe Calderon announced his intention to cap Mexican greenhouse gas emissions and allow GHG trading, beginning with state-owned energy producers. Mexico envisions eventually being part of a domestically regulated but internationally integrated North American GHG trading system. [104]

Mexico mainly promotes energy efficiency (including greater co-generation of heat and power by industrial sources) and renewable energy production, along with prevention of further deforestation, as its mitigation priorities. Principal instruments include Law for the Better Use of Renewable Energy and the Financing of Energy Transition (2007 or 2008) provide a number of legal energy reforms, including provisions that lay the groundwork for private investment in renewable electricity generation. The Law for the Sustainable Use of Energy created a three-stage program to 2050. It, *inter alia*, promotes renewable energy and energy efficiency. It also requires energy efficiency in all federal, state and local governments.

## 3. Covered Gases and Sectors

Six Kyoto Protocol gases. The cap-and-trade system under development is likely to cover energy production (oil and gas, refining, electricity), metals, chemicals, textiles, and cement. Analysis is underway to include a cap-and-trade program for vehicle fuel efficiencies as well.

## 4. Allocation of GHG Reductions to Various Sectors

Not yet determined.

## 5. Any Regulations or Exemptions Specific to Trade-sensitive Sectors

**Motor Vehicles:** The stringency of Mexico's vehicle efficiency standards was increased in 2004 to a mix of U.S. and European standards for different classes of vehicles.

**Oil and Gas Production, Refining and Distribution:** PEMEX, Mexico's state-owned petroleum company, has operated an internal carbon cap-and-trade system since 1998.

# RUSSIAN FEDERATION[105]

## 1. Overall GHG Emission Target, If Any, and Timing

The Russian Federation (hereafter "Russia") projects that its greenhouse gas (GHG) emissions in the year 2010 will be 28% below the 1990 level, which is Russia's GHG emissions cap (its "Assigned Amount") under the Kyoto Protocol.[106] Though GDP in 2006 was 3% below the 1990 level, Russia's GHG emissions were 34% below the 1990 level (inclusive of carbon uptake by forests and other vegetation, net GHG emissions were 74% below the 1990 level). Some four- fifths of the GHG reductions came from the energy sector. Russia's GHG emissions are thus below its Kyoto Protocol obligation, creating a large surplus of emission allowances (Assigned Amount Units, or AAUs, in the terminology of the Protocol). Under the rules of the Kyoto Protocol, Russia may sell its surplus AAUs to other Parties with GHG obligations.

A Presidential Decree[107] on measures for increasing the energy and environmental efficiency of the Russian economy was issued in 2008, setting a target to decrease the energy intensity of the economy by at least 40% by 2020, compared to the 2007 level. The government has also set a target to increase the share of renewable energy (excluding large hydroelectric production) in electricity generation to 4.5% by 2020, and to use 95% of associated natural gas (produced with oil) by 2014-2016.

In the Copenhagen negotiations, President Dmitry Medvedev has offered a GHG target for Russia's emissions of 10%-15% below 1990 levels by 2020.[108] With policies and measures in place, the Russian government has projected that its GHG emissions in 2010, 2015, and 2020 will be reductions of 28%, 21%, and 13%, respectively, of its 1990 emissions level. Other experts project them to be as much as 25% below 1990 levels in 2020 with current policies and economic outlooks.[109]

Although Russian leaders agreed in the G8 summit meeting of July 2008 to consider an 80% reduction from 1990 levels of GHG emissions from developed countries by 2050, they agreed only to a 50% reduction target for Russia.

## 2. Principal Policy Instrument(s)

Many observers contend that climate change has not attracted the interest of high level leaders in Russia and that, consequently, "[t]he government hardly has any official climate strategy, and little progress is occurring."[110] These claims persist in spite of apparent changes in the Russian leadership's diplomatic approach to the issue (e.g., an announcement of a climate "doctrine" accepting that GHG emissions would pose risks and would require actions to reduce emissions).[111] Many suspect that Russia's support for climate change actions is associated with expanding its export market for natural gas in Europe and, to a much smaller degree, the value of potentially selling its surplus AAUs to EU and other countries with GHG reduction obligations.

As noted above, Russia's reduced GHG emissions is due primarily to economic collapse, leading to steep drops in energy demand and production, as well as other activities (e.g., agriculture, waste) that lead to GHG emissions. Replacing old, inefficient manufacturing and other infrastructure has led to relatively slower increases in GHG emissions than in economic activity.

The government's strategy for economic and social development has relied on reform and expansion of the energy sector, in part because 50% of the central government's revenue comes from the oil and natural gas sector.[112] The export value of oil and natural gas has driven a policy emphasizing extraction of these resources for trade. However, many observers have noted a concomitant, low level of investment in new capacity. The 2006 Russian Energy Strategy to 2020 sought to increase reliance on nuclear and coal-fired electricity for domestic use in order to increase oil and natural gas available for export.[113] Investments are being made to back out natural gas use, for example, by investing in efficient, combined cycle gas turbine technologies. These energy initiatives have mixed effects on GHG trajectories.

In 2005, the government adopted the Complex Action Plan for Implementation of the Kyoto Protocol in the Russian Federation for 2004-2008. It gave coordinating authority to the Interdepartmental Commission on Implementation of the Kyoto Protocol in the Russia Federation. It established some sectoral targets for improving energy efficiency, although some commentators allege that no actions would be needed to achieve them.[114] The UNFCCC in-depth review concluded that these targets had been only partially met.

The Mid-term Social-economic Development Programme of the Russian Federation for 2003–2005 provided for economic incentives to modernize equipment and technologies, improving energy efficiency and thereby reducing GHG emissions. To supplement these initiatives, a Presidential Decree was issued in 2008 on measures for increasing the energy and environmental efficiency of the economy of Russia. Other reported actions include:

- Gazprom, Russia's state-owned natural gas enterprise, established an energy conservation program for 2001–2010.
- Gazprom is implementing measures to reduce $CH_4$ and $CO_2$ emissions through 2012 (the annual reductions expected are a 10% reduction in $CH_4$ emissions and a 2.5% reduction in $CO_2$ emissions); other measures to increase the efficiency of gas transport and decrease losses by Gazprom (emission reductions of 3 Mt $CO_2$ in the period 200 1–2004 through reconstruction of pump stations).
- A federal program for housing for 2002–2010 targets housing retrofit and modernization and includes energy efficiency measures and introduction of small-scale renewable energy generation in the residential and services sectors.

On November 12, 2009, President Medvedev addressed the Federal Assembly and outlined his proposal for Russia to "undergo comprehensive modernization." In this speech Medvedev announced that "increasing energy efficiency and making the transition to a rational resource consumption model is another of our economy's [five] modernization priorities."[115] To this end, he highlighted a number of new program proposals to:

- produce and install individual energy meters for households;
- transition to energy-saving light bulbs;
- introduce energy service contracts and introduce payment for consumption of services (and considering family incomes);
- increase efficiency in the public sector; and
- capture and sell natural gas co-produced with oil, instead of flaring gas.

President Medvedev also promoted developing waste-to-energy systems; super-conductors for electricity production, transmission, and use; and nuclear generation, including nuclear fusion.

The in-depth review of Russia's Fourth National Communication under the United Nations Framework Convention on Climate Change (UNFCCC) found that Russia did not report on its specific domestic measures to abate GHG emissions or detail on how they would contribute to meeting Russia's GHG commitments.[116] The review recommended that the government provide greater transparency of how Russia's policies and measures may be modifying long-term trends in anthropogenic GHG emissions and removals. According to the UNFCCC in-depth review,

> In the period 1990–1998, GHG emissions decreased almost in parallel with the economic decline. In the period 1998–2006, GDP growth was accompanied by a relatively slower increase in the level of GHG emissions, which was 9.9 per cent higher in 2006 than in 1998.
>
> The differences between GDP and the GHG emission trends are mainly driven by: shifts in the structure of the economy (particularly of non-energy intensive industries); shifts in the primary energy supply (the share of oil and coal has decreased and the share of natural gas and nuclear energy has increased); a decline in activities in the agriculture and transport sectors; the decrease in population (by 3.9 per cent); and the increase in energy efficiency. These trends resulted in a 31.9 per cent decrease in the Party's carbon intensity per GDP unit in 2006 compared with that in 1990.

Russia has not reported estimates of how government funding or financial incentives may influence GHG emissions.

Russia's latest energy strategy, as updated in August 2009, focuses in 2013-2015 on recovery from the current economic crisis. In its second phase, from 2015 to 2022, Russia would emphasize introducing new technologies and more efficiency into its energy sector. An expansion of renewable energy, including large hydroelectric plants, wind, and solar generation, would occur only in the third phase of the new strategy, from 2022 to 2030, along with continued development of hydrocarbon resources.

## 3. Covered Gases and Sectors

Russia's target under the Kyoto Protocol includes the six Kyoto Protocol gases.

## 4. Allocation of GHG Reductions to Various Sectors

None specified.

## 5. Any Regulations or Exemptions specific to Trade-sensitive Sectors

**Motor Vehicles:** In 2005, limits on motor vehicle pollutant emissions were introduced, including indicators of GHG emissions. These standards were comparable to the EURO 2–EURO 5 emission standards. (See Figure A-2 in the Appendix.)

# UNITED STATES

## 1. Overall GHG Emission Target, If Any, and Timing

The United States has not set legally binding targets to reduce its greenhouse gas emissions, neither under domestic law nor international treaty. President Barack Obama stated a policy to reduce U.S. GHG emissions to 14% below 2005 levels by 2020 (to approximately 1990 levels), and the Congress has been working on legislation (e.g., S. 1733 and H.R. 2454) that may set a comparable emission cap. Some Obama Administration officials have suggested that the U.S. and EU GHG targets are comparable, in that both parties[117] would reduce emissions approximately 1.4% annually through 2020.[118] (However, the EU's target is enacted into law.)

On November 25, the White House announced that President Obama would attend the international negotiations on an agreement to address climate change beyond the year 2012, stating that "he will take with him an emissions reduction target to drive progress...."[119] The numerical target was not given.

Had the United States become a Party to the Kyoto Protocol, it would have an obligation to reduce GHG emissions by 7% below 1990 levels during the first commitment period of 2008- 2012. In 2007, U.S. GHG emissions were about 16% above 1990 levels.[120]

Of the 50 States, 23 have set state-wide GHG mitigation targets, of which six are caps (maxima). While some are enforceable, others are not.

## 2. Principal Policy Instrument(s)

Current federal climate change policies provide incentives, but few requirements, explicitly to reduce GHG emissions; many programs exist, however, that contribute to limiting GHG emissions through energy efficiency standards, and technical assistance and financial incentives for renewable energy or other low-emitting technologies. For example, a number of tax incentives are in place to encourage investment in renewable energy, more efficient vehicles, and efficiency improvements to buildings. The White House identifies more than $80 billion of funding for clean energy provided under the American Recovery and Reinvestment Act of 2009 (P.L. 111-5), including the "largest-ever investment in renewable energy."[121] Other incentives induce agricultural producers to enhance soil carbon. While temporary financial incentives have been associated with greater investments, some stakeholders have indicated that longer duration of the incentives and combining with other market correction measures are important to effectiveness.

A suite of federal[122] programs, including the Energy Star, Climate Leaders, and Climate Challenge branded initiatives, provides information, technical assistance, and nominal awards to businesses, universities, and other consumers to quantify and reduce their GHG emissions; such programs generally are intended to encourage emission reductions that are already economical but do not occur because of market inefficiencies.

Some GHG reductions are achieved by existing or contemplated regulations. A major regulatory effort governs the energy efficiency of vehicles. For example, Corporate Average Fuel Economy (CAFE) standards will tighten for Model Year 2011 cars and trucks to approximately 27.3 miles per gallon (mpg). Again, these regulations have been put in place for reasons other than abating climate change. However, the Department of Transportation and the Environmental Protection Agency (EPA) are coordinating to propose new, joint CAFE and GHG emission standards for Model Years 2012-2016. The proposal would reach an estimated combined average of 34.1 mpg by 2016 (Table 1); combined with EPA's compliance credits for improving air conditioners of vehicles, the improvement could reach the GHG equivalent of 35.5 mpg. The proposed rules contain flexibilities for manufacturers to comply with the new standards by earning credits by over-complying, or by producing alternative or dual-fueled vehicles. Holders of credits may use them for compliance of other model years or classes, or trade them to another manufacturer. The agencies project that the new standards would reduce GHG emissions by about 900 million metric tons,[123] and reap net cost savings over the lifetimes of vehicles.

## Table 1 . Average Required Fuel Economies under Proposed Standards
## (in miles per gallon for model year vehicles)

|  | 2012 | 2013 | 2014 | 2015 | 2016 |
|---|---|---|---|---|---|
| Passenger Cars | 33.6 | 34.4 | 35.2 | 36.4 | 38.0 |
| Light Trucks | 25.0 | 25.6 | 26.2 | 27.1 | 28.3 |
| Combined | 29.8 | 30.6 | 31.4 | 32.6 | 34.1 |

Source: National Highway Traffic Safety Administration, "NHTSA and EPA Propose New national Program to Improve Fuel Economy and Reduce Greenhouse Gas Emissions for Passenger Cars and Light Trucks" fact sheet available at
http://www.nhtsa.dot.gov/portal/site/nhtsa/menuitem.d0b5a45b55bfbe582f57529cdba046a0/.

The United States has set minimum standards of energy efficiency for a wide variety of residential and commercial equipment since the 1970s, with updates by several more recent laws.[124] Efforts are currently underway to address a backlog of regulations, such as for residential water heaters, dishwashers, clothes dryers, and for commercial motors and lamps, and a number of new, more stringent standards were issues in 2009. About two dozen additional standards are planned over the next few years. In some instances, states may have set appliance efficiency standards more stringent than federal standards (e.g. television standards in California).

Methane emissions from landfills are controlled along with other air pollutants under the Clean Air Act. According to EPA, the regulation requires installation of gas collection and control systems for new and existing landfills and, generally, routing the gas to an energy recovery system. The gas control system must reduce collected landfill gas (LFG) emissions by 98%.[125]

Large programs are devoted to developing new technologies that would be necessary to reduce GHG emissions below current levels. Many experts contend that voluntary efforts (such as the U.S. Climate Leaders Program), research on technologies, and existing regulatory and tax incentives cannot achieve the GHG reductions necessary to avoid "dangerous" climate change.

Of the $6.4 billion in U.S. federal funding in FY2008 for climate change activities, almost all was for scientific and technological research and development. In addition, tax incentives that could help to reduce GHG emissions were equivalent to about $1.5 billion in FY2008. As mentioned above, more than $80 billion in funding was available in FY2009. Funding for regulatory, voluntary, and public education programs was a few percent of the total. President Obama has also pledged, along with leaders of more than 20 other countries, to seek to phase out subsidies for fossil fuels, reducing associated GHG emission by an estimated 10% or more by 2050.[126]

The 110th Congress enacted two broad pieces of legislation—an omnibus energy bill (P.L. 110-140) and a comprehensive appropriations act (P.L. 110-161)—that include climate change provisions. Both statutes increase climate change research efforts, and the energy act requires improvement in vehicle fuel economies, as well as other provisions that would reduce (or sometimes increase) GHG emissions. P.L. 110-161 directs the EPA to develop regulations that establish a mandatory GHG reporting program that applies "above appropriate thresholds in all sectors of the economy."

In the absence of a federal regulatory framework to address U.S. GHG emission reductions, a majority of states have established formal GHG mitigation policies, including targets for future reductions. Sixteen states[127] are regulating $CO_2$ emissions from electric utilities: 11 using a sectoral cap-and-trade approach, and five using emission performance standards. In several regions, including the Northeast, the Midwest and the West, states are working together to create regional schemes to cap GHG emissions and allow trading of emissions permits across borders. All states but four now support "net metering" to allow producers of renewably-generated electricity to sell what they don't use into the electric grid. Twenty-six states have set renewable portfolio standards and another four have set alternative energy portfolio standards; these standards require that a specified share of the state's electricity must be generated by renewable or alternative energy sources by a given date. An additional five states encourage renewable or alternative energy sources with non-binding goals.

In the transportation sector, 15 states, led by California, are adopting GHG emission standards for motor vehicles, and three additional states are poised to follow. Thirty-eight states offer tax exemptions, credits, and/or grants to promote biofuels, of which 13 have set regulations requiring a specified share of motor fuels to come from biomass. To address growth of traffic, 18 states have set "smart growth" policies. Arizona, for example, has enacted laws and required improved coordination of state agency spending to help communities address a variety of growth pressures. Three of these states have also set targets to reduce vehicle miles traveled in the state. For example, the State of Washington set a goal in 2008 to reduce annual per capita vehicle miles traveled by 18% by 2020, 30% by 2035, and 50% by 2050, compared to 1990 levels.

Building codes typically fall under local authorities, although a growing number of states have set performance standards that help to limit GHG emissions. Most states have set efficiency standards for state, commercial, and residential buildings. Twelve have set appliance efficiency standards as well.

Over the past five years, a proliferation of litigation relating to climate change also presses the federal government toward actions to reduce GHG emissions. For example, the Supreme Court ruled in 2007 that the EPA must consider regulating CO2 and other GHG emitted from motor vehicles as pollutants under the Clean Air Act.[128] The Obama Administration has made clear that it would prefer Congress to enact GHG-specific legislation but that it will move to regulate in the absence of such new law. Further litigation has been pursued, challenging the Executive Branch to action, using the Endangered Species Act, the Energy Policy and Conservation Act and the Outer Continental Shelf Lands Act. A few international-law claims have been filed against the United States as well.[129]

## 3. Covered Gases and Sectors

Only methane emissions currently are regulated directly, although CO2 has been proposed to be regulated from motor vehicles (in a joint rule with fuel economy standards) and is reduced through other regulatory measures.

## 4. Allocation of GHG Reductions to Various Sectors

Because no economy-wide reduction strategy is in place, there is no allocation among sectors.

## 5. Any Regulations or Exemptions Specific to Trade-sensitive Sectors

Because no economy-wide reduction strategy is in place, there are no regulations or exemptions in place specific to trade-sensitive sectors. H.R. 2454, which passed the House on June 26, 2009, includes two strategies to address possible shifts of GHG emissions from the United States to less regulated companies in other countries : (1) free allocation of allowances (similar to that of the EU), and (2) an international reserve allowance (IRA) scheme. The scheme would require importers of energy-intensive products from countries with insufficient carbon policies to submit a prescribed amount of "international reserve allowances," or IRAs, for their products to gain entry into the United States. Based on the GHG emissions generated in the production process, IRAs would be submitted on a per-unit basis for each category of covered goods from a covered country. Specifically, H.R. 2454 Section 768 requires EPA to promulgate rules establishing an international reserve allowance system for covered goods from the eligible industrial sector, including allowance trading, banking, pricing, and submission requirements.

(See also the Appendix, comparing U.S. efficiency standards for motor vehicles with those of other countries.)

## APPENDIX. COMPARISON OF VEHICLE EFFICIENCY STANDARDS INTERNATIONALLY (AS OF MID-2009)

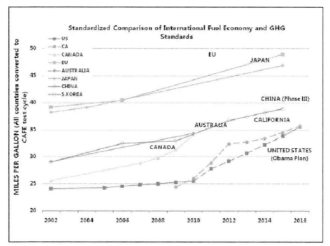

Source: Feng An, "Revised Chart for World Standards," Innovation Center for Energy and Transportation (iCET) (2009). Available at http://www.icet.org.cn.

Figure A-1. Comparison of International Fuel Economy and GHG Standards

| CAFE mpg | UNITED STATES | CALIFORNIA | CANADA | EUROPE | AUSTRALIA | JAPAN | CHINA |
|---|---|---|---|---|---|---|---|
| 2002 | 24.1 | | 25.5 | 39.2 | 29.1 | 38.2 | 29.0 |
| 2003 | | | | | | | |
| 2004 | | | | | | 39.1 | |
| 2005 | 24.3 | | | | | | |
| 2006 | 24.6 | | | 40.5 | | | 32.4 |
| 2007 | 24.9 | | 28.8 | | | | |
| 2008 | 25.0 | | 29.8 | | | | |
| 2009 | 25.3 | 24.4 | 31.3 | | | | 33.0 |
| 2010 | 25.5 | 26.0 | 34.1 | | 34.4 | | |
| 2011 | 27.8 | 28.9 | | | | | |
| 2012 | 29.2 | 32.4 | | 48.9 | | | 36.7 |
| 2013 | 30.7 | 32.7 | | | | | 37.4 |
| 2014 | 32.2 | 33.4 | | | | | 38.1 |
| 2015 | 33.8 | 34.5 | | | | 46.9 | 38.9 |
| 2016 | 35.5 | 35.7 | | | | | |

Note: all countries/regions normalized to US CAFÉ driving test cycle and converted to miles per gallon

Source: Feng An, "Revised Chart for World Standards," Innovation Center for Energy and Transportation (iCET) (2009). Available at http://www.icet.org.cn.

Figure A-2. Standardized Comparison of Select Vehicle Efficiency Standards Internationally (standards as of mid-2009)

# End Notes

[1] For a further discussion on trade-sensitivity issues, see CRS Report R40100, *"Carbon Leakage" and Trade: Issues and Approaches*, by Larry Parker and John Blodgett; and CRS Report R40914, *Climate Change: EU and Proposed U.S. Approaches to Carbon Leakage and WTO Implications*, by Larry Parker and Jeanne J. Grimmett.

[2] This section was prepared by Richard K. Lattanzio, Analyst in Environmental Policy (7-1754), with input from Larry Parker, Specialist in Energy and Environmental Policy (7-7238) and Jane A. Leggett, Specialist in Environmental and Energy Policy (7-9525).

[3] http://europa.eu/rapid/pressReleasesAction.do?reference=IP/09/1703&format=HTML&aged=0&language guiLanguage=en.

[4] See {COM(2008) 13 final}; {COM(2008) 16 final}; {COM(2008) 17 final}; {COM(2008) 18 final}; {COM(2008) 19 final} at http://ec.europa.eu/environment

[5] (2003/87/EC); see http://europa.eu/rapid/pressReleasesAction.do?reference=IP/09/628&format=HTML&aged=0&language=EN&guiLanguage=en. Also see CRS Report RL34 150, *Climate Change and the EU Emissions Trading Scheme (ETS): Kyoto and Beyond*, by Larry Parker.

[6] Each Member State is responsible for the implementation of Community law (adoption of implementing measures before a specified deadline, conformity and correct application) within its own legal system. Under the Treaties (Article 226 of the EC Treaty; Article 141 of the Euratom Treaty), the Commission of the European Communities is responsible for ensuring that Community law is correctly applied. Consequently, where a Member State fails to comply with Community law, the Commission has powers of its own (action for non-compliance) to try to bring the infringement to an end and, where necessary, may refer the case to the European Court of Justice. For additional information, see http://ec.europa.eu/community

[7] Directive 2009/28/EC of the European Parliament and of the Council of 23 April 2009.

[8] See, for example, Andres Cala, Europe Warming to Carbon Tax, Energy Tribune. "Spain and Ireland, which until recently were considered unlikely candidates to follow suit because of their high unemployment rates, are also weighing adding similar levies next year. Ireland's Finance Minister, Brian Lenihan, said recently that the government would not raise taxes to finance next year's budget, with the single exception of a carbon tax.... Spain's Prime Minister Jose Luis Rodriguez Zapatero, which has announced a fiscal reform to raise more

# An Overview of Greenhouse Gas (GHG) Control Policies in Various Countries 71

money to control a rampant deficit, called the carbon tax an "interesting" proposal and added carbon taxes will inevitably be applied by most countries." 23 Sept. 2009. http://www.energytribune.com/articles.cfm?aid=2354

[9] http://news.bna.com/deln/DELNWB/split_display.adp?fedfid=15354499&vname=dennotallissues&fn= 15354499&jd=a0c0y8h5r1&split=0;http://www.reuters.com/article/GCA-GreenBusiness/ idUSTRE59544A20091006.

[10] See CRS Report R40090, *Aviation and Climate Change*, by James E. McCarthy.

[11] http://www.consilium.europa.eu/uedocs/cms_data/docs/pressdata/en/misc/107136.pdf.

[12] Directive 2009/33/EC of the European Parliament and of the Council of 23 April 2009.

[13] http://www.consilium.europa.eu/uedocs/cms_data/docs/pressdata/en/misc/107136.pdf.

[14] If one or more countries requires carbon controls that add to production costs in businesses that compete internationally, it is possible for "carbon leakage" to occur if production in the controlled countries declines because purchasers instead buy increased supply from uncontrolled producers in other countries. Though emissions may decline from the controlled facilities, they may increase at uncontrolled facilities, thereby leading to "carbon leakage." This would offset the benefits of the emission controls.

[15] Of the 146 sectors, 117 have trade intensity > 30%; 27 have both estimated $CO_2$ costs >5% and trade intensity > 10%; and two sectors have $CO_2$ cost above 30% and trade intensity < 10%. Hans Bergman, "Sectors Deemed to be Exposed to a Significant Risk of Carbon Leakage—Outcome of the Assessment" presentation to Working Group 3 Meeting, 18 September 2009.

[16] http://ec.europa.eu/environment

[17] This section was prepared by Richard K. Lattanzio, Analyst in Environmental Policy (7-1754).

[18] Live market currency exchange rate for November 19, 2009 is listed as 1 Euro equivalent to 1.49 US$ (http://www.xe.com/). Currency rates are subject to fluctuation.

[19] Economist, Sept. 17 2009. http://www.economist.com/world/europe/displaystory.cfm?story_id=14460346.

[20] http://news.yahoo.com/s/afp/20090918/sc_afp/francegermanyclimateenvironmentuneu.

[21] This section was prepared by Richard K. Lattanzio, Analyst in Environmental Policy (7-1754).

[22] http://www.bmu.de/english/climate/doc/39945.php.

[23] See article at http://afp.google

[24] See article at http://www.eubusiness.com/news-eu/1200576720.98.

[25] This section was prepared by Richard K. Lattanzio, Analyst in Environmental Policy (7-1754).

[26] http://www.opsi.gov.uk/acts/acts2008/ukpga_20080027_en_2#pt1-pb2-l1g4

[27] For this and other policy descriptions, see the Department of Energy and Climate Change website: http://www.decc.gov.uk/en/content/cms/publications/lc_trans_plan/lc_trans_plan.aspx

[28] http://www.carbonreductioncommitment.info/carbon

[29] Live market currency exchange rate for November 19, 2009 is listed as 1UK£ equivalent to 1.67 US$ (http://www.xe.com/). Currency rates are subject to fluctuation.

[30] http://www.energysavingtrust.org.uk/Global-Data/Funding-Information/Carbon-Emissions-Reduction-Target-CERT.

[31] http://www.publications.parliament.uk/pa/cm200708/cmselect/cmenvaud/590/59003.htm.

[32] The Climate Change Levy revised and replaced a fossil fuel levy.

[33] Maria Pender, "UK Climate Change Programme: Business and Public Sector Economic Agreements."

[34] http://www.berr.gov.uk/energy

[35] DECC, *Investing in a Low Carbon Britain*, available at http://interactive.bis.gov.uk/lowcarbon/vision

[36] This section was prepared by Jane A. Leggett, Specialist in Environmental and Energy Policy (7-9525).

[37] http://www.environment

[38] Live market currency exchange rate for November 19, 2009 is listed as 1Aus$ equivalent to 0.92 US$ (http://www.xe.com/). Currency rates are subject to fluctuation.

[39] http://www.bloomberg.com/apps/news?pid=20601081&sid=aJXyEr9Kr_P4. As of November 18, 2009, the Australian Senate had just begun debate on the Labor Party's CPRS bill with the hope of passage before the year end break beginning November 25, 2009.

[40] Renewable Energy (Electricity) Amendment Act 2009, No. 78, 2009, C2009A00078; and Renewable Energy (Electricity) (Charge) Amendment Act 2009, No. 79, 2009, C2009A00079. http://www.comlaw.gov.au/ comlaw/Legislation/Act1.nsf/0/ 94CB90B9EED48B69CA25762D001B6F5F?OpenDocument.

[41] Australian Government, Carbon Pollution Reduction Scheme: Australia's Low Pollution Future: White Paper (December 2008).

[42] http://www.climatechange.gov.au/whitepaper/summary/index.html.

[43] http://www.environment

[44] Renewable Energy (Electricity) Amendment Act 2009, Schedule 2.

[45] This section was prepared by Richard K. Lattanzio, Analyst in Environmental Policy (7-1754).

[46] See http://www.reuters.com/article/marketsNews/idUSN1347815120091113.

[47] According to Brazil's National Institute of Space Research (INPE), Brazil's average rate of deforestation from 1996 to 2005 was 7,542 square miles annually, compared to averages of 6,574 annually from 1988 to 1995, and 4,974 from 2006 to 2008; http://www.mongabay.com/brazil.html. This target does not appear to include

forests, including open canopy forests, in other parts of Brazil, which may be cleared for agricultural production. Also, http://en.cop15.dk/ news/view+news?newsid=235 1, http://www.cmcc.it:8008/cmcc/blog

[48] http://www.mma.gov.br/estruturas/208/_arquivos/national_plan_208.pdf.

[49] Brazil received $100 million of the pledge on March 25, 2009. The remainder is pending. See http://inter.bndes.gov.br/english/news/not036_09.asp.

[50] See http://www.eenews.net/Greenwire/2009/11/13/4.

[51] http://www.redd-monitor.org/2009/01/23/brazils-national-plan-on-climate-change-and-the-amazon-fund-%E2%80%9Cthis-plan-does-not-create-any-carbon

[52] This section was prepared by Carl Ek, Specialist in International Relations (7-7286).

[53] Canada's New Government Announces Mandatory Industrial Targets to Tackle Climate Change and Reduce Air Pollution. News release. Environment Canada website. April 27, 2007. http://www.ec.gc.ca/default.asp?lang=En&n=714D9AAE-1&news=4F2292E9-3EFF-48D3-A7E4-CEFA05D70C21.

[54] No Clear Environmental Champion; Canada and the United States Have Shown Varied Levels of Aggressiveness in the Fight to Combat Climate Change. *Globe and Mail*. July 9, 2008. See also: Canada's Greenhouse Emissions Soaring Again: UN Report. *Canwest News Service*. April 21, 2009.

[55] Government Delivers Details of Greenhouse Gas Regulatory Framework. News release. Environment Canada website. March 10, 2008. http://www.ec.gc.ca/default.asp?lang=En&n=714D9AAE-1&news=B2B42466-B768-424C9A5B-6D59C2AE1C36.

[56] Notes for an address by the Honourable Jim Prentice, P.C., Q.C., M.P. Minister of the Environment on Canada's climate change plan. Speech. Environment Canada website. June 4, 2009. http://www.ec.gc.ca/default.asp?lang=En&n=6F2DE1CA-1&news=400A4566-DA85-4A0C-B9F4-BABE2DF555C7.

[57] CRS discussion with Canadian government official, September 10, 2009.

[58] Notes For an Address by the Honourable Jim Prentice, P.C., Q.C., M.P. Minister of the Environment on New Regulations To Limit Greenhouse Gas Emissions. Speech. Environment Canada website. April 1, 2009. http://www.ec.gc.ca/default.asp?lang=En&n=6F2DE1CA-1&news=D8C4903B-B406-4B70-8A4A-EDEF99B71D38.

[59] This section was prepared by Jane A. Leggett, Specialist in Environmental and Energy Policy (7-9525).

[60] Communications with CRS.

[61] http://www.ccchina.gov.cn/en/NewsInfo.asp?NewsId=19480.

[62] http://www.ccchina.gov.cn/en/NewsInfo.asp?NewsId=20325. This article also points to a study indicating that an 83% reduction of carbon intensity by 2050 would cost about 2.3% of GDP, while a 90% reduction of carbon intensity would cost about 7% of GDP. It is unclear whether this is a lost compared to the annual rate of GDP growth, or to cumulative GDP growth in 2050.

[63] Jing Li and Zhe Zhu, "Legislature Takes Urgent Action in Climate Change Fight," *China Daily*, August 28, 2009, http://www.chinadaily.com.cn/china/2009-08/28/content_8626140.htm.

[64] See, for example, http://news.xinhuanet.com/english/2007-06/20/content_6269732.htm; http://www.chinacsr.com/en/ 2009/06/18/5487-china-first-heavy-industries-fined-for-infringement-of-environmental-rules/; http://www.china.org.cn/ environment/2009-09/28/content_1 86191 89.htm; and http://www.china.org.cn /governmentcontent _1233 895 8.htm.

[65] http://www.reportbuyer.com/industry

[66] http://news.xinhuanet.com/english/2009-08/25/content_11942981_1.htm.

[67] Feng Gao et al., "Greenhouse gas emissions and reduction potential of primary aluminum production in China," *Science in China Series E: Technological Sciences* 52, no. 8 (2009): 2161-2166, doi:10. 1007/s1 1431-009-0165-6.

[68] http://experts.e-to-china.com/analysis/general_analysis/Taxation/2009/0728/58804.html.

[69] http://www.aluminum ContentDisplay.cfm.

[70] http://news.xinhuanet.com/english/2009-08/25/content_11942981_1.htm.

[71] http://china.lbl.gov/news/chinese-cement-companies-reduce-their-carbon

[72] Live market currency exchange rate for November 19, 2009 is listed as 1 CNY = 0.146 US$ (http://www.xe.com/). Currency rates are subject to fluctuation.

[73] This section was prepared by Jane A. Leggett, Specialist in Environmental and Energy Policy (7-9525).

[74] http://economictimes.indiatimes.com/articleshow/5007545.cms?prtpage=1; a published interview with Ramesh provides greater insights into the minister's thinking, at http://redgreenandblue.org/2009/07/06/no-funds-

[75] http://www.forbes.com/2009/09/23/jairam-ramesh-india-business-energy

[76] http://www.thaindian.com/newsportal/india-news/indian-forests-1.html.

[77] See, for example, http://in.reuters.com/article/oilRpt/idINDEL15998520090907?pageNumber=1& virtualBrandChannel=0.

[78] http://www.indg.in/rural-energy

[79] Ministry of New and Renewable Energy, "Statement of Dr. Farooq Abdullah on Jawaharial Nehru National Solar Mission – 'Solar India'" November 23, 3009.

[80] Pew Center, "Climate Change Mitigation Measures in India," International Brief 2, September 2008.

[81] http://online.wsj.com/article/SB125018657071529801.html.

[82] Among many sources: http://www.business-standard.com/india/news/23-thermal-plants-

[83] http://www.business-standard.com/india/news/govt-to-reduce-water-air-pollution

[84] http://www.dw-world.de/dw/article/0,,4707051,00.html.

[85] U.S. Census Bureau, Foreign Trade Statistics, http://www.census.gov/foreign-trade c5330.html.

[86] http://www.greencarcongress.com/2009/06/india-fe-20090603.html.

[87] This section was prepared by Jane A. Leggett, Specialist in Environmental and Energy Policy (7-9525).

[88] For a summary of the plan in English, see eneken.ieej.or.jp/data/en/data/pdf/443.pdf.

[89] Various press reports, including http://search

[90] Live market currency exchange rate for November 19, 2009 is listed as 1 JPY = 0.0112 USD (http://www.xe.com/). Currency rates are subject to fluctuation.

[91] http://www.planetark.com/enviro-news/item/54691.

[92] Law No.117 of 1998.

[93] 22 June 1979, Law No. 49. Revised in 10 December 1983, 31 March 1993, 12 November 1993, 9 April 1997, and 5 June 1998.

[94] Established by Nippon Keidanren, the Japan Business Federation. Negotiated environmental agreements in Japan have been used in lieu of legally binding regulation since the 1 990s, and are not comparable to "voluntary programs" in the United States or some other countries. For example, they may require inspections and there are few reported instances of non-compliance with set targets (Imura Hidefuri, "Building a Cooperative Relationship Between Industry and Regulatory Authorities," presented at OECD, Environmental Compliance Assurance: Trends and Good Practices Paris, 17-18 November 2008."

[95] http://www.jama-english.jp/asia/news/2009/vol36/index.html.

[96] Ibid.

[97] This section was prepared by Jane A. Leggett, Specialist in Environmental and Energy Policy (7-9525).

[98] http://www.korea.net/News/News/newsView.asp?serial_no=20091118002&part=101&SearchDay=&page=1.

[99] http://www.wwf.or.jp/activity/climate/lib/kyotoprotocol/20040928b.pdf.

[100] Mufson, "Asian Nations Could Outpace U.S. in Developing Clean Energy," *The Washington Post*, http://www.washingtonpost.com/wp-dyn/content/article/2009/07/15/AR2009071503731.html.

[101] Various press reports, including http://www.reuters.com/article/environmentNews/idUSTRE57308M20090804.

[102] http://www.wwf.or.jp/activity/climate/lib/kyotoprotocol/20040928b.pdf.

[103] This section was prepared by Jane A. Leggett, Specialist in Environmental and Energy Policy (7-9525).

[104] North American Leaders' Declaration on Climate Change and Clean Energy, August 10, 2009. Available at http://pm.gc.ca/eng/media

[105] This section was prepared by Jane A. Leggett, Specialist in Environmental and Energy Policy (7-9525).

[106] United Nations Framework Convention on Climate Change, *Report of the Centralized In-Depth Review of the Fourth National Communication of the Russian Federation* (Bonn, August 31, 2009), http://unfccc.int/documentation/ documents/advanced_search/items/3594.php?rec=j&priref=600005423.

[107] Decree 889, June 4, 2008.

[108] http://eng.kremlin.ru/speeches/2009/06/18/1241_type82916_218210.shtml.

[109] Anna Korppoo, "Linkages between Russian Energy and Climate Policies towards Copenhagen," October 16, 2009, http://docs.google. anna_korppoo.pdf+Russia+GHG+policies+measures&hl=en&gl=us&sig=AFQjCNEMV4xp1Ac-SYT7zh5Oh7v4UBit3Q.

[110] Anne Karin Saether, "Moscow Environmental Conference Places Climate Demands on Medvedev," Bellona, March 27, 2009, http://www.bellona.org/articles/articles_2009/environmentalists_ put_climate_changes_ to_medvedev; Simon Shuster, "Russia offers climate goal with no real bite," June 19, 2009, http://www.reuters.com/article/ environmentNews/idUSTRE55I3CP20090619; Ulkopoliittinen instituutti, "Russia's Post-2012 Climate Politics in the Context of Economic Growth," May 11, 2008, http://www.upi-fiia.fi/fi/event/195/; or, Simon Shuster, "Russia Still Dragging Its Feet on Climate Change," Time, October 8, 2009, http://www.time.com/time/specials/packages/article/ 0,28804,192907 1_1929070_1934785,00.html.

[111] Quirin Schiermeier, "Russia makes major shift in climate policy," *Nature -News* (May 26, 2009), http://www.nature.com/news/2009/090526/full/news.2009.506.html; Simon Shuster, "Russia offers climate goal with no real bite," June 19, 2009, http://www.reuters.com/article/environmentNews/ idUSTRE55I3 CP20090619; or 1. Oleg Shchedrov, "Russia's Medvedev warns of climate catastrophe," November 16, 2009, http://www.reuters.com/article/ environmentNews/idUSTRE5AF1 SU20091116.

[112] Jean Foglizzo, "Russia's New Energy Strategy Seems a Lot Like its Old One," The New York Times, March 30, 2008, http://www.nytimes.com/2008/03/30/business/worldbusiness/30iht-rnrgruss.1.11526942.html.

[113] Kevin Rosner, "Dirty Hands: Russian Coal, GHG Emissions & European Gas Demand," Journal of Energy Security (August 27, 2009), http://www.ensec.org/index.php?option=com_content&view= article&id=207: dirty-hands- :issuecontent0809&Itemid=349. The author raises, "The significant issue is whether it would be more advantageous, from an environmental-security perspective within the framework of Russia's coal paradigm, that the majority of new coal capacity is driven by comparatively more regulated OECD countries or whether it will revert back to Russia. Russia's environmental record is not exemplary in this regard."

[114] Ibid.

[115] Dimtry Medvedev, "Presidential Address to the Federal Assembly of the Russian Federation," http://www.kremlin.ru, November 12, 2009.

[116] UNFCCC, op. cit., p. 4.

[117] The European Union, as a regional economic integration organization, is a Party to the UNFCCC, as are its member countries.

[118] This is not the first quantitative GHG goal set for U.S. climate change policy: on April 21, 1993, President William J. Clinton "announce[d] our nation's commitment to reducing our emissions of greenhouse gases to their 1990 levels by the year 2000," consistent with the Article 4 aim of the UNFCCC. The challenge in meeting that aim with voluntary measures only led to agreement on mandatory GHG reduction obligations in the Kyoto Protocol. In 2002, President George W. Bush stated a goal of reducing carbon intensity – the amount of carbon dioxide emissions per unit of Gross Domestic Product (GDP) by 18% from 2002 to 2012, about a four percentage point improvement over business-asusual. At the time, the Administration projected GHG to increase to about 7,709 (MMTCO2e), or about 11% above 1990 levels. In April 2008, President George W. Bush announced a new national goal for climate policy—to halt increases in U.S. emissions of GHG by 2025.

[119] White House, "Combating Climate Change at Home and Around the World," November 25, 2009, http://www.whitehouse.gov/blog

[120] United States Environmental Protection Agency, *The U.S. Inventory of Greenhouse Gas Emissions and Sinks: 1990- 2007*, EPA 430-F-06-010 (Washington DC: Office of Atmospheric Programs, 2009).

[121] White House, 2009, op. cit.

[122] See http://www.epa.gov/climatechange/policy/neartermghgreduction.html, http://www.pi.energy and http://www.usda.gov/oce/climate_change/index.htm.

[123] White House, 2009, op. cit.

[124] Established by Part B of Title III of the Energy Policy and Conservation Act (EPCA), P.L. 94-163, as amended by the National Energy Conservation Policy Act, P.L. 95-6 19, by the National Appliance Energy Conservation Act, P.L. 100-12, by the National Appliance Energy Conservation Amendments of 1988, P.L. 100-357, and by the Energy Policy Act of 1992, P.L. 102-486, and by the Energy Policy of 2005, P.L. 109-58.

[125] http://www.epa.gov/reg3artd/airregulations/ap22/landfil2.htm.

[126] White House, 2009, op. cit.

[127] Data on state policies come from the Pew Center on Global Climate Change website, extracted November 20, 2009. http://www.pewclimate.org/states-regions.

[128] *Massachusetts v. EPA*, 127 S. Ct. 1438 (2007).

[129] See CRS Report RL32764, *Climate Change Litigation: A Survey*, by Robert Meltz.

In: Sources and Reduction of Greenhouse Gas Emissions
Editor: Steffen D. Saldana

ISBN: 978-1-61668-856-1
© 2010 Nova Science Publishers, Inc.

*Chapter 4*

# ANAEROBIC DIGESTION: GREENHOUSE GAS EMISSION REDUCTION AND ENERGY GENERATION

## *Kelsi Bracmort*

### ABSTRACT

Anaerobic digestion technology may help to address two congressional concerns that have some measure of interdependence: development of clean energy sources and reduction of greenhouse gas emissions. Anaerobic digestion technology breaks down a feedstock—usually manure from livestock operations—to produce a variety of outputs including methane. An anaerobic digestion system may reduce greenhouse gas emissions because it captures the methane from manure that might otherwise be released into the atmosphere as a potent greenhouse gas. The technology may contribute to the development of clean energy because the captured methane can be used as an energy source to produce heat or generate electricity.

Anaerobic digestion technology has been implemented sparingly, with 125 anaerobic digestion systems operating nationwide. Some barriers to adoption include high capital costs, questions about reliability, and varying payment rates for the electricity generated by anaerobic digestion systems. Two sources of federal financial assistance that may make the technology more attractive are the Section 9007 Rural Energy for America Program of the Food, Conservation, and Energy Act of 2008 (2008 farm bill, P.L. 110-246), and the Renewable Electricity Production Tax Credit (26 U.S.C. §45).

Congress could decide to encourage development and use of the technology by (1) identifying the primary technology benefit, so as to determine whether it should be pursued in the framework of greenhouse gas emission reduction or clean energy development; (2) determining if the captured methane will count as a carbon offset; and (3) considering additional financing options for the technology.

This chapter provides information on anaerobic digestion systems, technology adoption, challenges to widespread implementation, and policy interventions that could affect adoption of the technology.

## INTRODUCTION

An anaerobic digestion system (AD system) captures the methane that may otherwise be released from conventional manure handling methods, and has the potential to reduce greenhouse gas emissions and produce clean energy. Absent such technology, confined animal feeding operations typically collect and store manure (e.g., in a waste storage facility) under anaerobic conditions, which produces the potent greenhouse gas methane. The methane is released into the atmosphere as the manure decomposes while being stored. Livestock producers handle large quantities of manure on a daily basis that emit methane.[1] Some estimate that livestock operations produce more than a billion tons of manure yearly[2] and that 0.7% of total U.S. methane emissions are from managed manure.[3] Through methane capture, anaerobic digestion has the potential to reduce these emissions.

In addition, anaerobic digestion of manure produces biogas—a combination of methane, carbon dioxide, and trace amounts of other gases that can be used for renewable energy purposes or flared.[4] Recent legislation pertaining to agricultural sources of renewable energy has focused primarily on corn-based ethanol and cellulosic ethanol for liquid fuel purposes, and not biogas.[5] The economic environment is not currently favorable to profit from the investment required to compress biogas[6] produced from an AD system into a liquid fuel. There are, however, several successful cases where an engine-generator set has been used to generate electricity from the biogas or the biogas is burned in a boiler to produce heat. Biogas from an AD system could be used to assist livestock producers in an effort to have an energy self-sufficient operation as well as, potentially, to sell electricity to the local utility.

There are 125 AD systems operating nationally. Some factors that may be responsible for the low technology adoption rates are high capital costs, reliability concerns, and payment rates for the electricity generated. Congress may consider encouraging increased adoption of the technology by (1) identifying the primary technology benefit, so as to determine whether AD should be pursued in the framework of greenhouse gas emission reduction or clean energy development; (2) determining if the captured methane will count as a carbon offset; and (3) considering additional financing options for the technology. Moreover, Congress may receive insight on collaborative techniques by monitoring the first public-private partnership agreement between the U.S. Department of Agriculture and the Innovation Center for U.S. Dairy to reduce greenhouse gas emissions from dairy operations by 25% by 2020.[7] One goal of the agreement is to "accelerate and streamline the process for adopting anaerobic digesters by the United States dairy farm operators through various USDA programs."[8]

This chapter provides information on AD systems, technology adoption, and challenges to widespread technology implementation, and explores the issues facing Congress concerning adoption of the technology.

## WHAT IS AN ANAEROBIC DIGESTION SYSTEM?

Most manure management systems used on livestock operations store the manure in an open facility, allowing the manure to decompose naturally, which releases the potent greenhouse gas methane into the atmosphere. By contrast, an AD system feeds manure into a digester that breaks it down in a closed facility in the absence of oxygen (see Figure 1). The digested feedstock[9] is contained for a period of time[10] as anaerobic bacteria decompose the manure to produce several outputs, including biogas, liquid effluent, and dry matter. The captured biogas is flared (see footnote 4) or used for energy. The liquid effluent may be applied to the land as a fertilizer. The digested dry matter may be sold as a soil amendment product or used for animal bedding. AD systems have other benefits (e.g., odor reduction) that may curb negative impacts of livestock operations, including environmental pollution.[11]

An AD system is designed and constructed to suit the needs of an individual livestock operation and is typically selected based on the total solids (TS) content[12] of the manure and the manure handling system (see Figure 2). Other criteria taken into consideration when building an AD system include the feedstock quantity, feedstock quality, feedstock availability, feedstock handling, demand for effluent, use of captured biogas, and transportation logistics (e.g., feedstock may be transported to an AD system if it is not available on-site, or effluent transported to a receiving entity if not used on-site).

AD systems in the United States are in use on dairy cow, swine, and poultry operations. More than 75% of the operating AD systems are located on dairy cow operations. Some argue that less complex AD systems could be constructed on dairy cow operations at a lower cost and that the manure could be easily transported to the AD system, thus making them more economically appealing. Others contend that fewer AD systems are installed on swine operations because many swine operations currently store liquid manure in pits beneath the livestock; the producer would need to redesign the swine operation to incorporate an AD system. Installation of an AD system on a swine operation may also be more expensive if a storage facility must be constructed to contain the digested feedstock. Some assert that AD systems are not a favorable addition to a poultry operation partly because the litter is dry and may require more resources for transport to an AD system and additional inputs to digest.

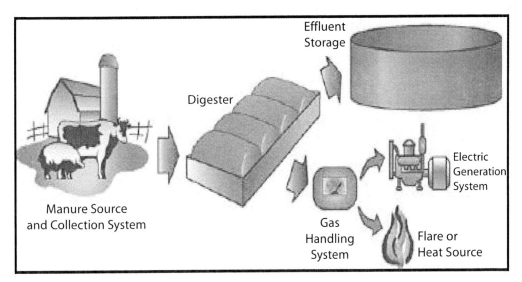

Source: U.S. EPA, The AgSTAR Program.

Figure 1. Anaerobic Digestion System Schematic

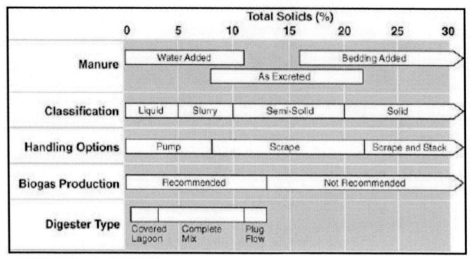

Source: US EPA, The AgSTAR Program.

Figure 2. Manure Characteristics and Handling Systems for Specific Types of Anaerobic Digestion Systems

Source: EPA, The AgSTAR Program.

Figure 3. Plug Flow AD System

# THREE COMMON AD SYSTEM TYPES FOUND IN THE UNITED STATES

*Covered lagoon AD system*:

- Usually an earthen structure containing manure affixed with a flexible geosynthetic cover (e.g., high density polyethylene).
- Typically operated at ambient temperature.
- Biogas production may occur seasonally when weather is warmer.

- Ideal for manure with 0.5%-2% TS content.
- Time required to treat manure ranges from approximately 30 to 60 days.
- Low capital costs.

*Plug flow AD system* (see Figure 3):

- Usually a rectangular concrete tank affixed with a flexible geosynthetic cover.
- Digestion occurs in a plug fashion where the digested manure exits the system as raw manure enters the system.
- Heated structure.
- Ideal for manure with 1 1%-13% TS content.
- Time required to treat manure ranges from approximately 18 to 20 days.

*Complete mix AD system* (see Figure 4):

- Usually a round concrete or steel tank.
- Heated structure.
- Capable of digesting a range of feedstock with varying total solids content.
- Ideal for manure with 3%-1 0% TS content.
- Time required to treat manure ranges from approximately 5 to 20 days.
- High capital costs.

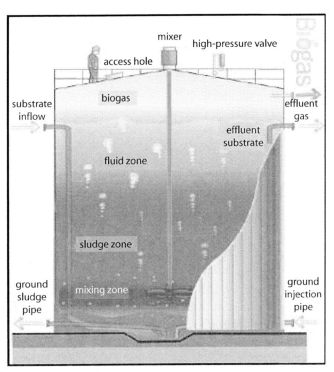

Source: Renewable Energy Association.

Figure 4. Complete Mix AD System Schematic

**Table 1. Methane Emissions from Animal Agriculture (Tg $CO_2$e)**

| Source | 2005 | 2006 | 2007 |
|---|---|---|---|
| Enteric fermentation | 136.0 | 138.2 | 139.0 |
| Manure management | 41.8 | 41.9 | 44.0 |

Source: US EPA, 2009 U.S. Greenhouse Gas Inventory Report

Notes: Select values obtained from EPA, 2009 U.S. Greenhouse Gas Inventory Report, Table 6-1. A teragram (Tg) is equivalent to 1 trillion grams. A Tg $CO_2$e (teragram of carbon dioxide equivalent) is a principal unit of measurement across greenhouse gases. Agricultural activities and livestock emit an assortment of greenhouse gases in various quantities given numerous factors. $CO_2$e is used to compare greenhouse gases emitted from different sources on the same basis.

**Table 2. Methane Emissions from Manure Management (Tg $CO_2$e)**

| Animal Type | 2005 | 2006 | 2007 |
|---|---|---|---|
| Swine | 18.6 | 18.3 | 19.7 |
| Dairy cattle | 17.2 | 17.5 | 18.1 |
| Poultry | 2.7 | 2.7 | 2.7 |
| Beef cattle | 2.4 | 2.5 | 2.4 |
| Horses | 0.8 | 0.8 | 0.8 |
| Sheep | 0.1 | 0.1 | 0.1 |
| Goats | + | + | + |

Source: US EPA, 2009 U.S. Greenhouse Gas Inventory Report

Notes: Select values obtained from EPA, 2009 U.S. Greenhouse Gas Inventory Report, Table 6-6. The + symbol denotes a value that does not exceed 0.05 Tg $CO_2$e. A Tg $CO_2$e (teragram of carbon dioxide equivalent) is a principal unit of measurement across greenhouse gases. Agricultural activities and livestock emit an assortment of greenhouse gases in various quantities given numerous factors. $CO_2$e is used to compare greenhouse gases emitted from different sources on the same basis.

## CHARACTERISTICS OF ANAEROBIC DIGESTION SYSTEMS

### Methane Capture

Methane ($CH_4$) is one of the primary greenhouse gases associated with the agricultural sector.[13] The odorless, colorless, flammable gas is potent because it is 21 times more effective at trapping heat in the atmosphere than carbon dioxide ($CO_2$) over a 100-year timeframe.[14] In other words, it takes 21 tons of $CO_2$ to equal the effect of 1 ton of CH4. Methane has a relatively short atmospheric lifetime (approximately 12 years) when compared to the atmospheric lifetime for carbon dioxide, which has a half-life of roughly 100 years; thus some argue that efforts to capture methane from anthropogenic sources may provide near-term climate change abatement.

Major sources of methane emissions from animal agriculture are enteric fermentation[15] and manure management (see Table 1). While enteric fermentation emissions are much larger than manure management emissions, they are also much more difficult (nearly impossible) to control. Emission factors[16] affiliated with methane released from enteric fermentation, and conversion factors[17] affiliated with methane released from a manure management system, are used to estimate overall methane emissions from animal agriculture. Estimation incorporates a multi-step process that takes into account livestock population data, waste characteristics, waste management system data, and other variables.

Capturing methane with an AD system is beneficial because it reduces emissions of a harmful greenhouse gas from an agricultural source of methane: managed manure (e.g., manure stored in pit storage, an anaerobic lagoon, or a storage facility).[18] Swine and dairy cattle are the two dominant livestock emitters of methane for managed manure (see Table 2). Emissions from managed manure vary on a statewide basis depending on the livestock population and the manure handling systems in place (see Figure 5).

Captured methane may qualify as a carbon offset because the methane would no longer be released directly into the atmosphere. Carbon offsets[19] are a facet of the climate change cap-andtrade program debate currently underway in the 111th Congress. Accurate quantification and verification of the methane captured from an AD system requires robust data, observed or inferred, to ensure that actual reductions are occurring as projected.

## Biogas Quality and Use

Another AD system benefit is the production of biogas, which can be used as a renewable energy source. Biogas consists of 60%-70% methane, 30%-40% carbon dioxide, and trace amounts of other gases (e.g., hydrogen sulfide, ammonia, hydrogen, nitrogen gas, carbon monoxide). Biogas can be explosive if exposed to air, depending on the concentration of methane in a confined space. The quality (i.e., heat value) and amount of biogas produced varies based on the hydraulic retention time of the AD system, the manure total solids content, and temperature.

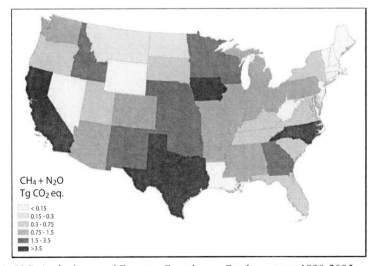

Source: USDA, U.S. Agriculture and Forestry Greenhouse Gas Inventory: 1990-2005.

Figure 5. Greenhouse Gas Emissions (Methane and Nitrous Oxide) from Managed Waste in 2005

Biogas can be used to produce heat or generate electricity. The biogas may be burned in a boiler for space heating or water heating on-site. Another option is to use an engine-generator set to create electricity from the biogas. Biogas that is upgraded to pipeline quality[20] may be sold to a natural gas utility. Typically a producer will decide on only one use for the biogas— either generating electricity or producing heat—due to the expense associated with energy generation and boiler equipment. Few choose to sell the biogas to a natural gas utility.[21]

If electricity is generated from the captured biogas, it may be sold to a utility in addition to its on-site use.[22] A rule of thumb is that 7-10 cows supply fuel for 1 kilowatt capacity.[23] Sales to a utility require a contractual agreement between the producer and the local utility provider that typically outlines safety, reliability, and performance standards. Additionally, any federal and state environmental requirements must be met (e.g., AD system generators may have to meet Best Available Control Technology[24] requirements). Net metering,[25] a preferred form of billing for many renewable electricity generators, is an option for producers selling the electricity generated to electric utilities in 40 states.[26]

Biogas may be used as a fuel if it contains at a minimum a 50% concentration of methane (CH4).[27] The biogas must be cleaned, upgraded, and compressed if it is to be used in a mobile engine. Cleaning biogas removes hydrogen sulfide. Removing any moisture and carbon dioxide upgrades the biogas and increases the British thermal unit (Btu) value.[28] With an energy content of 600 Btu/ft$^3$ for biogas with approximately a 60% methane concentration, biogas is a low-Btu fuel compared to other fuels (see Table 3). A high-Btu fuel is necessary for energy applications requiring greater amounts of power.

Biogas is flared if not used for energy purposes. Flaring the biogas destroys the methane and yields the greenhouse gas carbon dioxide ($CO_2$) and water.[29] Carbon dioxide has a longer atmospheric lifetime (~100 years) and is less effective at trapping heat in the atmosphere when compared to methane, which has an approximate atmospheric lifetime of 12 years. Some view the release of the carbon dioxide due to biogas flaring as more environmentally acceptable than releasing unused biogas into the atmosphere.[30] Environmental impacts of flaring methane require further investigation to determine flare efficiency under varying wind speeds and various biogas compositions.

**Table 3. Energy Content for Select Fuels**

| Fuel | Energy Content |
|---|---|
| Biogas[a] | 600 Btu/ft$^3$ |
| Natural gas | 1,000 Btu/ft$^3$ |
| Propane | 92,000 Btu/gallon |
| Diesel fuel | 138,000 Btu/gallon |
| Coal | 25,000,000 Btu/ton |

Source: James C. Barker. *Methane Fuel Gas from Livestock Wastes: A Summary*. North Carolina State University Cooperative Extension Service, EBAE 071-80, 2001.

a. Assumes a 60% methane concentration.

## U.S. Adoption of Anaerobic Digestion Technology

Federal funding (discussed in the following section) has supported the installation of 125 AD systems nationwide operating with a total annual energy production of approximately 244,000 megawatt-hours (MWh).[31] Approximately 90% of the AD systems are currently generating electricity. These AD systems restrict roughly 36,000 metric tons of methane, or 756,000 metric tons of carbon dioxide equivalent ($CO_2e$), from entering the atmosphere, which is comparable to the annual greenhouse gas emissions from approximately 138,000 passenger vehicles.[32]

## Startup Financing

Congress as well as states have enacted legislation that provides financial assistance for AD system installation. Loans, grants, tax credits, tax exemptions, and production incentives are common financial assistance tools available at the federal and state levels.[33] One principal source of federal funding is the Section 9007 Rural Energy for America Program (REAP) of the Food, Conservation, and Energy Act of 2008 (2008 farm bill, P.L. 110-246).[34] Section 9007 authorizes $255 million in mandatory funding for 2009-2012, with an additional $25 million per year in discretionary funding. For FY2009, $55 million is available, part of which is to be distributed to eligible applicants for AD system projects. Grants dispensed to applicants are not to exceed 25% of the cost of the activity. Loan guarantees dispensed to applicants are not to exceed $25 million. The maximum amount of a combined loan and grant can be no more than 75% of the cost of the activity. Approximately $40 million in loans and grants from Section 9006 of the 2002 farm bill was allocated for the construction of 121 AD system projects. A minimal share of the 121 AD systems are operational, while most are still under development.[35]

Another source of financial assistance offered for anaerobic digestion projects is the Renewable Electricity Production Tax Credit (REPTC; 26 U.S.C. §45).[36] The REPTC grants a one cent per kilowatt-hour tax credit for electricity generated from open-loop biomass (e.g., agricultural livestock waste nutrients).[37] The tax credit period is five years for an open-loop biomass facility using agricultural livestock waste placed in service after October 22, 2004, and before August 9, 2005. The tax credit period is 10 years for a facility placed in service after August 8, 2005.

## Anaerobic Digestion Obstacles

Some technology deployment and adoption barriers exist because of the complexity involved in designing and operating an AD system. Thus far, a significant number of successful AD systems are operated by producers who have sophisticated anaerobic digestion technology knowledge. Some assert that the technology will reach its full potential when an array of concerns are addressed. Some challenges include:[38]

- *Lack of economic return.* AD systems are capital-intensive. The cost fluctuates depending upon the system type, system size, livestock operation type, and factors specific to the site. Capital costs generally include the AD system cost, the engine-generator set, the engineering design process, and installation. Extra costs are incurred for additional elements (e.g., a post-digestion solids separator, utility and interconnection fees). System costs range from a few hundred thousand dollars to a few million dollars.[39] A covered lagoon AD system could cost at a minimum a few hundred thousand dollars.[40] A complete mix AD system or plug flow AD system can cost a few million dollars. An analysis of 38 AD systems indicates that approximately one-third of the total system cost is estimated to be spent on the electrical generation equipment.[41] A producer may find it feasible to forgo producing electricity to save money and to use the biogas produced on-site for heating purposes, which still requires a boiler. A general approximation is that 300-500 head for a dairy cow operation and 2,000 head for a swine operation is the minimum number of head necessary to produce electricity with an AD system at a possible profit.[42]

- *Reliability.* Some argue that producers hesitate to adopt anaerobic digestion technology due to the poor performance rate observed in the 1970s. AD system performance has improved over time due to better engineering, construction materials, and management. Communicating current performance rates may provide producers with the information needed to gain more confidence in the improved technology. AD system performance data may be expanded with mandatory reporting on a periodic basis by an independent third party, which might bolster performance claims made by some AD system construction companies. Information (e.g., demonstration projects, long-term performance records) that communicates recent performance may mitigate producers' doubts about technology reliability and may verify energy generation and greenhouse gas emission reduction data reported.[43] Some producers and construction companies are opposed to a national reporting program because of the release of potentially proprietary information.

- *Lack of an engineering practice standard.* A national practice standard that lists performance criteria, safety precautions, technical components, and design elements and has undergone review from a standards developing organization is not available for anaerobic digestion technology. Some producers may be reluctant to make a financial investment in a technology that may or may not meet future environmental and technical requisites. The USDA Natural Resources Conservation Service (NRCS) issued an anaerobic digester conservation practice standard in 2003 that is currently undergoing public review and comment.[44]

- *Utility collaboration.* Rates paid by utilities for the electricity generated from AD systems vary by state and within each utility. It may not be economically attractive for a producer to sell the renewable energy generated depending on the cost per kilowatt-hour offered. Some net-metering agreements pay wholesale rates instead of retail rates for the electricity generated, thus limiting potential profitability. Some gas utility companies may be reluctant to accept a renewable energy fuel originating from a technology with no official standard that may contaminate an otherwise clean energy source.

- *Unquantifiable co-benefits.* Certain benefits (e.g., odor reduction) may not be quantifiable in dollar terms but may add to the value of an AD system.

- *Operation and maintenance.* AD systems perform optimally when they are well maintained. Some estimate that daily AD system operation and maintenance may require 30 minutes to an hour. The producer may have to acquire a level of technical expertise not previously necessary.

## ISSUES FOR CONGRESS

The 111th Congress is faced with proposed legislation centered on clean energy and climate change mitigation. There may be an opportunity for the agricultural community, particularly the livestock industry, to participate in forthcoming efforts for energy generation and greenhouse gas emission reduction with anaerobic digestion technology. Prior to incorporating anaerobic digestion technology into legislation, Congress may choose to consider:

- *Identifying the primary benefit offered by an AD system.* Selecting a primary benefit (renewable energy generation or greenhouse gas emission reduction) may assist with determining which policy vehicle could support technology deployment (e.g. energy legislation, climate change legislation, agricultural legislation). A single message regarding the technology benefit may encourage producers to adopt the technology to achieve the policy goal.

- *Determining if the methane captured from the technology will be included as a carbon offset.* The climate change debate underway in the 111th Congress includes carbon offsets as a potential greenhouse gas emission reduction strategy. Producers may find it economically worthwhile to invest in AD technology if it produces an additional revenue source in the form of carbon offsets. If the methane captured and combusted is not treated as a carbon offset, will the 111th Congress consider regulating the methane captured as a pollutant?

- *Identifying whether alternate sources of financial support for technology implementation are appropriate.* Most of the federal financial assistance available comes in loans and grants for AD system construction. A shorter payback period for an AD system may occur if producers receive a more substantial monetary sum for the energy generated and transferred to a utility company via a federal electricity rate premium. Additional tax credits may also improve the economic return for AD technology (e.g., raising the tax credit value for agricultural livestock waste nutrients to that of closed-loop biomass for the Renewable Electricity, Refined Coal, and Indian Coal Production Credit).

# End Notes

[1] For example, the amount of manure excreted from one lactating dairy cow is estimated at 150 pounds per day based on the American Society of Agricultural and Biological Engineers (ASABE) Manure Production and Characteristics Standard D384.2, March 2005.

[2] Amanda D. Cuellar and Michael E. Webber, "Cow Power: The Energy and Emissions Benefits of Converting Manure to Biogas," *Environmental Research Letters*, vol. 3 (2008).

[3] The agriculture sector was responsible for roughly 6% of all U.S. greenhouse gas emissions in 2007, approximately 11% of which is attributable to methane emissions from managed manure. U.S. Environmental Protection Agency, *2009 U.S. Greenhouse Gas Inventory Report*, Inventory of U.S. Greenhouse Gas Emissions and Sinks: 1990-2007, 2009.

[4] Flaring is the combustion of gas without commercial purposes. Flaring emits fewer greenhouse gases than simply releasing the biogas as is into the atmosphere. See the "Biogas Quality and Use" section for additional details.

[5] For more information on agriculture-based renewable energy, see CRS Report RL34130, *Renewable Energy Programs in the 2008 Farm Bill*, by Megan Stubbs, CRS Report RL34738, *Cellulosic Biofuels: Analysis of Policy Issues for Congress*, by Kelsi Bracmort et al., and CRS Report RL32712, *Agriculture-Based Renewable Energy Production*, by Randy Schnepf.

[6] Biogas consists of 60%-70% methane, 30%-40% carbon dioxide, and trace amounts of other gases.

[7] U.S. Department of Agriculture, "Agriculture Secretary Vilsack, Dairy Producers Sign Historic Agreement to Cut Greenhouse Gas Emissions by 25% by 2020," press release, December 15, 2009, http://www.usda.gov/wps/portal/ !ut/p/_s.7_0_A/7_0_1OB?contentidonly=true&contentid=2009/12/0613.xml.

[8] *Memorandum of Understanding between the United States Department of Agriculture and The Innovation Center for U.S. Dairy*, December 15, 2009, http://www.usda.gov/documents/FINAL_USDA

[9] The feedstock is usually manure, but can also include other organic matter. The digestion of two or more types of organic matter is referred to as co-digestion.

[10] The time required for the manure to be treated is referred to as the hydraulic retention time (HRT). HRT is normally expressed in days.

[11] For more information on environmental issues facing the livestock community, see CRS Report RL32948, *Air Quality Issues and Animal Agriculture: A Primer*, by Claudia Copeland, and CRS Report RL33691, *Animal Waste and Hazardous Substances: Current Laws and Legislative Issues*, by Claudia Copeland.

[12] Total solids content for manure is the amount of solid material remaining after all moisture has been removed from a sample.

[13] For additional information concerning agricultural sources of methane, see CRS Report RL33898, *Climate Change: The Role of the U.S. Agriculture Sector*, by Renée Johnson

[14] The IPCC Second Assessment Report issued in 1996 assigned methane a Global Warming Potential of 21. Global Warming Potential is an estimate of how much a greenhouse gas affects climate change over a quantity of time relative to $CO_2$, which has a GWP value of 1.

[15] Enteric fermentation is the production and release of methane via eructation (burping) and flatulence as ruminant animals digest their feed.

[16] The amount of methane produced by an animal expressed per mass unit for one year (kg $CH_4$/head/year).

[17] The potential amount of methane produced from a manure management system for a given animal expressed in percent. The factor for dry manure management systems may vary according to climate.

[18] Manure that is not managed (e.g., manure deposited in a pasture from livestock grazing) has low methane emissions, but relatively high nitrous oxide emissions. The greenhouse gas nitrous oxide is 310 times more effective at trapping heat in the atmosphere than carbon dioxide over a 100-year timeframe.

[19] A carbon offset is a measurable reduction, avoidance, or sequestration of GHG emissions from a source not covered by an emission reduction program. For more information on carbon offsets, see CRS Report RL34705, *Potential Offset Supply in a Cap-and-Trade Program*, by Jonathan L. Ramseur, and CRS Report RL34436, *The Role of Offsets in a Greenhouse Gas Emissions Cap-and-Trade Program: Potential Benefits and Concerns*, by Jonathan L. Ramseur.

[20] Pipeline quality is achieved by the removal of carbon dioxide and other contaminants so that only the methane is sold to the natural gas utility.

[21] Elizabeth R. Leuer, Jeffrey Hyde, and Tom L. Richard, "Investing in Methane Digesters on Pennsylvania Dairy Farms: Implications of Scale Economics and Environmental Programs," *Agricultural and Resources Economics Review*, vol. 37, no. 2 (October 2008), pp. 188-203.

[22] The quality of the biogas, measured by its Btu value, necessary to generate electricity varies depending on the engine.

[23] Wisconsin Focus on energy, *Farm Energy from Manure*, REN2003-0602, 2002, http://focusonenergy.com/files/ Document_Management_System/Renewables/farmenergyfrommanure_factsheet.pdf.

[24] Best Available Control Technology (BACT) is a pollution control standard mandated by the Clean Air Act.

[25] Net metering is an energy metering method that uses a bidirectional meter, thus allowing the meter to run backwards if a customer generates more electricity than being consumed.

[26] Interstate Renewable Energy Council , *Map of state net metering rules*, 2009, http://www.dsireusa.org.

[27] Jenifer Beddoes, Kelsi Bracmort, and Robert Burns et al., *An analysis of energy production costs from anaerobic digestion systems on U.S. livestock production facilities*, USDA Natural Resources Conservation Service, October 2007.

[28] A Btu (British thermal unit) is a unit of energy used to express the heating value of fuels.

[29] Stoichiometric equation for biogas combustion: $CH_4 + 2O_2 \rightarrow CO_2 + 2H_2O$.

[30] According to the California Climate Action Registry, *Livestock Project Reporting Protocol Version 2.1*, August 2008, the $CO_2$ released from flaring the biogas is considered biogenic and therefore more environmentally acceptable. See http://www.climateregistry.org/resources CCARLivestockProjectReportingProtocol2. 1.pdf.

[31] U.S. Environmental Protection Agency, The AgSTAR Program. Online AgSTAR Digest: Spring 2009. The AgSTAR Program is a voluntary effort jointly sponsored by the U.S. Environmental Protection Agency, the U.S. Department of Agriculture, and the U.S. Department of Energy. The program encourages biogas capture and utilization at animal feeding operations that manage manures as liquids and slurries. One megawatt-hour (MWh) is equivalent to 1,000 kWh. The energy generated could provide approximately 21,724 average homes with electricity for one year assuming the average residential home uses 936 kWh on a monthly basis (DOE).

[32] Passenger vehicle estimate computed using the EPA Greenhouse Gas Equivalencies Calculator available at http://www.epa.gov/solar/energy. Calculation based on 2005 average fuel economy assumptions and average vehicle miles traveled at 19.7 mpg and 11,856 miles per year, respectively. Additional information on passenger vehicle calculation available at website provided.

[33] For a comprehensive list of AD system financing tools, visit the EPA AgSTAR Funding Database website at http://www.epa.gov/agstar/resources and consult the EPA AgSTAR document *Funding On-Farm Biogas Recovery Systems: A Guide to Federal and State Resources*, at http://www.epa.gov/agstar/pdf/ag_fund_doc.pdf.

[34] REAP is an extension of the Farm Security and Rural Investment Act of 2002 (2002 farm bill, P.L. 107-171).

[35] William F. Lazarus, *Farm-Based Anaerobic Digesters as an Energy and Odor Control Technology: Background and Policy Issues*, USDA Office of the Chief Economist Office of Energy Policy and New Uses, Agricultural Economic Report Number 843, February 2008, http://www.usda.gov/oce/reports/energy

[36] For more information on renewable energy policy, see CRS Report RL33578, *Energy Tax Policy: History and Current Issues*, by Salvatore Lazzari, and CRS Report R40412, *Energy Provisions in the American Recovery and Reinvestment Act of 2009 (P.L. 111-5)*, coordinated by Fred Sissine.

[37] For more information about the Renewable Electricity, Refined Coal, and Indian Coal production credit visit http://www.irs.gov/pub/irs-pdf/f8835.pdf

[38] No relative importance is intended by the order in which challenges are listed.

[39] AgSTAR Program, *Estimating Anaerobic Digestion Capital Costs for Dairy Farms*, February 2009, http://www.epa.gov/agstar/pdf/conf09/crenshaw_digester_cost.pdf.q

[40] Assuming the producer is only purchasing the cover and the biogas recovery equipment to add to an existing lagoon.

[41] Jenifer Beddoes, Kelsi Bracmort, and Robert Burns et al., *An analysis of energy production costs from anaerobic digestion systems on U.S. livestock production facilities*, USDA Natural Resources Conservation Service, October 2007.

[42] U.S. Environmental Protection Agency, *Market Opportunities for Biogas Recovery Systems*, EPA-430-8-06-004; The Minnesota Project, *Profits from Manure Power*, http://www.mnproject.org/pdf/AD%20economics.pdf.; Wisconsin Focus on Energy, *Farm Energy from Manure*, http://focusonenergy.com/files/Document_Management_System/Renewables/farmenergyfrommanure_factsheet.pdf.

[43] For an example of reporting criteria to communicate technology performance, see John H. Martin, *A Protocol for Quantifying and Reporting the Performance of Anaerobic Digestion Systems for Livestock Manures*, January 2007.

[44] The conservation practice standard must be adhered to for AD systems that will be constructed with financial or technical support from USDA NRCS. Conservation practice standards are provided at http://www.nrcs.usda.gov/ technical/standards/nhcp.html.

In: Sources and Reduction of Greenhouse Gas Emissions
Editor: Steffen D. Saldana

ISBN: 978-1-61668-856-1
© 2010 Nova Science Publishers, Inc.

*Chapter 5*

# FARM-BASED ANAEROBIC DIGESTERS AS AN ENERGY AND ODOR CONTROL TECHNOLOGY: BACKGROUND AND POLICY ISSUES

## *William F. Lazarus*

## ABSTRACT

This chapter summarizes the existing literature and analytical perspectives on farm-based digesters, highlights major efforts in the United States and Europe to expand digester usage, and discusses key policy issues affecting digester economics. The study was largely a review of the "gray literature" on digesters, and it serves as a snapshot overview of the industry. Digesters are fairly capital-intensive when viewed primarily as an energy source. On a strictly market basis, current U.S. average electricity prices do not appear to provide sufficient economic justification for digesters to move beyond a fairly limited niche. Digesters make the most sense today where the odor and nutrient management benefits are important, or where the electricity or heat has a higher-than-average value. Digester biogas is mainly methane, which is destroyed when flared or used for electricity. This methane destruction is beneficial in terms of climate change. The associated carbon credits may become a more significant farm revenue source in the future.

## ACKNOWLEDGMENTS

This chapter was prepared while I was a visiting scholar with the USDA Office of Energy Policy and New Uses in 2006-07. The report grew out of a December 2006 seminar at USDA. The report would not have been possible without the help and encouragement of Roger Conway and the Office of Energy Policy and New Uses (OEPNU) staff. Jim Duffield, Kelsi Bracmort, William Hohenstein, and Marvin Duncan provided helpful comments on the manuscript. My interest in digesters grew out of earlier work on an Environmental Quality Incentives Program-funded project from the USDA Natural Resource Conservation Service.

That project and follow- up work were coordinated by staff of the Minnesota Project, Inc., nonprofit in St. Paul, Minnesota, and involved a number of University of Minnesota staff as well as other agency and utility staff. Deborah Allen, Fantu Bachewe, Amanda Bilek, Roger Becker, Carl Nelson, Margot Rudstrom, David Schmidt, and Mike Schmitt have been terrific collaborators in that work. Phil Goodrich deserves a special note of thanks for sharing his insights on digester engineering and for putting up with me on our European digester tour. Finally, Dennis and Marsha Haubenschild have hosted us at their farm many times and shared their views on what it is like to actually own and operate a digester. I owe them my thanks for their cooperation and enthusiasm.

## SUMMARY

Renewable energy is currently viewed as key to the energy security of the United States and as an economic opportunity for rural communities. Biofuels such as ethanol and biodiesel have been receiving most of the attention, but other renewable energy sources can also play an important role. Concern about climate change and greenhouse gases is growing, with methane from livestock production being one of the key areas where mitigation may be feasible. This chapter summarizes the outlook for farm-based digesters, highlights major efforts in the United States and Europe to expand digester usage, and discusses key policy issues affecting digester economics.

## What Is the Issue?

Digesters are of interest with respect to climate change, energy, air quality, water quality, and land use. However, digesters are capital-intensive and difficult to maintain. Many of the digesters installed in the 1970s went out of business. Have the bugs been worked out sufficiently to improve success rates in the future? What Government policies currently help facilitate adoption of this technology? Do those policies need fine-tuning to speed adoption further? How significant can digesters be as a source of renewable energy?

## What Did the Study Find?

Farm-based anaerobic digesters can make a significant contribution to U.S. energy security as well as help to minimize livestock odors. Digester technology has progressed as a result of a number of active development efforts in North America and Europe. The European countries have shown that biogas can supply a significant percentage of national electricity needs or can even serve as a transportation fuel if need be. However, digesters are fairly capital-intensive when viewed primarily as an energy source. On a strictly market basis, current U.S. average electricity prices do not appear to provide sufficient economic justification for digesters to move beyond a fairly limited niche. Digesters make the most sense today where the odor and nutrient management benefits are important, or where the electricity or heat has a higher-than-average value. Continued high fossil fuel prices and/or public sector support could accelerate digester adoption.

## How Was the Study Conducted?

The study was largely a review of the "gray literature" on digesters. By "gray literature" we mean publications other than peer-reviewed journal articles, such as extension monographs, slide sets, Web sites of public entities and private firms, consultant feasibility analyses, government information bulletins, and university research project reports. A substantial body of literature exists on digesters but is somewhat scattered and reflects a number of different analytical perspectives. Some of the material in this chapter was presented in a seminar at the U.S. Department of Agriculture in December 2006. A May 2006 study tour of digesters in Denmark, Sweden, and Germany helped shape the perspective taken in this chapter. The initial impetus for the study dates back to an interdisciplinary project monitoring a Minnesota dairy farm digester over several years. Followup projects looking at the feasibility of centralized digesters, digesters for smaller farms, and various biogas utilization options have also provided insights for this chapter.

## INTRODUCTION AND OVERVIEW

Anaerobic digestion converts volatile acids in livestock manure into biogas consisting of methane (55–70 percent), carbon dioxide (30–45 percent), and small amounts of water and other compounds.[1] The organic matter that can be processed in anaerobic digesters includes manure from dairy, swine, beef, and poultry farms; wastewater sludge; municipal solid waste; food industry wastes; grain industry and crop residues; paper and pulp industry wastes; or any other biodegradable matter. The methane produced by this process can be used to generate electricity or for heating. Under favorable circumstances, there is also a potential for purifying the methane into a marketable, naturalgas-grade biogas suitable for household and industrial use. If we move to a "hydrogen economy," biogas can be an excellent source of hydrogen. In addition to generating renewable energy, anaerobic digestion leads to reduced odor pollution, fewer pathogens, and reduced biochemical oxygen demand. There is little change in the nutrient value of the manure and organic matter that passes through the process, which can then be used as fertilizer.

Anaerobic digesters are relevant to concerns about climate change, energy supplies, air quality, and land use. Farm-based digesters are an attractive technology for addressing climate change because they reduce livestock- related emissions of methane, a potent greenhouse gas.[2] Evaluation of the greenhouse gas reduction benefits of a farm-based digester involves three considerations: how much methane is generated in the digester and then burned in the engine or flare; how much methane would have been generated otherwise if the farm had used a manure handling system that did not include a digester; and how much $CO_2$ would otherwise have been generated in producing the electricity that is replaced by the electricity generated by the digester system. Because digesters are designed to maximize methane production, they typically generate more methane than manure handling systems that do not include digesters. Calculating the greenhouse benefits based on digester output rather than output of the alternative manure handling system can result in an overestimate of the benefits. See the Winter 2006 AgSTAR Digest for more details on calculating methane destruction benefits [U.S. AgSTAR, 2006a].

The interest in digesters was initially driven by energy concerns, with biogas viewed as a source of electricity or a substitute for natural gas. Digestion converts volatile organic compounds in manure to more stable forms that can be land-applied with fewer objectionable odors; so many farm digesters have been installed to address neighbors' complaints. Municipal sewage treatment plants tend to use digesters to reduce the volume of solids and minimize the land required for spreading sludge.

## DIGESTERS IN EUROPE AND NORTH AMERICA

Europe faced energy reductions during and after World War II and is still more dependent on imports of oil and natural gas than the United States [Lusk, 1998]. For example, Sweden has no domestic sources of oil or natural gas, and in 2004 Germany was dependent on imports for 89 percent of its natural gas consumption [U.S. Energy Information Administration, 2007a; WestStart-CALSTART, 2004]. Thus, it is not surprising that Europe has moved more aggressively to develop digesters, along with other renewable energy sources, than has the United States.

Of the European countries, Denmark, Sweden, and Germany have the best known biogas industries. Many European digesters have been designed as centralized units to serve the typically small farms. Denmark currently has around 20 centralized digesters. The number of individual farm digesters in Denmark was not reported, but the newer digesters are individual ones on larger swine operations [Al Seadi, 2000; Hjort-Gregersen, 1999]. The Al Seadi report states that: "The first biogas plants were only designed for generating energy. Later it occurred that the plants made a significant contribution to solve a range of problems in the fields of agriculture, energy and environment. Consequently, increased attention has been paid to these issues, and centralized biogas plants are now considered as integrated energy production, manure and organic waste treatment and nutrient redistribu-tion facilities" (see p. 3 of Al Seadi). The centralized digesters described in that report all co-digest materials such as food processing waste along with livestock manure, which increases biogas output as well as solving disposal problems for other industries. The biogas from those digesters is utilized in combined heat-and-power plants, which in some cases supply heat to district heating systems in nearby villages.

Danish legislation has encouraged the centralized digesters, which are common in that country [Al Seadi et al., undated]. Manure storage is required, and is often integrated with digesters. There are limits on manure application rates based on nitrogen (N) levels (digestion increases N availability, providing more immediate benefit from the N that is allowed). Industrial processors have incentives to deliver wastes to digesters and pay tipping fees. Power companies are required to purchase digester electricity at regulated prices. Biogas is exempted from energy taxes. Grants, subsidized loans, and production subsidies are also available for digesters.

The Swedish biogas industry is unique in its emphasis on the use of biogas as a fuel for vehicles designed to run on compressed natural gas [WestStartCALSTART, 2004; Krich et al., 2005]. The biogas production and distribution system was developed with the involvement of all the players—farmers, waste haulers, technology providers, national and municipal governments, transit authorities, energy providers, vehicle manufacturers, and consumers. Transit buses are the "anchor customers," and Volvo bi-fuel vehicles also use the biogas, which is upgraded to as high as 97-percent methane. The Krich et al. report states that four plants in Sweden are currently upgrading biogas to biomethane.[3] The WestStart-CALSTART report states that only the Laholm plant injects its biomethane directly into the national natural gas distribution grid. The other plants distribute the gas through dedicated pipelines to biogas refueling stations or inject "partially cleaned" biogas into "town gas" pipeline networks for residential use.

Germany currently has at least 2,700 digesters, operating with a combined electrical generating capacity of 650 megawatts [Effenberger, 2006]. Some observers have estimated the number of German digesters in late 2007 to be as high as 4,000. They supply 0.8 percent of Germany's electricity needs (all renewable fuels combined provide 9.4 percent). Energy crops such as corn silage are a more common digester feedstock in Germany than in Denmark or Sweden. Effenberger cites one study that suggests energy crops may supply almost half of the total feedstock of digesters in the German State of Thuringia.

Farm-based digesters first became popular in the United States during the 1970s oil crisis. One of the first American digesters was installed in Iowa in 1972 [Mattocks and Wilson, 2004]. Around 70 digesters were installed on U.S. farms between 1970 and 1990, but most of those early designs failed or were taken out of service [Mattocks and Wilson, 2004; Roos, 2002; Lusk, 1998]. Energy prices declined in the late 1980s and 1990s, and it is unclear whether the success rate might have been higher if prices had remained at the levels of the early 1980s. In year-2000 dollar terms, West Texas crude oil prices reached $50/barrel in 1982 and then declined to near $10 in 1999 before climbing past the $50 level again in 2006 [U.S. Energy Information Administration, 2007b]. Digesters are not the only renewable energy technology that has seen large numbers of systems cease to operate. For example, nearly 30 percent of biomass-burning power plants built since 1985 are currently nonoperational [Peters, 2007].

A total of 111 farm digesters were operating in the United States in 2007. Most of these digesters generate electricity, with an estimated output of 215 million kilowatt-hours equivalent of useable energy. Others simply flare off the biogas for odor control, use it for heating, or upgrade the gas for injection into the natural gas pipeline [Ball, 2007]. AgSTAR estimates that around 7,000 large dairy and swine operations could operate profitable biogas systems, with a generating potential of 722 megawatts—0.1 percent of total U.S. electrical generating capacity, or enough to supply almost 1 million homes [U.S. AgSTAR, 2006a].

The number of operating digesters in the United States is somewhat difficult to track. When a digester is installed, it is often publicized, especially if public support was provided. So, new digesters are generally identified and counted. When a digester ceases to operate, however, that status change may not become known until such time as someone does a comprehensive survey. Another potential source of confusion is that some large farms have installed several digesters, so when a count is done one must be clear about whether the number of farms or the number of digesters was counted.

Two surveys of digesters by Kramer in seven Midwestern States 2 years apart, in 2002 and 2004, show the dynamic nature of digester installations [Kramer, 2004]. Of 23 digesters on 19 farms with digesters in 2002, 18 digesters on 14 farms were still operating in 2004, while 5 had ceased to operate. Ten more digesters on six farms were under construction or in startup. In many cases, systems that are no longer operational did not fail (defined as ceasing to operate) because of technology shortcomings, but because the farmer was unwilling to continue operating the Anaerobic Digestion (AD) system given the operation and maintenance costs.

Dairy producers who have digesters are reported to be almost universally pleased with them, although the motivation for installing them varies. A survey of 64 producers across the United States and 10 in California found that reducing odor and improving air and water quality were the main motivations for digester installation [Hettinga, 2007; U.S. AgSTAR, 2007b]. Electricity generation was viewed as a secondary benefit. Negotiating with the local electrical utility was the biggest challenge faced by these producers and, in many cases, discouraged them from installing digesters that had been planned. The national sample consisted of the producers who had received USDA Section 9006 grants in 2003 or 2004. By the time of the survey, 27 (40%) of the digesters were operational or under construction. Twenty eight (45%) were still in the planning stages or the producers were undecided whether to install them, while nine (15%) had decided not to move forward with installation. Financing also proved to be difficult for many in the national group and obtaining permits was more difficult for the California group.

Canada is also developing a digester industry, which was initiated largely by health concerns. One activity of the Canadian government to address the environmental problems posed by livestock manure is the ManureNet Web site, a collection of over 23,000 links on digesters, manure, and nutrient management [Agriculture and Agrifood Canada, 2005]. ManureNet includes information on digesters, organized into topics such as government programs, technology providers and consultants, digester designs, electricity net metering, impacts of antibiotics on digester performance, pathogen reduction, and economic assessments.

# U.S. POLICIES RELATED TO DIGESTERS

Federal, State and local policies have been instrumental in digester development in the United States. Policies relating to digesters include educational and technical assistance, grants and production incentives, distributed power generation policy, environmental policy, and support for research and development.

## Educational and Technical Assistance

One Federal digester support program is AgSTAR, a voluntary effort jointly sponsored by the U.S. Environmental Protection Agency (EPA), the U.S. Department of Agriculture (USDA), and the U.S. Department of Energy [U.S. AgSTAR, 2007a]. The AgSTAR program provides publications and technical tools, including a handbook and software for doing initial economic evaluations. The program helps interested livestock producers to identify potential project developers, suppliers, and partners, and conducts workshops and conferences.

One recent AgSTAR product is a protocol for rigorous evaluations of digester designs. The protocol was used to compare manure use in two typical upstate New York dairy farms, one with a digester and one without [Martin, 2004]. While the two-farm sample is obviously small, this side-by-side comparison provides insights into how a digester changes the manure effluent. For example, digestion reduced volatile solids, chemical oxygen demand, and pathogen levels. Ammonia N was increased in the effluent, with the addition presumably coming from the mineralization of the organic N . Martin did not report ammonia emissions into the atmosphere from the effluent, although total N losses from the manure storage structure were measured as less than from the undigested manure. Ammonia emitted into the atmosphere is a concern because it contributes to eutrophication of surface waters and nitrate contamination of groundwaters, and is a precursor of fine particulate matter (PM2.5) air pollution [U.S. Environmental Protection Agency, 2004; Shih et al., 2006].

The USDA Natural Resources Conservation Service also provides technical assistance on digesters, including conservation practice standards and a technical note on energy production costs [USDA Natural Resources Conservation Service, 2004; USDA Natural Resources Conservation Service, 2007; Beddoes et al., 2007].

## Support for Research and Feasibility Studies

Other Federal support for digesters includes basic research by the USDA Agricultural Research Service (ARS). The ARS facility at Beltsville, MD, recently installed a set of small digesters designed for replicated studies of digester feedstocks such as food wastes [U.S. AgSTAR, 2004]. State and local governments have also provided funding for basic research, as well as feasibility studies, such as the California biomethane study [Krich et al., 2005] and centralized digester studies in King County, Washington [Environmental Resource Recovery Group, LLC, 2003], Michigan [Frazier, Barnes & Associates, LLC, 2006], and Minnesota [Sebesta Blomberg & Associates, Inc., 2005].

## Grants, Loans, and Production Incentives

The main Federal sources of financial support for digesters are (1) the 2002 Farm Bill section 9006 grants (up to 25 percent of project costs) and guaranteed loans (up to 50 percent), and (2) the Renewable Electricity Production Credit (REPC) [USDA, 2007; Office of the Law Revision Counsel, U.S. House of Representatives, 2004]. Environmental Quality Improvement Program (EQIP) cost-share funds are also available for digesters that address pollution problems (see, for example, USDA Natural Resources Conservation Service, 2003). Since 2003, the 9006 grants for digesters have amounted to around $26 million and have leveraged $123 million in private investment. Ninety-one digesters have been funded. Of those, 19 are operational. Six more are nearing completion, and the other 66 are still under development [Lusk, 1995]. A number of other funding sources are discussed in a feasibility study of a Minnesota digester by Sebesta Blomberg & Associates, Inc. (2005).

Digesters are considered "open-loop biomass" for the purpose of the REPC, which is available at 0.9 cents/kwh for the first 5 years of digester operation. The REPC is also reduced by up to half the amount of Government grants (such as section 9006), subsidized financing, and other credits, so a livestock producer receiving a section 9006 grant for 25 percent of the project cost would need to consider that tradeoff. The REPC is currently set to expire on December 31, 2008 [North Carolina Solar Center, 2007a]. Some States, such as Minnesota, offer their own per-kwh production incentives and low-interest loans for digesters.

## Environmental Policy

The U.S. Environmental Protection Agency is currently monitoring air emissions from the livestock industry under a 2005 consent agreement, to help determine how to regulate those emissions [U.S. Environmental Protection Agency, 2005]. While the regulatory scheme is not yet determined, digesters may well be one of the technologies that could be installed to reduce emissions of volatile organic compounds. Hydrogen sulfide, ammonia, and particulate matter are also being monitored.

Farm-based digesters may play an important role in the Methane to Markets Partnership, a voluntary, nonbinding framework for international cooperation to reduce methane emissions [Methane to Markets Partnership, 2007]. The partnership will focus on key technologies, market assessment, project financing, country-specific needs, cooperative opportunities, and communication and outreach. Agriculture, coal mines, landfills, and oil and gas systems are being targeted. Livestock waste management is the main focus of the agricultural work.

## Distributed Power Pricing and Interconnection Policy

While distributed power policy is too complex to describe fully here, a few of the key aspects relating to digester electricity will be noted. A longstanding Federal law regulating distributed electrical generators such as digesters is the 1978 Public Utility Regulatory Practices Act (PURPA), which requires utilities to interconnect with qualifying facilities and to buy the electricity at the utility's avoided cost. The definition of a "qualifying facility" relates to size, which generally may not exceed 80 megawatts, and fuel use, which must be mainly biomass, waste, or renewable or geothermal resources. "Avoided costs" means the incremental costs to an electric utility of electric energy or capacity or both which, but for the purchase from the qualifying facility or qualifying facilities, such utility would generate itself or purchase from another source [Regulations Under Sections 201 and 210 of the Public Utility Regulatory Policies Act of 1978 With Regard to Small Power Production and Cogeneration, 18 U.S.C. 292.101 et seq., undated; Cogeneration and small power production, 16 U.S.C. 824a–3]. Derivation of avoided costs involves an energy component and a capacity component. Capacity planning must consider the 10 succeeding years [Availability of electric utility system cost data, 18 U.S.C. 292.302]. The Energy Policy Act of 2005 loosened the requirements on utilities to purchase electricity from small producers who have nondiscriminatory access to wholesale electricity markets. There is a rebuttable presumption that qualifying facilities below 20 megawatts do not have nondiscriminatory access to the market [Termination of obligation to purchase from qualifying facilities, 18 U.S .C. 292.309].

The value of digester electricity is an important factor in economic feasibility, and negotiating favorable terms with local utilities for valuing avoided purchases and excess sales has long been a concern. The United States and the European countries have followed somewhat different paths in encouraging renewable sources of electricity. In the United States, the Federal Government and the States have tended to set quantitative mandates or targets, while European countries have tended to set minimum purchase prices ("feed-in tariffs") that utilities must pay suppliers of renewable electricity. Kildegaard argues that the minimum-price approach is more efficient [Kildegaard, 2006].

## STATE POLICIES AND ELECTRIC RATE DIFFERENTIALS

Farm-based digesters can be found in most of the States with significant dairy or swine industries. California, New York, Pennsylvania, and Wisconsin are the only States with more than five digesters each, however [Ball, 2007]. What explains the popularity of digesters in those States?

Digesters are a technology that produces electricity, requires large volumes of livestock manure, and minimizes livestock odors, so it could be expected that States with high electricity rates and/or large livestock operations would be likely to explore digester development. Data are not available on the value of the electricity generated at most farm-based digesters, except for a few situations discussed in the economic assessment section below. Most States have adopted rules implementing PURPA [National Rural Electric Cooperative Association, 2005]. For example, Minnesota's implementation language is in Minnesota Statutes section 216B. 164 [Cogeneration and Small Power Production, Minnesota Statutes 2 16B.164, 2006].

The Kildegaard paper describes the rulemaking procedure that the Minnesota Public Utility Commission has followed to set rates and interconnection standards for distributed electricity [Kildegaard, 2006]. The most difficult problem has been to arrive at the value of the backup generating capacity that the utility must provide in case the distributed facility shuts down, in particular, how the distributed facilities affect the utilities' need to meet projected future demand growth and over what timeframe. A white paper by the National Rural Electric Cooperative Association provides a rural electric cooperative's perspective on the question of how to value distributed generation [National Rural Electric Cooperative Association, 2005]. That paper argues that distributed generation is in danger of being oversold, and that it poses genuine safety and reliability risks, and can pose economic risks to some incumbent utilities and their consumers.

State incentives include renewable electricity portfolio minimums that may induce utilities to offer attractive prices for digester electricity, particularly in States where wind resources are not good enough to make wind farms feasible. For example, Minnesota provides a 1 .5-cent/kwh production incentive and low-interest construction loans for digesters [Minnesota Department of Agriculture, 2007; Minnesota Department of Commerce, 2007]. The California Energy Commission's Dairy Power Production Program was mentioned above [Marsh and LaMendola, 2006]. A description of every State's programs is beyond the scope of this analysis—see the Database of State Incentives for Renewable Energy for State program details [North Carolina Solar Center, 2007b]. An analysis of experiences in 11 States with tradable renewable credits, also known as green tags or renewable energy credits, is provided in Fitzgerald et al. (2003).

**Table 1. State Electricity Rates Compared with the U.S., Danish, and German Averages, with Numbers of Milk Cows, Pigs, and Digesters**

| | Average retail electricity rates, cents/kwh[1] | | | Dairy cows (000) | Pig crop (000) | Number of digesters in 2007 |
|---|---|---|---|---|---|---|
| | All sectors | Residential | Industrial | | | |
| U.S. total | 8.83 | 10.55 | 6.12 | 9,129 | 105,259 | 111 |
| California | 12.15 | 12.20 | 9.71 | 1,790 | 330 | 15 |
| Connecticut | 14.79 | 17.27 | 11.70 | 19 | 5 | 2 |
| Florida | 10.50 | 11.43 | 7.72 | 130 | 56 | 1 |
| Idaho | 4.66 | 6.10 | 3.10 | 502 | 50 | 1 |
| Illinois | 7.19 | 8.79 | 4.73 | 103 | 7,377 | 4 |
| Indiana | 6.45 | 8.69 | 4.86 | 166 | 4,858 | 3 |
| Iowa | 6.71 | 9.58 | 4.87 | 210 | 16,583 | 3 |
| Maryland | 10.78 | 10.47 | 12.05 | 60 | 64 | 1 |
| Michigan | 8.24 | 9.83 | 6.13 | 324 | 1,765 | 3 |
| Minnesota | 6.81 | 8.66 | 5.31 | 455 | 10,209 | 2 |
| Mississippi | 7.55 | 8.83 | 5.60 | 22 | 695 | 1 |
| Nebraska | 5.83 | 7.32 | 4.41 | 60 | 6,514 | 1 |
| New York | 14.33 | 16.92 | 9.17 | 628 | 149 | 16 |
| North Carolina | 7.68 | 9.80 | 5.35 | 48 | 20,048 | 3 |
| Oregon | 6.71 | 7.62 | 4.89 | 115 | 52 | 5 |
| Pennsylvania | 8.48 | 10.54 | 6.38 | 550 | 1,676 | 16 |
| Texas | 10.16 | 12.89 | 7.51 | 347 | 1,651 | 3 |
| Utah | 6.06 | 7.43 | 4.40 | 86 | 1,365 | 2 |
| Vermont | 11.48 | 13.84 | 8.10 | 140 | 5 | 4 |
| Virginia | 6.78 | 8.75 | 4.81 | 100 | 548 | 1 |
| Washington | 6.51 | 6.99 | 5.17 | 235 | 48 | 2 |
| Wisconsin | 8.09 | 10.59 | 5.90 | 1245 | 935 | 20 |
| Wyoming | 5.49 | 8.27 | 4.37 | 7 | 377 | 2 |
| Denmark, 2004 | | 28.30 | 9.60 | | | |
| Germany, 2004 | | 19.80 | 7.70 | | | |

[1]The U.S. electricity prices are for October 2006. The latest Danish and German rates available through the EIA are for 2004.

Sources: The electricity rates are from the U.S. Energy Information Administration. Milk cow and pig crop numbers are from the USDA National Agricultural Statistics Service. The digester numbers are from the AgSTAR 2007 Update.

Local governments sometimes require new or expanded livestock facilities to install digesters for odor control. Electricity may then be generated as a side benefit, or the farm may decide to just flare the gas. Local policies vary widely, but in Minnesota large new feedlots typically must have public hearings before they are permitted (see, for example, the feedlot ordinance for Nicollet County, Minnesota [Department of Environmental Services, Nicollet County, Minnesota, 2007]). Digesters are not explicitly required in that ordinance, but might be one of a range of technologies that the feedlot operator and officials could consider in addressing concerns raised by the public.

The avoided costs that PURPA required utilities to use as a basis for payment rates generally are not public information, but retail electric rates may serve as a proxy for avoided costs. Retail electric rates are a function of utility costs for generation, transmission, and distribution. Assuming that the main costs avoided are for generation and that transmission and distribution costs are relatively constant across States, average retail electricity rates are likely to provide at least a rough estimate of the variation in avoided costs across States. Table 1 gives a breakdown of the 2006 average retail electricity rates in the 23 States that had a total of 111 digesters [Ball, 2007]. The most recent electricity rates available for Denmark and Germany (for 2004) are shown for comparison. The numbers of milk cows as of January 2007 and the 2006 pig crops are also indicated. Figure 1 shows how the digester numbers relate to the electricity rates.

California, Florida, Maryland, Texas, and Vermont, like New York and Connecticut, have double-digit electricity rates. California also has the most dairy cows, concentrated on quite large operations in an area where air quality is a concern. Dairy farm digesters are one part of a broader energy and waste management project in the "Inland Empire" region of California that lies east of Los Angeles [Inland Empire Utilities Agency, 2006]. The California Energy Commission's Dairy Power Production Program provides construction buydown grants and electrical generation incentive grants; to date, 14 projects have been approved for grants totaling $5,792,370. The projects have an estimated generating capacity of 3.5 megawatts. Five dairy farm digesters funded by this program have been operating long enough to be described in 90-day evaluation reports in 2006, with others evaluated in 2005 [Marsh and LaMendola, 2006].

Wisconsin is interesting in that the State has more digesters than California but cheaper electricity (8.09 cents compared to 12.15 cents/kwh). A comprehensive discussion of each State's situation is beyond the scope of this paper, but a few possible reasons can be identified for the popularity of digesters in Wisconsin. One factor is that Wisconsin was one of the first States to enact a renewable electricity portfolio standard [North Carolina Solar Center, 2007b]. States further west, such as Minnesota, are relying mainly on wind as a source of renewable electricity, but Wisconsin's wind resources are not as good, so there is more of a need for digesters to meet that requirement. Also, Wisconsin is the second-ranked dairy State in the United States. The Wisconsin dairy industry has been consolidating into larger operations with new facilities, so the addition of a digester to a facility construction plan may be a logical step. Many of the Wisconsin digesters are plug-flow designs, which will work with dairy manure and are less costly than the complete- mixed digester designs that are generally used with swine manure, so that investment requirements are somewhat less than for States where swine is the predominant livestock type.

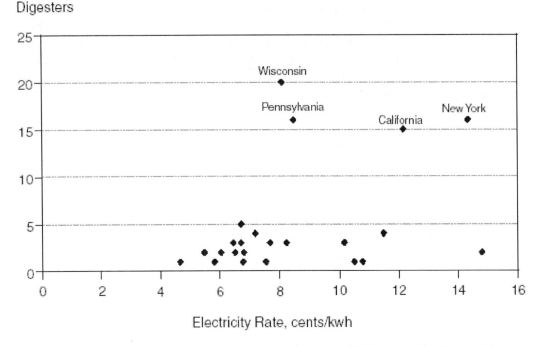

Figure 1. U.S. Digesters by State in 2007 Compared With October 2006 Average Retail Electricity Rates

Aside from the retail electricity rates, State policies on distributed power pricing and interconnection also differ among States and, some authors argue, can greatly affect the economic viability of digesters and other distributed power generation sources. Contractual arrangements for digesters operating in parallel with electric utilities tend to be of three types: "buy all-sell all," "surplus sale," and "net metering" [U.S. AgSTAR, 2006b, chapter 5]. In a buy all-sell all arrangement, the utility usually offers an avoided cost rate for all electricity generated, whether the amount generated exceeds usage on the farm or not. In a surplus-sale arrangement, excess generation is sold at the avoided cost rate, while excess usage is purchased at the retail rate. This allows the farm to realize the retail value of the amount used on the farm. However, some utilities impose "demand" or "standby" charges to pay for the availability of electricity to the farm when the generator is not running. At least one digester operator avoids standby charges by purchasing electricity on an interruptible basis. He runs a backup generator (separate from the digester generator) for 80 to 100 hours of peak demand per year when requested by the utility, and so also has the backup generator available when the digester generator needs servicing [U.S. AgSTAR, 2007b].

Net metering allows customer/generators, such as farm digester operators, to spin their meters backwards, in effect paying the customer/generator the retail rate for the electricity they generate but do not immediately consume. If customers generate more electricity than they use over a specified period, they are typically paid for the net excess generation at the utility's avoided cost, or at the wholesale rate. In some cases, however, they are paid at the retail rate, or their excess electricity may be granted to the utility with no payment [Chapman, 2006; New Rules Project, 2007]. Lack of net metering may not necessarily be a barrier to digester installation if the other terms of the electricity sales agreement are favorable, but availability of net metering with a favorable payment for excess generation might help stimulate digester adoption. The Energy Policy Act of 2005 requires States to offer some type of net metering (see 109th Congress, 2005, Title XII, Subtitle E, section 1251), but the specifics are regulated by the States. Maximum sizes of generators that qualify for net metering, and maximum percentage of total utility capacity vary by State; digesters qualify in some States and not in others. Most digesters have generators of at least 100 kilowatts, so a net metering maximum below that level is unlikely to help promote digesters. Of the four States with more than five digesters, California and New York have net metering caps greater than 100 kilowatts. California has a 1-megawatt maximum, whereas New York's maximum for digesters is 400 kilowatts [Chapman, 2006]. Wisconsin's and Pennsylvania's caps are too low to be beneficial to digesters—Wisconsin's limit is 20 kilowatts, whereas Pennsylvania has a 10- kilowatt cap [North Carolina Solar Center, 2007b; Chapman, 2006].

New York and Connecticut have the highest electricity rates of the States with significant digester activity. The New York State Legislature formed the New York State Energy Research and Development Authority (NYSERDA) in 1975 with funding from an assessment on utilities [New York State Energy Research Development Agency, 2007]. Since that time, Cornell University has had an active research and extension program on digesters. The program is responsible for developing the plug-flow digester design, which has lower capital costs and is easier to operate than the early complete-mix digesters [Lusk, 1998]. The Cornell program has issued papers on a variety of topics, including biogas processing, use of manure solids for dairy cow bedding, and operating experience and economic comparisons of digesters on a number of New York dairy farms [Cornell Manure Management Program, 2007].

The Central Vermont Public Service Corporation is encouraging dairy farm digesters by offering financial incentives, along with a 4-cent/kwh premium on digester electricity. The 4-cent premium is a voluntary addition that over 2,500 consumers have chosen to pay on their electric bills [Dunn, 2007]. So far, one Vermont farm is operating a digester under the program.

As a possible adjunct to the electricity being generated by digesters in California, the potential for producing biomethane from dairy manure and other wastes has been studied [Krich et al., 2005]. Krich includes biogasupgrading cost estimates, based on information from four Swedish upgrading plants, and a discussion of prospects for siting farm digesters to feed the biomethane into existing compressed natural gas outlets. The feasibility of operating centralized digesters in California has also been analyzed [Hurley, Ahern, and Williams, 2006].

Pennsylvania also has relatively cheap electricity, but it has 16 digesters. Pennsylvania has a relatively large number of rural residences interspersed with its livestock operations, so urban sprawl-related odor concerns may have been a more significant motivation for digesters there than in the more rural States further west. Illinois and Iowa have low electricity rates, but large swine industries, with odor concerns related to urban encroachment that were the motivation for early digesters [Lusk, 1998]. While North Carolina's electricity rates are relatively low, North Carolina State University has an active livestock waste management research program, funded via an agreement between the North Carolina Attorney General and six large swine operations to preserve and enhance water quality [North Carolina State University, 2007]. This research focuses on innovative digester designs along with other technologies for air and water quality preservation and enhancement. Oregon and Washington have low-cost electricity, but they have water quality concerns that have motivated digester development [Environmental Resource Recovery Group, LLC, 2003; Port of Tillamook Bay, 2006].

## TECHNOLOGY AND ECONOMICS OF FARM-BASED DIGESTERS

Digester technology is discussed in detail in many of the references mentioned above. This section will highlight only a few common themes and issues particularly relevant to energy policy. For more details, a reader might start with the AgSTAR handbook for general background; the Martin protocol description, with and without comparison on performance measures; Lusk for historical perspective; Krich et al. on biomethane; and the Marsh and LaMendola reports on startup issues [U.S. AgSTAR, 2006b; Krich et al., 2005; Martin, 2006; Martin, 2004; Lusk, 1998; Marsh and LaMendola, 2006].

The three main designs for farm-based digesters are the covered anaerobic lagoon, plug-flow, and complete mix (or continually stirred tank reactor) (see, for example, Krich et al., 2005). Digester temperature must be regulated, so anaerobic digesters are also classified by working temperature. Those that work at temperatures between 95 and 105 degrees Fahrenheit. are called mesophilic, whereas those that work between 120 and 140 degrees are known as thermophilic. Covered lagoons operate at "psycrophilic" temperatures lower than 95 degrees. They are lower in cost and are commonly used where odor control is the main objective. However, in some warmer locations covered lagoon digesters are successfully used to produce energy. Digestion also occurs naturally at temperatures lower than 95 degrees in ponds, swamps, and open lagoons.

The solids content of the material to be digested is an important criterion in the choice of digester design. Plug-flow digesters work best at a solids content of 11–13 percent, so they work well with dairy manure from operations that collect it by scraping or other methods that do not add much additional water. Complete-mix digesters work at a wider range of 2–10 percent solids, which makes them suitable for a greater variety of materials, including swine manure and processing wastes as well as dairy manure. For more dilute wastes such as those in municipal sewage treatment plants or flush manure systems, "fixed-film" or "filter" digesters are designed to retain the bacteria on some type of medium long enough to break down the waste rather than allowing it to be immediately flushed out of the system. See the Sebesta report for a discussion of several newer digester designs [Sebesta Blomberg & Associates, Inc., 2005].

It is not practical to run the manure from all livestock through digesters. The potential for methane production from livestock waste depends on: (1) size of the farm operation, (2) freshness of the waste, and (3) concentration of digestible materials in the manure. EPA AgSTAR staff have identified 500 dairy cows or 2,000 head of swine as the minimum for which a digester is likely to provide positive financial returns [U.S. AgSTAR, 2007]. Forty percent of the dairy cows and approximately 75 percent of the swine are on operations larger than the AgSTAR figures. Free-stall dairy operations with daily-scraped alleys work well with digesters because the manure does not get mixed with dirt or stones and is moved into the digester while fresh. Drylot dairies, beef, sheep, and poultry operations work less well because the manure may decompose before it is scraped. Flushed manure collection systems also produce less gas because the digestible materials are diluted. Deep-pitted swine-finishing buildings would need to be modified to remove manure more frequently before a digester would be practical.

## Biogas Use

Most farm-based digesters use the biogas output to fuel internal combustion engines that generate electricity. The electricity is most valuable when used on-farm, but many digester systems have the capacity to generate more than the farm needs. The economic feasibility of selling the extra electricity to the local utility varies widely.

If the number of digesters with internal combustion engine generators were greatly increased in the United States, emissions of the regulated air pollutant NOx could become an issue. Lean-burn engines with relatively low NOx emissions are currently available in sizes of 350 kilowatts or more, but that is larger than most individual farm digesters currently in use (see Krich et al. (2005), pp. 36-37). Microturbines are another widely discussed option, but they have not been used successfully on digesters because biogas impurities corrode the engines. Fuel cells would be cleaner, if they could be developed to the point of commercial feasibility.

Mesophilic and thermophilic digesters typically use engine heat to heat the digester. Engine heat is also often used for heating farm buildings, barn alleys, and water. The biogas can also be burned directly for space heating, which is most common at small digesters where an engine would be uneconomical.

As mentioned earlier, in Sweden biogas is being upgraded, injected into the natural gas grid, and used as a transportation fuel [WestStart-CALSTART, 2004]. Biogas is also being used in combined heat and power plants in villages in Denmark. Krich et al. evaluated the feasibility of this use in the United States.

Two digester installations, in Texas and Wisconsin, began marketing biogas that had been upgraded to biomethane or "renewable natural gas" in 2007 [Smith, 2007; Agri-Waste Energy, Inc., 2007]. The Texas facility, Huckabay Ridge, is located at a dairy manure composting plant. This location provided a large volume of manure, but has added complications in material handling and maintaining the desired solids levels compared to locating at a dairy farm. Huckabay Ridge has also experienced startup problems with gas cleanup and compression equipment and problems with the digester chemical balance that were caused by delays in acquiring a permit for land application of separated liquids, but those problems have now apparently been solved [USDA Natural Resources Conservation Service, 2007].

## Feedstocks

Livestock manure is the main feedstock for farm-based digesters, but there are other important feedstocks, including organic wastes from nonfarm sources like food processing industries, as well as energy crops that are digested without going through livestock first. Almost half of the 2005 biogas production in Thuringia State, Germany, was from energy crops [Effenberger, 2006]. The biogas potential of different feedstocks varies widely [Effenberger, 2006; Steffen et al., 1998].

Nonfarm industries that have organic wastes to dispose of will sometimes pay tipping fees to a farm digester to accept the waste. For the digester enterprise, the tipping fees can be an important side benefit of accepting this feedstock, making the difference between profit and loss (see, for example, the comparison of New York State farm digesters by Wright et al. (2004) and the analysis of centralized Danish digesters by Nielsen and HjortGregersen [Wright et al., 2004; Nielsen and Hjort-Gregersen (undated)]. Industrial organic wastes are an important digester feedstock in Europe [Al Seadi, 2000; WestStart-CALSTART, 2004]. European digesters that accept industrial waste are required to either operate at thermophilic temperatures or pasteurize the wastes at high temperatures before digestion to ensure the destruction of pathogens [Bendixen, undated]. Pasteurization is reported to have the side benefit of speeding up gas production. The required pasteurization tanks and heat exchangers likely add to capital investment and operating costs, but the magnitude of the added cost is not detailed in any of the publicly available economic analyses identified for this chapter. Another potential concern is that on livestock farms with small land bases, the livestock manure alone may already have too much nitrogen and phosphorus for the cropland available. Imported nonfarm organic wastes would contain additional nutrients, which could exacerbate the cropland nutrient imbalance. The tipping fees and added gas output need to be weighed against potentially greater manure hauling costs to take the effluent to more distant cropland, where the nutrients can be utilized.

## Digestate Utilization and Benefits Relative to Raw Manure

Digestion is generally considered to reduce odor during land application of the digestate because the odor-causing volatile organic compounds are converted to more stable forms during digestion. (See pages H- 119 through H-123 of Jacobson et al. for principles and empirical studies of the impact of digestion on odor [Jacobson, Moon, Bicudo, Janni, and Noll, 2007]). The reduced odor potential may have economic value to the livestock operation when spread on the operation's own cropland if it minimizes the chance of neighbors' complaints or nuisance lawsuits. The reduced odor may also make the digestate more marketable to crop farms. The economic analyses of U.S. digesters reviewed for this chapter have not included estimates of the economic value of reduced odor, however.

Pathogen reduction is another frequently cited benefit of digestion. Martin's two-farm comparison found a 99.9-percent reduction in fecal coliforms and *Mycobacterium avium paratuberculosis* [Martin, 2004]. The California spinach recall in 2006 demonstrates the potential food supply disruption and costs that pathogens can cause. No studies have been identified that have quantified the economic benefit of pathogen reduction, however, or the percentage reduction that would be required in order to provide an economic benefit in a given situation.

Many of the farm digester systems described in the literature include separators that extract manure solids that may be used as a bedding material in dairy farm free-stall barns, or sold as a soil amendment for landscaping or gardening. Separation also may help reduce manure application costs because the remaining liquid can be applied at higher per acre rates on cropland close to the livestock facility, since separation removes some of the nutrients from the liquid. The solids are nutrient-dense but have less volume, so they can be hauled to fertilize distant fields at less cost than hauling the original manure. There appear to be few empirical economic assessments of solids separators that document the impact on land-application costs.

These benefits of solids separators can be achieved without incorporating a digester in the system. For that reason, the AgSTAR digester evaluation protocol recommends setting boundary conditions for digester evaluations that leave out the separator part of the system [Martin, 2006]. One reason to include the separator in digester evaluations is that many dairy farms that do not have digesters use sand bedding, and sand tends to settle out in a digester and reduce capacity until it is cleaned out, which can be an expensive process [Moser and Langerwerf, 2004]. A farm operator who might not otherwise consider a separator may decide to install one as part of a digester system so that the solids are available to replace the sand that can no longer be used.

Manure solids bedding is controversial because of concerns that it might increase mastitis problems in dairy cows. The concern is greatest in warm and moist conditions. Determining the impact on mastitis is complicated by a number of factors, including different ways of measuring the concentration of bacteria in bedding (by weight on a wet or dry basis or by volume); changes in bacterial levels in bedding during the time it sits in the stall; the relationship between bacteria in the bedding and on teat ends; and the impact of bacteria in bedding and on teat ends on the occurrence of mastitis and on milk quality. These aspects of solids bedding for dairy animals are summarized in a recent literature review by the Cornell Waste Management Institute [Cornell Waste Management Institute, 2006]. The mastitis associated with factors such as bedding is termed "environmental mastitis," as opposed to "contagious mastitis." A common way of detecting mastitis is to estimate the somatic cell count (SCC) of a milk sample, which consists mainly of white blood cells, or leucocytes. A high SCC count is correlated with reduced milk production, and reduces the milk's value for uses such as cheesemaking. A high bulk tank SCC count generally indicates a problem with contagious mastitis. Herds with lower bulk tank SCC counts have lower levels of subclinical mastitis and better udder health. However, leucocytes in the udder help protect it from other sources of mastitis, so low SCC counts may predispose cows to environmental mastitis. The literature review suggests that more research is needed to clarify the impact of bedding type on mastitis, in the context of the many other management factors on a typical dairy farm.

The economic value of separated and composted manure solids as an off-farm soil amendment appears to vary widely, depending on the seller's marketing expertise and location. Compost may have a higher market value if sold as a replacement for fertilizer, fungicide, or sand-based golf course topdressing material rather than as a substitute for peat moss. In an online publication, Alexander (2004) provides some example prices for these products that might be useful for anyone who is considering their use [Alexander, 2007].

A centralized digester was recently proposed to serve an estimated 15 dairy farms in King County, Oregon, with a total of 6,075 cows [Environmental Resource Recovery Group, LLC, 2003]. A followup study examined the feasibility of marketing the digested manure solids from that digester [Terre- Source LLC, 2003]. The Terre-Source study found that a number of the original study's assumptions were not realistic. It found that the cost of handling the solids, which had been estimated at $5 per ton, was actually $ 10–$ 15 per ton at a similar operation. It also found that equipment and storage space may also have been underestimated.

The followup study further concluded that a 3–5 year ramp-up period would be needed to develop the solids market before the projected price could be achieved. Marketing staff would be needed. The solids volume would be reduced due to solids degradation and moisture loss, there might be odor problems, and sulfides in the solids might need to be dealt with (see Terre-Source, p. 22).

Discussion of solids marketing raises a question about the potential size of the U.S. market for digested manure solids, assuming that the number of digesters becomes greatly expanded and that it proves profitable to market the solids throughout the United States (obviously a long way from the current situation.) The Canadian Sphagnum Peat Moss Association claims to supply over 98 percent of the peat moss used in the United States and to have sold 10.3 million m3 to the United States in 1999 for almost $170 million [Canadian Sphagnum Peat Moss Association, 2007]. That works out to $ 16.50/m3. A study by Hurley et al. (2006) assumes a solids production rate of around 17 yd3/cow/year or 13 m3/cow/year [Hurley, Ahern, and Williams, 2006]. If the market for manure solids could be expanded to replace, say, 25 percent of that peat moss, that would be around 2.5 million m3. This would equal the supply from digesters for around 200,000 dairy cows, assuming that all of the digesters were for dairy rather than swine or poultry. There were 9 million dairy cows in the United States in 2005, so that would be 2 percent of the total dairy herd.

## Economic Assessments of Farm-Based Digesters

A number of digesters have been described in case study reports and other publications over the past two decades (see, for example, Kramer, 2004; Lusk, 1998; Lusk, 1995; Martin, 2005; Martin, 2004; Wright et al., 2004). The ManureNet Web site lists 32 other North American and European studies under the "Economic Assessments" heading [Agriculture and Agrifood Canada, 2005]. The capital costs from 38 digesters described in Kramer, Wright et al., and the two Lusk reports have been summarized in Beddoes et al. The latter report found that on average 36 percent of total investment on these digesters was for the electrical generation equipment, suggesting that substantial cost savings may be possible in situations where the biogas can be used for heating rather than to produce electricity.

Methodologies and the amount of economic information on revenues and operating costs vary across these studies, however, making an overall assessment difficult. Martin (2006) suggests that, "... the decision to construct and operate manure-based biogas systems depends ultimately on the anticipated ability to at least recover any internally derived capital investment with a reasonable rate of return and service any debt financing over the life of the system. Otherwise, other investment opportunities become more attractive unless the need for environmental quality benefits, such as odor control, justify the net cost of system operation." It is fairly common for these studies to provide the size and capital cost of the digester. Data are published less often on biogas output, electricity prices, and maintenance costs. It is axiomatic that any technology can be profitable if subsidized heavily enough, so information on grants and operating subsidies received for digesters is important, but it is not always provided.

Few of the published economic assessments appear to be peer reviewed, so accuracy and any bias toward excessively optimistic or conservative assumptions are difficult to evaluate.

Even when assessments include comprehensive sets of actual operating performance data, the assessments are usually done fairly early in the expected useful life. Future costs for engine overhauls, flexible cover replacements, and other maintenance, along with gas output declines as the digester fills up with sludge, are difficult to predict, as are future electrical rates.

A few overall conclusions can be drawn from the studies that are relatively recent and complete. First, under current economic conditions digester profitability appears marginal when manure is the primary feedstock, when electricity is the primary source of value, when not all of the electricity can be used onsite, when electricity retail rates (as a proxy for utility avoided costs) are around the national average, and when subsidies are minimal or are left out of the analysis. Second, there are digesters that appear to be operating profitably or that may not be quite covering financial costs, but that are viewed as successful because they are providing nonmonetary benefits such as odor control.

Livestock operations often expand over time depending on profitability, so determining the optimal size for a digester that is expected to operate for 10 years or more can be difficult. Martin discusses one digester that was oversized in anticipation of a dairy herd expansion that did not happen. In this situation, the profitability of a digester can be analyzed two ways: (1) as operating at less than full capacity, or (2) as if it were operating at full capacity. The second approach gives an indication of the potential of the technology, but it may be worth also looking at the as-operating performance, given that future farms may not always be able to anticipate expansion with certainty.

A digester located at the ML dairy farm in New York is profitable [Wright et al., 2004]. The estimated annual revenues are reported as $287,685, which is over half the total capital cost of $490,269. Profitability in this case is largely the result of tipping fees and expanded gas production from off-farm food processing wastes that the farm is accepting for digestion. The profit- ability projections are based on a relatively favorable electricity price of 10 cents/kwh. Wright also mentions that this farm has received grants, but the amounts were not specified. Another New York digester, at AA Dairy, was analyzed by both Wright et al. and Martin [Martin, 2004]. Wright and colleagues base their analysis on a 10-cent/kwh electricity price, but Martin notes that around a third of the electricity was excess sold to the utility for a lower price, which averaged around 5.25 cents/kwh. That digester is profitable in the Wright calculations (at the 10-cent/kwh price) and would have a 7.5-year payback in the Martin calculations as operated, or 2.8 years if it could be operated at full capacity. In the author's calculations, AA Dairy does not quite cover the digester costs if the revenue is recalculated based on the electricity price information from Martin. The Wright article describes three other New York digesters that were losing money. The AA Dairy digester was also described as receiving a grant, although the amount was not specified.

The Haubenschild Farms digester in Minnesota has also been profitable [Lazarus and Rudstrom, 2007]. This digester received grants and subsidized financing, and received a 7.3-cent electricity price for its sales in the first 5 years of operation. In contrast, the Tillamook, Oregon, centralized digester has had financial difficulties during its first few years of operation, due partly to higher-than-expected manure transportation costs and lower-than-expected revenue from solids sales [DeVore, 2006].

The investment required to generate electricity via a farm-based digester is somewhat higher than for two other non-fossil-fuel-based electrical generation technologies, a wind generator or an advanced nuclear power plant (table 2). The nuclear costs, from the U.S. Energy Information Administration, are obviously speculative since no nuclear plants have been built in the United States for many years. The investment amounts are shown per cow (for the digesters), per kilowatt of generating capacity. In interpreting the measures of the wind generator's per kilowatt production capacity, it is important to remember that a wind generator produces only when the wind is blowing. The last column of the table adjusts the wind amount for an assumed 35-percent capacity factor to make it more comparable with the digesters. The digester investment/kw of output numbers is calculated by taking the estimated annual electricity output from the reports, divided by 365 days and 24 hours.

Table 3 shows several profitability measures calculated for the four digesters compared with the wind generator, with the electricity price assumed as an average of the electricity purchases avoided and the excess sales. The profit- ability of the Haubenschild digester considers the investment buydown from the grant as well as the benefit of a zero-interest loan. The amounts of grants received by AA and ML Dairies are not included.

Figure 2 is a graphical representation of the costs and benefits for these four digesters versus the wind generator. The investments, costs, and benefits were annualized and then divided by annualized total cost for comparability, using the assumptions described in Lazarus and Rudstrom [Lazarus and Rudstrom, 2007]. The bottom line on the graph shows annualized operating costs, with annualized total costs as the top line. The stacked bars show the various revenue sources. The ML and AA benefits were provided only as undifferentiated totals, rather than by breaking out co-generated heat and solids or digestate value as was done for the Haubenschild and Tillamook digesters. Finally, carbon credit trading may offer a way to internalize the greenhouse gas reduction externality presented by farm digesters. Haubenschild Farms has also recently begun selling carbon credits through an intermediary to the Chicago Climate Exchange, although the annualized value of the credits appears negligible in Figure 2 partly because the sale was not arranged until the digester had been operating for most of its estimated useful life [Bilek, 2006; Haubenschild, 2006]. In present value terms, the production tax credits for which the wind generator is eligible are larger relative to the electrical market value than are the grants and interest subsidy that the Haubenschild digester received.

Digester biogas and electricity, like other renewable energy sources, compete with fossil fuels. Hence, the economic feasibility of farm-based digesters will be dictated to a large extent by the direction of fossil fuel prices in the future. The oil price swing from $50 per barrel in the early 1980s to $10 in the late 90s and up to over $100 in 2007 was accompanied by a rash of digester terminations, followed by a recent resurgence of interest in digesters.

Looking ahead, prospects seem good for digesters given that further fossil fuel price increases seem likely, but history suggests that the downside risk cannot be discounted entirely.

The renewable electricity portfolio minimums enacted in many States are stimulating the interest of utilities in all forms of renewable electricity, including that from digesters, but here wind electricity is a competitor. For example, competition from wind electricity appears to be placing an implicit ceiling on digester electricity rates in Minnesota, where wind generators are being installed at a rapid rate. A common rate for wind electricity is 3.3 cents/kwh on a 20-year flat-price contract [Tiffany, 2005].

**Table 2. Capital Investments Made for Selected Dairy Farm Digesters Compared with a Wind Generator and an Advanced Nuclear Plant**

| | Cows | Generating capacity, kw | Investment/cow | Investment/ kw capacity | Investment/ kw output |
|---|---|---|---|---|---|
| AA Dairy, NY, 1998 | 950 | 140 | $374 | $2,536 | $2,941 |
| ML Dairy, NY, 2001 | 550 | 130 | $446 | $1,886 | $4,118 |
| Haubenschild Dairy, MN, 1999 | 740 | 135 | $663 | $3,632 | $3,632 |
| Tillamook (central), OR, 2003 | 4,000 | 500 | $425 | $3,400 | $3,723 |
| Four California dairies average, 2005[1] | 3,176 | 267 | $561 | $3,774 | NA |
| Wind generator, 2005 | NA | 1,650 | NA | $1,003 | $2,866 |
| Advanced nuclear, 2002 | NA | 1,000,000 | NA | $2,117 | NA |

[1]The average herd size for the four California dairies includes heifers and bulls, but not calves. Marsh and LaMendola also prepared a report on the centralized digester at the Inland Empire facility, which was an upgrade to an existing digester. The investment at that facility was somewhat cheaper on a per cow basis but more expensive per kilowatt of generating capacity. Investment per kw of output was not calculated for these four digesters because they had only been operating for 90 days, so the output numbers provided may not be indicative of longer-run performance.

Sources: The AA Dairy is described by both Wright et al., (2004) and Martin, (2004). The ML Dairy data are from Wright et al. The Haubenschild data are from Lazarus and Rudstrom (2007). The Tillamook data are from DeVore (2006). The California dairy data are from Marsh and LaMendola (2006). The wind generator data are from Tiffany (2005). The nuclear plant data are from the Energy Information Administration (undated).

**Table 3. Electricity Price and Profitability Indicators for Selected Dairy Farm Digesters Compared with a Wind Generator**

| | Electricity price, $/kwh | Payback, years | Return on assets | NPV, annualized |
|---|---|---|---|---|
| ML Dairy, NY, 2001 | 0.100 | 2 | 39% | $134,463 |
| AA Dairy, NY, 1998 | $0.086 | 8 | 3% | $(3,495) |
| Haubenschild Dairy, MN, 1999 | $ 0.058 | 4 | 12% | $9,392 |
| Tillamook (central), OR, 2003 | $0.047 | 10+ | <-10% | $(247,666) |
| Wind generator, 2005 | $ 0.033 | 7 | 9% | $31,034 |

Sources: The AA Dairy is described by both Wright et al., (2004) and Martin, (2004). The ML Dairy data are from Wright et al. The Haubenschild data are from Lazarus and Rudstrom, 2007. The Tillamook data are from DeVore (2006). The California dairy data are from Marsh and LaMendola (2006). The wind generator data are from Tiffany (2005).

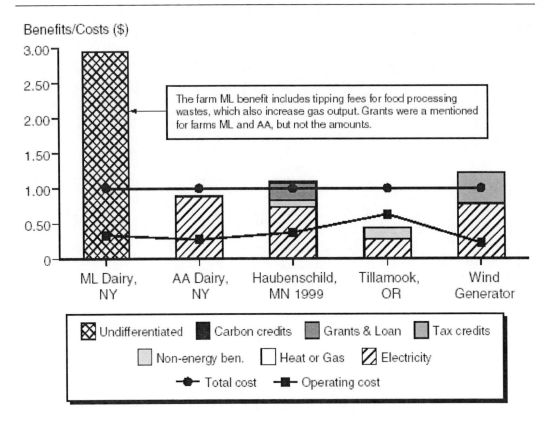

Figure 2. Comparison of Economic Assessments of Selected Dairy Farm Digesters Compared with a Wind Generator Under Minnesota Conditions

## Digester Economies of Size

The relationship between digester size and capital cost is difficult to estimate at present due to the small number of operating digesters, differences in digester design, and inflation over time. AgSTAR staff have estimated the cost-size relationship for 15 dairy plug-flow digesters with flexible covers and internal combustion engine-generator sets, based on dairy herd size [U.S. AgSTAR, 2006a]. They arrived at the regression equation:

Plug-Flow/Flexible Cover/Dairy:
$$y = 226.69x + 288{,}936 \ (R^2 = 0.76)$$

The individual farm data are provided only as a scatterplot, so specifics are not available but, based on the scatterplot, 14 of the 15 farms appear to be between 100 and 2,200 cows. The other farm has 7,000 cows. The total capital cost and per cow cost for some representative herd sizes implied by this relationship are:

**Table 4. Cost-Size Relationship for Dairy Farm Plug-Flow Digesters**

| Dairy herd size | Capital cost - total | Per cow cost |
|:---:|:---:|:---:|
| x | y | y / x |
| 100 | $311,605 | $3,116 |
| 500 | 402,281 | 805 |
| 2,000 | 742,316 | 371 |
| 7,000 | 1,875,766 | 268 |

**Table 5. Cost-Size Relationship for Large Dairy Farm Complete-Mix Digesters**

| Dairy herd size | Capital cost - total | Per cow cost |
|:---:|:---:|:---:|
| 9,100 | $5,364,712 | $590 |
| 27,120 | 14,135,387 | 521 |
| 50,774 | 26,216,241 | 516 |

They also estimated cost-size relationships for dairy and swine covered lagoon digesters, but the observations were more limited—five dairy and three swine lagoons. Those regressions were:

$$\text{Lagoon, dairy: } y = 233.43x + 38,056 \ (R^2 = 0.967)$$

$$\text{Lagoon, swine: } y = 63.863x + 35,990 \ (R^2 = 9792)$$

They did not estimate the cost-size relationship for complete-mix digesters, which tend to be larger and more costly than plug-flow designs.

A feasibility analysis of three hypothetical, centralized, scraped-manure, complete-mix digesters for large California dairies provides some information on the cost-size relationship at larger sizes [Hurley, Ahern, and Williams, 2006]. The 1.5-megawatt digester required 9,000–10,600 cows with the manure trucked up to 1 mile (one way), while the 10-megawatt size required 48,200–63,500 cows with a haul distance of up to 4 miles. Hurley' s per cow capital cost estimates for the 100 percent participation scenario are roughly double the AgSTAR equation estimate for the largest plug-flow digester (see their table 4-5):

Part of Hurley et al.' s higher costs may be for the trucks and loading/ unloading facilities that would not be required for an individual farm digester. They found the difference in required electricity prices between a 1.5-megawatt digester and a 10-megawatt one to be 0.5–1.25 cents/kwh depending on assumptions about financing terms and farm participation rate. For example, with 100 percent of the farms participating near the digester location, and a 9-percent interest rate, 9.75 cents/kwh would be required for a 1.5-megawatt digester and 8.50 cents for a 10-megawatt one.

## CONCLUSIONS AND SUGGESTIONS FOR FUTURE RESEARCH

One conclusion of this analysis is that farm-based digesters are a multifaceted technology that offers a range of benefits. The benefits tend to appeal to different policy constituencies, which can become confusing. Digesters are a source of renewable energy. They destroy methane, a greenhouse gas that contributes to global warming. Future digester installations can help address the Nation's energy situation, but their contribution is likely to be small. Odor concerns have been the main motivation for many of the existing digesters. Aside from odor, other factors conducive to economic viability are where the biogas can replace large, onsite retail purchases of electricity or heat; where the digester electricity is sold to the grid in a region with higherthan-average electricity prices; or where offsite organic wastes are available. Evidence so far suggests that digesters do not reduce ammonia emissions, and may increase them slightly. There are other technologies and practices that can address ammonia emissions, such as biofilters and soil injection of manure. Digesters have been included in some large manure handling systems that have been constructed to address water quality concerns; however, it appears that other components of those systems, such as solids separators, are the components that offer most of the water quality benefits. Digesters do not significantly reduce total nutrients in the effluent.

This analysis of existing literature indicates that public funding support and technical assistance along with private entrepreneurship are resulting in advances in digester technologies. Continued growth in the industry can be expected. The expanded application of biogas, in particular conversion to natural gas that can be put on existing pipelines, is already underway at facilities in Wisconsin and Texas.

The concept of integrating farm-based digesters with ethanol plants in a relatively closed system, where the ethanol co-products supply livestock, which then supply biogas to heat the ethanol plant, seems promising. Indications so far are that the associated livestock operation needs to be relatively large and/ or the ethanol plant fairly small compared to industry norms for such a design to work, so it will be interesting to see how this possibility affects overall economies of size and structure of the livestock and ethanol industries.

Two other questions for future research are: 1) How much is digester technology improving over time? and 2) Will those improvements along with policy and market shifts lead to more rapid digester adoption in the future? Regarding the first question, in a certain sense basic digester technology has not changed much since the 1970s. The digester is still a covered tank into which manure or other wastes react to produce biogas, which is then often burned in an internal combustion engine to generate electricity. However, changes are apparent in some areas of digester technology. Hydrogen sulfide removal from biogas is one such area of active academic research and commercial development (see, for example, an evaluation of micro-aeration [Duangmanee, 2007]. New digester designs are also being adopted, such as induced blanket digesters and biogas-mixed plug-flow digesters [Martin and Roos, 2007; Sebesta Blomberg & Associates, Inc., 2005]. Digester monitoring and control systems have also improved, in part due to availability of the Internet and cellular phone technology [Goodrich, 2007].

One suggestion for future research is to consider applying the concept of an "experience curve" to digester technology. The idea is to quantify technological learning over an extended period of time and to identify the drivers underlying the improvements. If there is learning, production costs will tend to decline by a fixed percentage with each doubling in cumulative production. An analysis of the dry mill ethanol industry found that cumulative dry grind ethanol production doubled 7.2 times over the period 1983–2005, while ethanol processing costs (without corn and capital costs) declined by 45% after adjusting for inflation [Hettinga, 2007]. The progress ratio of this curve was estimated as $0.87\pm0.01$, indicating that ethanol processing costs declined 13 percent per doubling in cumulative production. As profit- ability data become available for more digesters, it might be possible to test for a change in costs over time. Such an analysis would be complicated by the range of motivating factors for digester installations and the different sources of value involved.

## REFERENCES CITED

Cogeneration and Small Power Production, Minnesota Statutes 2 16B.164.

Availability of electric utility system cost data, 18 U.S.C. 292.302. Accessed 7/16/2007.

Cogeneration and small power production, 16 U.S.C. 824a–3. Accessed 7/16/2007.

Termination of obligation to purchase from qualifying facilities, 18 U.S.C. 292.309. Accessed 7/16/2007.

Regulations Under Sections 201 and 210 of The Public Utility Regulatory Policies Act of 1978 With Regard to Small Power Production and Cogeneration, 18 U.S.C. 292.10 1 et seq. Accessed 7/16/2007.

109th Congress. Energy Policy Act of 2005, P.L. 109-58. Accessed 7/18/2007.

Agri-Waste Energy. (2007). Inc., *Turning Animal Waste into Energy* (Web page), http://www.agriwasteenergy.com/home.html, accessed 11/9/2007.

Agriculture and Agrifood Canada. (2005). *ManureNet Canada: Manure Digesters Background Information* (Web page), http://res2.agr.ca/initiatives/ manurenet/en/man_ digesters_back.html, accessed 6/28/2005.

Al Seadi, T., Hjort-Gregersen, K. & Holm-Nielsen, J. B. Impact of the Legislative Framework on the Implementation and Development of Manure Based, Centralised Co-Digestion Systems in Denmark. undated.

Al Seadi, Teodorita. Danish Centralized Biogas Plant Descriptions. 2000. Bioenergy Department, University of Southern Denmark. http://websrv5. sdu.dk/bio/pdf/rap2.pdf.

Alexander, Ron. (2004). *Strategies for Increasing the Value of Biosolids Compost* (Web page), www.alexassoc.net, accessed 2/5/2007.

Ball, Jeffrey. Gas Leak: Kyoto's Caps On Emissionism Hit Snag In Marketplace. *Wall Street Journal,* A1. 12/3/2007.

Beddoes, Jenifer C., Bracmort, Kelsi S., Burns, Robert T. & Lazarus, William F. An Analysis of Energy Production Costs from Anaerobic Digestion Systems on U.S. Livestock Production Facilities, Technical Note No. 1. 10/2007. USDA Natural Resources Conservation Service. http://directives

Bendixen, H. J. Hygiene & Sanitation Requirements in Danish Biogas Plants, unpublished mimeo. undated.

Bilek, Amanda. Trading Carbon Credits From Methane Digester. *In Business 28(4),* 16. 7/2006-8/31/2006.

Canadian Sphagnum Peat Moss Association, *Welcome To The Canadian Sphagnum Peat Moss Association* (Web page), www.peatmoss.com, undated, accessed 1/12/2007.

Chapman, Shaun. Freeing the Grid: How Effective State Net Metering Laws Can Revolutionize U.S. Energy Policy. 11/2006. Network for New Energy Choices. http://www.newenergychoices.org/uploads/netMetering.pdf.

Cornell Manure Management Program, *Manure Treatment: Anaerobic Digestion* (Web page), http://www.manuremanagement.cornell.edu/ HTMLs/AnaerobicDigestion.htm, undated, accessed 1/31/2007.

Cornell Waste Management Institute. Using Manure Solids as Bedding: Literature Review. 12/2006. http://cwmi.css.cornell.edu/beddinglitreview.pdf.

Department of Environmental Services, Nicollet County, Minnesota, *Nicollet County Zoning Ordinance – Feedlot Sections* (Web page), http://www. co.nicollet.mn.us/dept/9/N icollet%20County%20Feedlot%20Ordinance. pdf, undated, accessed 2/2/2007.

DeVore, George. Tillamook, Oregon Methane Energy Development Program. 4/24/2006. http://www.epa.gov/agstar/pdf/conf06/devore.pdf.

Duangmanee, Thanapong Jack. Micro-Aeration for Sulfide Removal in Anaerobic Treatment of High-Solid Wastewater. AgSTAR National Conference. 11/27/2007.

Dunn, David J., (2006). *Diversifying Vermont Farms though Renewable Generation* (Web page), http://www.epa.gov/agstar/pdf/conf06/dunn.pdf, accessed 2/1/2007.

Effenberger, Mathias. Biogas Opportunities—Pathways and Wrong Tracks to Success. Organic Inputs and Renewable Energy: Learning from the European Biogas Model. 4/6/2006. http://www.omafra.gov.on.ca/english/ engineer/facts/bg_pres4.pdf.

Environmental Resource Recovery Group, LLC. Anaerobic Digesters for King County Dairies, Final Report for King County, Washington. 6/2003. http://dnr.metrokc.gov/ wlr/lands/2003-anaerobic-digesters-report.htm.

Fitzgerald, Garrett, Wiser, Ryan, and Bolinger, Mark. The Experience of State Clean Energy Funds with Tradable Renewable Certificates. 11/2003. http://eetd.lbl.gov/ea/ ems/cases/TRC_Case_Study.pdf.

Frazier, Barnes & Associates, LLC. Feasibility Study: West Michigan Regional Liquid Livestock Manure Processing Center (LLMPC). 4/25/2006. http://www.michigan.go v/documents/CISMichigan_Centralized_AD_ Feasibility_Study2006_165007_7.pdf.

Goodrich, P. Professor, Department of Bioproducts and Biosystems Engineering, University of Minnesota. Personal communication about experiences with remote monitoring and control of digester operation. 11/20/2007.

Haubenschild, D. personal communication about carbon credit sales. 12/2006.

Hettinga, Willem. Technological Learning in U.S. Ethanol Production: Quantifying Reductions in Production Costs and Energy Use, Master Thesis. 5/2007. The Netherlands, Energy & Resources Track, Utrecht University.

Hjort-Gregersen, Kurt. (1999). Centralized Biogas Plants Integrated Energy Production, Waste Treatment and Nutrient Redistribution Facilities. Danish Institute of Agricultural and Fisheries Economics. http://websrv4. sdu.dk/bio/pdf/centra.pdf.

Hurley, S., Ahern, J. & Williams, D., *Clustering of Independent Dairy Operators for Generation of Bio-Renewable Energy: A Feasibility Analysis, A Report Prepared for the California Institute for the Study of Specialty of Crops* (Web page), http://cissc.calpoly.edu/research/ dairy-grant-final-report-7-3 1-06-exec-summary-2.pdf, 2006, accessed 12/4/2006.

Inland Empire Utilities Agency. (2006) *Inland Empire Dairy Manure to Energy "Cow Power" Renewable Energy Program* (Web page), http://www.epa. gov/agstar/pdf /conf06/clifton.pdf, accessed 12/4/2006.

Jacobson, Larry D., Moon, Roger, Bicudo, Jose, Janni, Kevin, and Noll, Sally, *Generic Environmental Impact Statement on Animal Agriculture: A Summary of the Literature Related to Air Quality and Odor* (Web page), http://www.eqb.state.mn.us/ geis/LS_AirQuality.pdf, 2005, accessed 6/11/2007.

Kildegaard, A. (2006). "Renewable Energy Policy in Minnesota: Can We Change the Subject?" *CURA Reporter*,3-8.

Kramer, Joseph M. (2004). Agricultural Biogas Casebook Update. 9/2004. Resource Strategies, Inc., 22 North Carroll Street, Suite 300, Madison, WI 53703. http://www.rs-inc.com/deliverables.asp.

Krich, Ken, Augenstein, Don, Batmale, J. P, Benamann, John, Rutledge, Brad & Salour, Dara. Biomethane from Dairy Waste: A Sourcebook for the Production and Use of Renewable Natural Gas in California. 7/2005. Sustainable Conservation. http://www.suscon.org/news/biomethane_ report/Full_Report.pdf.

Lazarus, W. & Rudstrom. M. (2007). "The Economics of Methane Digester Operations on Minnesota Dairy Farms." *Review of Agricultural Economics 2*,349–364.

Lusk, P. Methane Recovery from Animal Manures: A Current Opportunities Casebook. 1995. Washington, DC, U.S. Department of Energy, Regional Biomass Energy Program, National Renewable Energy Laboratory.

Lusk, P. Methane Recovery from Animal Manures: The Current Opportunities Casebook, 3rd Edition. 9/1998. National Renewable Energy Laboratory. http://www.nrel.gov/ docs/fy99osti/25145.pdf.

Marsh, Michael L. H. & LaMendola, Tiffany. Dairy Methane Digester System 90-Day Evaluation Reports—Eden-Vale, Hilarides, Koetsier, and van Ommering Dairies, and Inland Empire Utilities Agency. 12/2006. Modesto, California, prepared by: Western United Resource Development, Inc., for the California Energy Commission. http://www. energy

Martin, J. H. Jr. & Roos, K. F. Comparison of the Performance of a Conventional and a Modified Plug-Flow Digester for Scraped Dairy Manure. International Symposium on Air Quality and Waste Management for Agriculture, ASABE Publication Number 701P0907cd. 9/16/2007.

Martin, John H. A Comparison of Dairy Cattle Manure Management With and Without Anaerobic Digestion And Biogas Utilization. 6/17/2004. Washington, DC, AgSTAR Program, U.S. Environmental Protection Agency.
http://www.epa.gov/agstar/pdf/nydairy2003.pdf.

Martin, John H. An Evaluation of a Mesophilic, Modified Plug-Flow Anaerobic Digester for Dairy Cattle Manure. 7/20/2005. Washington, DC, AgSTAR Program, U.S. Environmental Protection Agency.
http://focusonenergy.com/data/common/dmsFiles/W_RW_REFR_ GordondaleFeasReport.pdf.

Martin, John H. A Protocol for Quantifying and Reporting the Performance of Anaerobic Digestion Systems for Livestock Manures. 9/2006. Washington, DC, AgSTAR Program, U.S. Environmental Protection Agency.
http://www.epa.gov/agstar/pdf/assessing_digester_perform_ protocol.pdf.

Mattocks, Richard & Wilson, Richard. Latest Trends in Anaerobic Digestion in North America. Biocycle Fourth Annual Conference. 11/2004.

Methane to Markets Partnership, *Methane to Markets* (Web page), http://www.methanetomarkets.org/, undated, accessed 2/26/2007.

Minnesota Department of Agriculture, *FAQs about the MDA Methane Digester Loan Program* (Web page), http://www.mda.state.mn.us/feedlots/digester.htm, undated, accessed 2/2/2007.

Minnesota Department of Commerce, *Renewable and Efficiency Incentives* (Web page), http://www.state.mn.us/portal/mn/jsp/content.do?id=- 536881 350&subchannel=-53688 1511&sc2=null&sc3=null&contentid=536885915&contenttype= EDITORIAL&programid=536885394&agency=C ommerce, undated, accessed 2/2/2007.

Moser, M. & Langerwerf, L., *Langerwerf Dairy Digester Facelift: What We Found When We Took Apart a 16 Year-old Dairy Plug Flow Digester* (Web page), http://www.epa.gov/agstar/resources undated, accessed 12/20/2004.

National Rural Electric Cooperative Association. White Paper on Distributed Generation. 12/2/2005. http://www.nreca.org/Documents/PublicPolicy/ DGWhitepaper.pdf.

New Rules Project, *Net Metering of Electricity* (Web page), http://www.newrules.org/el ectricity undated, accessed 1/5/2007.

New York State Energy Research Development Agency, *About NYSERDA* (Web page), http://www.nyserda.org/About/default.asp, 2007, accessed 1/31/2007.

Nielsen, L. H. & Hjort-Gregersen, K. Socio-Economic Analysis of Centralised Biogas Plants. undated. Danish Research Institute of Food Economics, C/o University of South Denmark, Niels Bohrsvej 9, DK6700 Esbjerg, Denmark.

North Carolina Solar Center, *Renewable Electricity Production Tax Credit* (Web page), http://www.dsireusa.org/library/includes/incentive2. cfm?Incentive_Code=US 1 3F&State=federal&currentpageid=1&ee=0&re =1, 2007a, accessed 3/13/2007a.

North Carolina Solar Center. (2007b). *DSIRE Database of State Incentives for Renewables & Efficiency* (Web page), dsireusa.org,, accessed 2/2 1/2007b.

North Carolina State University, *Waste Management Programs* (Web page), http://www.cals.ncsu.edu/waste_mgt/, undated, accessed 1/31/2007.

Office of the Law Revision Counsel, U.S. House of Representatives. Internal Revenue Code, Electricity produced from certain renewable resources.

Peters, Jerome. Debt Financing. Biomass Finance & Investment Summit. 1/18/2007-1/19/2007.

Port of Tillamook Bay, *Bio-Gas Methane Facility—Hooley Digester* (Web page), http://www.potb.org/methane-energy undated, accessed 12/4/2006.

Roos, K. Status & Issues Facing the Anaerobic Digester Industry Relative to the U.S. Livestock Market. Wisconsin Manure Biogas Symposium, (Proceedings available on CD from the University of Wisconsin Center for Dairy Profitability, Madison, Wisconsin). 4/4/2002.

Sebesta Blomberg & Associates, Inc. Methane Digester Economic & Technical Feasibility Study, West River Dairy & City of Morris. 10/2005. Agricultural Utilization Research Institute (AURI) Center for Producer- Owned Energy.

Shih, Jhih-Shyang, Burtraw, Dallas, Palmer, Karen & Siikamaki, Juha. (2006). Air Emissions of Ammonia and Methane from Livestock Operations: Valuation and Policy Options,

Discussion Paper 06-11. Resources for the Future. http://agecon.lib.umn.edu/cgi-bin/pdf_view. pl?paperid=21066.

Smith, J. B. Manure-to-gas plant promises relief for Texas' dairy country. Waco Tribune-Herald. 7/9/2007. http://www.news-journal.com/green/content/shared/news/stories/dairy_070907.html.

Steffen, R., Szolar, O. & Braun, R. (1998). Feedstocks for Anaerobic Digestion. Austria. http://res2.agr. ca/initiatives/manurenet/download/feedstocks_AD.pdf.

Terre-Source LLC. (2003). Study to Evaluate the Price and Markets for Residual Solids from a Dairy Cow Manure Anaerobic Digester. ftp://dnr. metrokc.gov/dnr/library/2003/kcr1540.pdf.

Tiffany, Douglas G. (2005). Economic Analysis: Co-generation Using Wind and Biodiesel-Powered Generators, Staff Paper P05-10. University of Minnesota, Department of Applied Economics. http://agecon.lib.umn.edu/ cgi-bin/pdf_view.pl?paperid= 18 123&ftype=.pdf.

U.S. AgSTAR. Guide to Operational Systems. 6/30/2004. http://www.epa. gov/agstar/operational.html.

U.S. AgSTAR. (2004). *Market Opportunities for Biogas Recovery Systems: A Guide to Identifying Candidates for On-Farm and Centralized Systems* (Web page), http://www.epa.gov/agstar/pdf/biogas%20recovery%20systems_ screenres.pdf, accessed 6/11/2007.

U.S. AgSTAR (2006). Winter AgSTAR Digest. 2006a. http://www.epa. gov/agstar/.

U.S. AgSTAR. AgSTAR Handbook and Software, Second Edition. 11/3/2006b. Washington, D.C. http://www.epa.gov/agstar/resources/

U.S. AgSTAR. (2007a). *The AgSTAR Program* (Web page), http://www.epa.gov/ agstar/ , accessed 2/22/2007a.

U.S. AgSTAR. Anaerobic Digesters Continue Growth in U.S. Livestock Market. 11/2007b. http://www.epa.gov/agstar/pdf/2007_digester_update. pdf.

U.S. Energy Information Administration. (2006a). *International Data, Natural Gas* (Web page), http://www.eia.doe.gov/emeu/international/contents.html, accessed 1/30/2007a.

U.S. Energy Information Administration. (2007b). *Petroleum Spot Prices* (Web page), http://tonto.eia.doe.gov/dnav/pet/pet_pri_spt_s1_d.htm, accessed 1/30/2007b.

U.S. Energy Information Administration, NEMS (*Renewable Energy Modeling Summit Series*) (Web page), http://www.epa.gov/cleanrgy/ pdf/model_sum_nems_100 202.pdf#search=%22nems%20model%22, undated, accessed 2/9/2007c.

U.S. Environmental Protection Agency. National Emission Inventory— Ammonia Emissions from Animal Husbandry Operations, Draft Report. 1/30/2004. http://www.epa.gov/ttn/chief/ap42/ch09/related/nh3inventorydraft_jan2004.pdf.

U.S. Environmental Protection Agency. Consent Agreement And Final Order (Afo Air Agreement). 1/21/2005. http://www.epa.gov/compliance resources/agreements/caa/cafo-agr-0604.html.

USDA, *Section (9006). Renewable Energy and Energy Efficiency Program* (Web page), http://www.rurdev.usda.gov/rbs/farmbill/, undated, accessed 1/9/2007.

USDA Natural Resources Conservation Service. (2003) Minnesota EQIP Conservation Practice Payment Docket(440-V-CPM, First Edition, Amend. MN-27, May, 2003) MN515-230(48). 2003. http://www.mn.nrcs. usda.gov/programs/eqip/2003mn_eqi p_docket3.pdf.

USDA Natural Resources Conservation Service. Conservation Practice Standard, Anaerobic Digester Controlled Temperature, Code 366. 8/2004. http://efotg.nrcs.usda.gov/references/public/pa/pa366_std.pdf.

USDA Natural Resources Conservation Service. Conservation Practice Standard, Anaerobic Digester Ambient Temperature, Code 365. 2/2007. http://efotg.nrcs.usda.gov references/public/WY/Anerobic_Digester_Ambient_ Temperature__365_Standard.pdf.

WestStart-CALSTART. (2004). Swedish Biogas Industry Education Tour: Observations and Findings. 11/12/2004 . http://www.calstart.org/info/ publications/Swedish_Biogas_ %20Tour_2004/Swedish_Biogas_Tour_ 2004.pdf.

Wright, Peter, Inglis, Scott, Ma, Jianguo, Gooch, Curt, Aldrich, Brian, Meister, Alex & Scott, Norman. (2004). Preliminary Comparison of Five Anaerobic Digestion Systems on Dairy Farms in New York State. ASAE/CSAE Annual International Meeting. 8/1/2004-8/4/2004.
http://www.manuremanagement.cornell.edu/Docs/ASAE%20paper%20044032 %20Final.pdf.

## End Notes

[1] The methane content of biogas varies depending on the amounts of carbohydrate, protein, and fats in the digestate (material being digested), and on $CO_2$ dissolution in the digester water (see Appendix A of Krich et al. (2005)).

[2] Discussions of greenhouse gases are often expressed in terms of carbon dioxide ($CO_2$) equivalents, so some people are confused by the fact that a digester doesn't reduce $CO_2$. The methane from a digester is destroyed through combustion in an engine, flare, or other device. Combustion actu- ally produces $CO_2$ and water ($H_2O$). Methane ($CH_4$) is considered to be around 23 times as powerful as $CO_2$ in its effect on global warming, however, so the overall impact of converting $CH_4$ to $CO_2$ is considered beneficial.

[3] "Biomethane" refers to biogas that has been upgraded to natural gas quality.

In: Sources and Reduction of Greenhouse Gas Emissions　　ISBN: 978-1-61668-856-1
Editor: Steffen D. Saldana　　　　　　　　　　　© 2010 Nova Science Publishers, Inc.

*Chapter 6*

# THE COSTS OF REDUCING GREENHOUSE-GAS EMISSIONS

## *Congressional Budget Office*

Human activities around the world are producing increasingly large quantities of greenhouse gases, particularly carbon dioxide ($CO_2$) resulting from the consumption of fossil fuels and deforestation. Most experts expect that the accumulation of such gases in the atmosphere will result in a variety of environmental changes over time, including a gradual warming of the global climate, extensive changes in regional weather patterns, and significant shifts in the chemistry of the oceans.[1] Although the magnitude and consequences of such developments are highly uncertain, researchers generally conclude that a continued increase in atmospheric concentrations of greenhouse gases would have serious and costly effects.[2]

A comprehensive response to that problem would include a collection of strategies: research to better understand the scientific processes at work and to develop technologies to address them; measures to help the economy and society adapt to the projected warming and other expected changes; and efforts to reduce emissions, averting at least some of the potential damage to the environment and attendant economic losses. Those strategies would all present technological challenges and entail economic costs.

Reducing emissions would impose a burden on the economy because it would require lessening the use of fossil fuels and altering patterns of land use. This issue brief discusses the economic costs of reducing greenhouse-gas emissions in the United States, describing the main determinants of costs, how analysts estimate those costs, and the magnitude of estimated costs. The brief also illustrates the uncertainty surrounding such estimates using studies of a recent legislative proposal, H.R. 2454, the American Clean Energy and Security Act of 2009.

## WHAT DETERMINES THE COSTS OF REDUCING EMISSIONS?

The costs of reducing emissions would depend on several factors: the growth of emissions in the absence of policy changes; the types of policies used to restrict emissions;

120                          Congressional Budget Office

the magnitude of the reductions achieved by those policies; the extent to which producers and consumers could moderate emission-intensive activities without reducing their material well-being; and the policies pursued by other countries. (For a discussion of different concepts of cost, see Box 1.)

## Emissions in the Absence of Policy Changes

In 2006, the United States emitted roughly 7 billion metric tons (MT) of greenhouse gases, measured in $CO_2$ equivalents ($CO_2e$, or the amount of $CO_2$ that would cause an equivalent amount of warming).[3] Eighty percent of domestic emissions consisted of $CO_2$ from the burning of fossil fuels in activities such as manufacturing, electricity generation, transportation, agriculture, and the heating and cooling of buildings. The remaining 20 percent—consisting of $CO_2$ emitted from sources other than fossil fuels, methane ($CH_4$), nitrous oxide ($N_2O$), and a variety of fluorinated gases—were produced by myriad processes and activities throughout the economy. Under current land-use patterns in the United States, forests and soils absorb nearly 900 million MT $CO_2$ every year, putting net U.S. emissions in 2006 at about 6 billion MT $CO_2e$. U.S. emissions of $CO_2$ from the burning of fossil fuels accounted for one-fifth of global $CO_2$ emissions from such activities. However, net U.S. emissions of all greenhouse gases accounted for only about 12 percent of net global emissions.

---

### BOX 1. MEASURING THE COSTS OF REDUCING GREENHOUSE-GAS EMISSIONS

Economists characterize and measure the costs of reducing emissions in a number of related ways, each of which provides a useful perspective on how a program to control emissions would affect the economy.

#### Incremental Cost

Not all emissions of greenhouse gases are equally expensive to reduce. If emissions were restricted in a manner that allowed firms and households flexibility in how they met the restriction, the least costly reductions would tend to be undertaken first. If the restriction was tightened and people had to forgo increasingly highly valued activities to further reduce their emissions, each additional ton of reductions would be more expensive. For any amount of reduction, the incremental cost of cutting the last ton necessary—what economists call the marginal cost—reflects how much consumption of goods and services people would have to forgo to eliminate that last ton of emissions. If emissions were restricted using a cap-and-trade system, the incremental cost would determine the market price of allowances, even if most of the reductions were less expensive than the very last, incremental ton. If, instead, a tax was imposed on emissions, the incremental cost would equal the tax rate: People would make reductions that cost less than the tax; for the rest, they would simply pay the tax. Traditional command-and-control approaches are unlikely to mandate the least expensive ways of restricting emissions and therefore are likely to achieve any given reduction in emissions at greater cost than market-based approaches such as taxes or cap-and-trade programs.

## Aggregate Cost

Researchers measure and report the aggregate cost of restrictions in a variety of ways. One common measure, the direct resource cost of restrictions on emissions, is simply the sum of all resources that society must give up to meet the restrictions—that is, the sum of all the incremental costs discussed above. However, significant restrictions would impose costs on the economy over and above the direct cost as economic effects rippled beyond markets directly related to emissions and into other parts of the economy. Such adjustments would include changes in aggregate spending, saving, investment, and nonpaid activities, such as childrearing, production of goods and services in the home, and leisure activities. Those broader impacts would depend not only on how emission restrictions were implemented but also on whether the program raised revenues for the government and how the revenues were used. An ideal measure of the overall effect of the policy on households' welfare would account for all such aspects of the policy as well as all the changes in behavior that might be triggered by the policy. Unfortunately, little information is available regarding the likely magnitude of many of those changes, so estimates of the overall economic burden of policies are highly uncertain. Most researchers avoid the difficulties associated with estimating that burden by focusing on the impacts of policies on less comprehensive but more readily estimated measures of aggregate economic activity, such as gross domestic product, personal consumption expenditures, or employment.

## The Distribution of Costs

Some studies analyze how the costs of policies to reduce emissions would be distributed among households and among industries. Most policies would not only generate a cost to the economy as a whole but would also cause significant shifts of income among households. Such shifts would not involve net costs to the economy, since one household's gain would be another's loss, but the gains and losses clearly would matter for the well-being of the affected households. For example, if fossil-fuel producers were required to pay a tax per ton of emissions resulting from the burning of their products, the tax burden would be passed on to consumers in the form of higher prices and to some owners of capital in the form of reduced profits. The distribution of the tax revenues would depend on how the government used them either to reduce other taxes or to increase spending. By contrast, if the government instituted a cap-and-trade policy and gave away emission allowances, the recipients would receive the full value of the allowances. Such transfers would probably be significantly larger than the aggregate costs associated with a given policy, particularly in the initial years of the policy.

Distributional studies typically find that, ignoring the allocation of tax revenue or allowance value flowing from the policy, low-income households would bear lower costs than high-income households in absolute terms. However, the former group would bear higher costs measured as a fraction of their income because they consume a larger share of their income and because their consumption is more energy-intensive. Studies also conclude that the owners of fossil-fuel-producing or energy-intensive industries would bear a disproportionate share of producers' costs (ignoring any effects from the allocation of allowances or tax revenues). However, the allocation of allowances or revenues from the policy could greatly affect those distributional impacts.

Experts generally expect that, in the absence of policy changes to reduce them, domestic greenhouse-gas emissions will grow substantially in the next few decades, totaling roughly 330 billion MT $CO_2e$ between now and 2050. However, long-term trends in emissions are notoriously difficult to project because they will be influenced by population and income growth, by advances in technology, and by the availability and price of fossil fuels; total emissions, therefore, could be substantially higher or lower than that central estimate.[4] The more rapidly that emissions are projected to grow without policy changes, the greater the changes that would be required and the greater the mitigation costs that would be incurred to keep emissions below any specific level.

### *The Types of Policies Adopted*

The costs of reducing emissions would depend critically on the approach that policymakers adopted to achieve that goal. In particular, costs would depend on whether the policy worked primarily through conventional regulation or market-based approaches, on the stringency of emission reductions, and on other policy choices.

### Conventional Regulation versus Market-Based Approaches

A basic choice facing policymakers is whether to adopt conventional regulatory approaches, such as setting standards for machinery, equipment, and appliances, or to employ market-based approaches, such as imposing taxes on emissions or establishing cap-andtrade programs. Experts generally conclude that market-based approaches would reduce emissions to a specified level at significantly lower cost than conventional regulations.[5] Whereas conventional regulatory approaches impose specific requirements that may not be the least costly means of reducing emissions, market-based approaches would provide much more latitude for firms and households to determine the most cost-effective means of accomplishing that goal.

### Alternative Market-Based Approaches

If a tax was imposed on emissions, people would make reductions that cost less than the tax, and the incremental cost would equal the tax rate. Proposals for such taxes generally specify rates that gradually rise year by year, with the aim of making emission-producing activities increasingly expensive. Cap-and-trade proposals, by contrast, explicitly restrict the quantity of emissions that can be produced over any given period. Under such programs, allowances to emit greenhouse gases would be allocated or sold, and then could be traded. Market forces would yield an allowance price equal to the incremental cost of meeting the cap. Cap-and-trade proposals generally specify caps that gradually decrease over time in absolute terms, so households and firms would incur gradually rising incremental costs to reduce emissions.

If policymakers had full and accurate information about the costs of reducing emissions, either taxes or caps could be used to achieve a given goal for emissions. Policy- makers could set a cap and know what allowance price it would yield; or they could set a tax at that same allowance price and achieve the same reduction in emissions. Thus, policymakers could use either approach to balance the incremental costs of reducing emissions against the incremental benefits of doing so, thereby achieving the greatest possible net benefit.

However, policymakers face great uncertainty about the cost of reducing emissions. The two approaches therefore are likely to yield different outcomes: A tax on emissions would leave the resulting amount of emissions uncertain, whereas a cap would leave the resulting allowance price uncertain.

Most experts conclude that, in the face of such uncertainty, policies that set the year-by-year price of emissions to be consistent with the projected incremental benefits of reducing emissions (as with a tax) would probably yield higher net benefits than policies that specified yearby-year caps on emissions or even a cap on cumulative emissions over many years.[6] The cost of meeting a fixed emission cap is likely to vary substantially from year to year— depending on the weather, economic activity, and the price of fossil fuels. A tax would ensure that firms and households had an incentive to make all reductions that cost less to achieve than that expected incremental benefit. By contrast, a cap could easily generate incremental reductions that cost substantially more or less than the expected benefit.

Even if policymakers chose a target measured in terms of average global temperature or atmospheric concentrations (for example, staying under 450 parts per million $CO_2e$), meeting rigid year-to-year (or even cumulative) targets for emissions would be relatively unimportant. The uncertainty about how a given quantity of emissions would ultimately affect concentrations or temperatures is so great that little additional certainty would be gained from choosing fixed emission goals over price paths that were expected to achieve the same goal.

If policymakers chose cumulative emission targets, policies that offered firms and households greater flexibility as to when they reduced their emissions would achieve such targets at a lower cost than policies that afforded them less flexibility. Various provisions of a cap-and-trade program could provide greater flexibility in the timing of emission reductions.[7] One important provision of this sort would permit regulated entities to "bank" emission allowances in any given year for use in future years. Such provisions would tend to moderate the overall costs of the policy by giving firms flexibility to make larger emissions reductions when such reductions were less expensive. A related "borrowing" provision would enable firms to use allowances from future years (to be repaid with interest) during periods when particularly high demand led to spikes in allowance prices. A variant would create a "reserve pool" of allowances from future years that could be used only under certain circumstances (for instance, when allowance prices rose above a designated threshold). Another widely discussed provision—referred to as a "safety valve"—would allow firms to exceed annual caps if the market price for allowances rose above some specified value. That value, typically specified to rise over time, would determine the maximum incremental cost of reducing greenhouse-gas emissions in any given period. Proposals could also keep incremental costs above a minimum by specifying a "price floor." A "price collar" would specify both a ceiling and a floor. Unlike banking, borrowing, and a reserve pool, a safety valve or a price collar

124 Congressional Budget Office

would not guarantee that a given cumulative cap would be achieved. Policymakers might have to shift the price collar in order to ensure that a desired cap was met.[8]

**Other Policy Approaches**

Some proposals would augment basic cap-and-trade or tax provisions with subsidies for activities that reduced emissions or with regulations such as standards for machinery, equipment, and appliances. Some such approaches—subsidies for basic energy research, for example—would probably be useful and effective supplements to market-based approaches. Standards might also be effective when the nature of the emission-producing activities made caps difficult to implement and taxes difficult to levy (as would be the case with emissions associated with agricultural practices) or where market forces did not convey appropriate incentives (for example, when a tax on energy would not spur landlords to make efficiency improvements if renters were responsible for paying the electricity bills and rents were not adjusted to reflect the energy efficiency of apartments). However, standards would tend to increase the costs of a cap-and-trade program if they supplanted the effective reliance on market forces—even though they would also tend to reduce the allowance price in the program by reducing emissions covered under the program.

Other types of government activity could also affect the costs of restricting emissions. Many experts believe that nuclear power could displace a significant amount of fossil-fuel use, but only if the regulatory framework was adjusted to allow for the greater use of nuclear power to generate electricity. Similarly, generators would be unlikely to adopt technologies for the capture of $CO_2$ and its sequestration in the ground unless an extensive regulatory structure was put in place to address issues involving property rights, rights-of-way for pipelines, and liability for emissions that escape from the ground.

Governmental activities not immediately related to the energy sector could affect the costs of reducing emissions as well. For example, the tax treatment of investment could influence the cost and availability of particular technologies. Similarly, existing land-use regulations and the continued construction of highways might hinder efforts to increase urban density and to foster the development of public transportation networks.

**Coverage, Timing, and Stringency**

Policymakers also face decisions about which types of emissions to control, and when and how much to reduce them.[9] Coverage (that is, the types of emissions subject to control) could sharply affect costs. Most recent policy proposals would control all or nearly all $CO_2$ emissions produced by the burning of fossil fuels and would cover at least some other emissions of $CO_2$ and non-$CO_2$ greenhouse gases. Because monitoring and measuring emissions from some sources would be difficult, no proposals include all emissions from all sources. Nevertheless, many cap-and-trade proposals provide incentives for reducing emissions from sources not covered under the cap-and-trade program. For example, many proposals would allow landowners to earn credits by planting trees that absorb $CO_2$ from the atmosphere—credits that could then be sold to entities covered by the cap-and-trade program, who might submit those credits in lieu of allowances. Proposals often limit the use of such "offsets" (which, in some proposals, could be purchased in foreign countries as well) to a fixed annual amount or a fixed fraction of the total quantity of emissions allowed. Greater latitude for the use of offsets could help moderate the costs of achieving a given emission target because inexpensive reductions by uncovered sources could then substitute for a larger share of costlier reductions that covered sources would otherwise have to achieve. However, difficulties in ensuring the credibility and permanence of offsets could at least partially undermine the effectiveness of policies at achieving their stated emission goals.[10]

Proposals vary as to where in the chain of production and consumption the restrictions on emissions would apply. For example, emissions from petroleum products could be restricted by limiting production at refineries or by limiting consumers' purchases at the gas pump; emissions from coal could be restricted by limiting shipments of coal from mines or by limiting emissions from coal-fired power plants. Such differences, however, would have only small effects on incremental costs or on the distribution of costs among producers or consumers: No matter where the restrictions applied, costs would be determined mainly by the characteristics of supply and demand in the markets for fossil fuels and other goods and services.[11]

Most recent proposals call for gradual reductions in emissions through 2050, a significant shift in focus from that in the 1990s, when policy proposals usually focused on establishing targets for the 2007–2011 period to be incorporated in the 1997 Kyoto Amendment to the Framework Convention on Climate Change. Recent proposals typically would impose relatively modest tax rates or restrictions on emissions in the early years and then gradually raise rates or tighten caps over time. Gradually increasing levels of stringency would provide firms and households time to anticipate and adapt to rising costs by slowly replacing long-lived energy-using structures and equipment with newly developed, more energy-efficient or non-fossil-fuel substitutes. Such policies would cost a good deal less than policies that imposed severe restrictions all at once—even if they resulted in equivalent amounts of total emissions over time. Some recent proposals would limit cumulative emissions through 2050 (which are projected to total roughly 330 billion MT $CO_2e$) to less than 200 billion MT $CO_2e$, leading to annual emissions in 2050 that would be a fraction of today's. The more stringent the cuts, the higher the incremental and total costs eventually would be, other things being equal.

## Allocation of Allowances and Use of Additional Government Revenue

If policymakers decided to adopt market mechanisms to control emissions, they would also face decisions about how to allocate allowances under a cap-and-trade program or how to use the revenues generated by taxes on emissions. No matter which approach they adopted, the resulting policy would almost certainly involve shifts of resources among households, in their capacity as consumers, workers, and owners or shareholders of firms. For example, a program that distributed allowances to firms for free would be giving those firms a valuable right to emit gases—a right they could then exercise or sell—while consumers would pay more for fossil fuels and fossil-fuel-intensive goods and services. Such a program and allocation would effectively shift income from consumers of fossil-fuel-intensive goods and services to recipients of allowances. Alternatively, the government could auction allowances to firms and use the revenue to provide new tax credits or rebates, reduce existing taxes, or finance government spending. (Revenue raised from a tax on emissions could be used in any of those ways as well.) Revenue-raising approaches would effectively shift income from consumers of fossil-fuelintensive goods and services to the people who would benefit from the resulting spending increases or tax reductions.[12]

In a cap-and-trade system, policymakers' choices about the allocation of allowances and the uses of additional government revenue would affect both the overall economic cost of the program and the distribution of that cost. If allowances were given away in a manner that offset the increases in prices of fossil fuels (increases that would otherwise encourage households and businesses to reduce their consumption of such fuels), then a larger share of the overall reduction in carbon emissions would need to occur elsewhere in the economy. For example, some recent proposals would direct a portion of tax revenues or proceeds from the sale of allowances to households or businesses to help compensate for some of the loss in purchasing power they would experience under a climate policy. To the extent that such an allocation masked the price signals needed to alter behavior, it would raise the overall economic cost of meeting the cap. However, to the extent that the allocation did not mask price signals, the overall cost would not be affected by who received the allowances— although the distribution of that cost might be greatly affected.

If allowances were auctioned rather than freely allocated, and the revenue was used to reduce marginal rates on taxes on labor and capital, the economic cost of a capand-trade program would be reduced.[13] If the revenue from auctions was not used to reduce marginal rates, then decisions about who received that revenue would have little effect on the program's overall economic cost— although they could have important effects on the distribution of that cost.

## The Response of the Economy

By gradually increasing the prices of fossil fuels and other goods and services associated with greenhouse-gas emissions, market-based policies would induce firms and households to change their practices—in the short run, by driving slightly less, adjusting thermostats, and switching fuels in the power sector; and in the long run, by buying more-efficient vehicles and equipment, building more-energy-efficient buildings in denser neighborhoods, and building power plants that used less (or no) fossil fuel or that captured $CO_2$ and sequestered it in the ground. Depending on the specifics of the policy, people might also plant trees and change agricultural practices to absorb and sequester $CO_2$.

Rising costs of emission-intensive activities would tend to dampen overall economic activity by reducing the productive capacity of existing capital and labor; by reducing households' income (which, in turn, would tend to reduce consumption and saving); by reducing real (inflation-adjusted) wages and, thereby, the supply of labor; and by discouraging investment through increasing the costs of producing capital goods (which is a relatively energy-intensive process) and through diverting investment and research toward the production of less emission-intensive but more expensive sources of energy.[14] Taken together, those changes would affect the levels and composition of both gross domestic product (GDP) and employment and would thus influence households' economic well-being, although the effect on overall output would be modest compared with expected future economic growth. The more easily that producers and consumers can respond to price changes by altering their production techniques and behavior and by bringing low- emission fuels and technologies to market, the lower the costs of reducing emissions would be.

However, analysts have only a limited understanding of how easily such responses can occur—especially over an extended period—because energy use is important in so many economic activities and because such a wide range of activities besides energy use generate emissions. Uncertainty about how the economy would respond to price changes contributes importantly to the wide variation in estimates of the cost of achieving any particular emission target. For example, expert opinion varies considerably about which types of technologies are likely to be available at different points in the future or how emissions restrictions might shift the pace of their development and deployment. Some experts argue that nuclear energy is likely to dominate other alternatives to fossil fuels; others believe that $CO_2$ capture and storage shows greater promise; and still others believe that renewable energy sources could be most promising of all.

## Efforts by Other Countries

The stringency of other nations' efforts to reduce emissions could strongly influence the costs of reducing them in the United States. As long as a significant percentage of the world's economy did not restrict greenhouse-gas emissions, a portion of any reductions achieved in the United States would probably be offset by increases in emissions elsewhere. For example, as U.S. consumption of oil declined, pushing down international oil prices, foreign consumption of oil would rise. In addition, energy-intensive production overseas (and exports of such products to the United States) would most likely grow as U.S. manufacturing costs rose relative to foreign costs. Such emissions "leakage" would lead countries that were controlling emissions to achieve smaller net reductions in global emissions and to incur greater costs than they would if all countries were controlling emissions simultaneously.

Leakage could be avoided if most countries restricted emissions at the same time. Even so, the policies used in other countries would influence costs in the United States. For example, demand for biofuels from many countries would raise biofuel prices more than would demand from the United States alone. Moreover, if a domestic cap-and-trade system was linked to similar systems in other countries, the United States might benefit from being able to buy low-cost foreign allowances—or it could find that prices for domestic allowances were driven up by foreign demand.

## HOW ARE COSTS ESTIMATED?

Researchers estimate the costs of reducing emissions using a variety of techniques. Those techniques range from detailed analyses of specific technologies—which can be termed "engineering estimates"—to highly aggregated analyses—which can be termed "economic estimates."

## Engineering Estimates

Often referred to as bottom-up studies, engineering estimates typically analyze the net direct costs of installing and operating many specific types of equipment.[15] Engineering studies frequently yield low estimates of the cost of achieving particular emission targets and can yield estimates of negative costs for some applications—that is, energy savings appear to more than outweigh the capital costs of investing in certain types of equipment. Most economists conclude that such estimates generally do not represent the full costs of adopting the specific technologies, partly because they may not adequately consider all of the costs of installing new equipment or the costs of various impediments—such as borrowing constraints or the effort to acquire needed information about alternatives—that deter firms and households from making certain investments. Moreover, such estimates may not sufficiently account for all of the characteristics that users value in the equipment.

## Economic Estimates

Economic estimates typically rely on aggregate data to project the cost of reducing energy use across broad classes of activities. Such studies often simulate the effects of policies using large-scale economic models that attempt to measure not only the full range of emission- producing activities that would be directly affected by restrictions on carbon emissions but myriad other economic activities that might be indirectly affected as firms and households adjust their behavior in response to the restrictions.[16] Most studies use so-called *general- equilibrium* models that simulate how idealized house-holds and businesses would respond to policies by adjusting their consumption, saving, hours worked, and production behavior; simulations using those models often run for decades into the future. In many of the models used to analyze climate policies, however, households are assumed to have perfect foresight—that is, to take into account fully and accurately events far into the future. As a consequence, the modeled economies respond rather quickly to changes in policy. A smaller set of studies uses so-called *macroeconometric models*, which are designed to simulate gradual responses of consumer spending and investment to unexpected economic shocks over shorter time periods. Those models show how adjustments in behavior might result in lower aggregate economic activity.

Some economic estimates also include considerable engineering detail about different technologies used in the energy sector. One well-known example is the Energy Information Administration's National Energy Modeling System (NEMS), which integrates a highly detailed representation of the energy sector with a macroeconometric model of the U.S. economy. Some models include only cursory representations of international linkages, while others include the entire global economy. Nearly all modeling efforts in the United States rely on the EIA for baseline (reference) forecasts, and they draw on engineering estimates of abatement costs by the Environmental Protection Agency for nearly all types of greenhouse-gas emissions other than $CO_2$ resulting from the combustion of fossil fuels.

## How Large Are Estimated Costs?

In recent years, a few legislative proposals for long-term emission reductions have been analyzed using several different models, providing an opportunity to compare cost estimates and to understand the sources of differences in estimates. Most recently, several groups have released estimates of the economic impact of H.R. 2454, the American Clean Energy and Security Act of 2009.[17] (For more information on those estimates, see Box 2.) That bill would create two cap-and-trade programs for greenhouse-gas emissions—a large one applying to $CO_2$ and most other greenhouse gases, and a much smaller one applying to hydrofluorocarbons—and would make further significant changes in climate and energy policy. The larger cap-and-trade program, on which most analyses focus, would restrict greenhouse-gas emissions from covered entities by requiring them to hold allowances or off-set credits for their emissions. The annual allocation of allowances would fall to 83 percent of 2005 emission levels by 2020 and to 17 percent of 2005 levels by 2050. Fully phased in by 2016, the cap would cover roughly 85 percent of projected total U.S. greenhouse-gas emissions. The bill would allow for unlimited banking, and limited borrowing, of allowances. The bill also would allow covered entities to meet a significant portion of their compliance obligations by purchasing offset credits from domestic and international providers; in total, entities could use offset credits in lieu of reducing up to 2 billion tons of greenhouse-gas emissions annually (which would represent more than half of the annual emission reductions projected until about 2030).

---

### Box 2. Recent Analyses of H.R. 2454, the American Clean Energy and Security Act of 2009

A number of analyses of different versions of the American Clean Energy and Security Act of 2009 have been published (some at the request of Members of Congress) by government agencies, academic institutions, and private organizations. The Congressional Budget Office published cost estimates for two versions of the bill, one as ordered reported by the House Committee on Energy and Commerce (available online at www.cbo.gov/ftpdocs/102xx/ doc10262/hr2454.pdf) and another as amended and reported by the House Committee on Rules (available at www.cbo.gov/ftpdocs/103xx /doc10376/hr2998Waxman Ltr.pdf). In addition, CBO reported on other aspects of the bill in the following publications: *The Estimated Costs to Households from the Cap-and-Trade Provisions of H.R. 2454* (June 19, 2009), www.cbo.gov/ftpdocs/103xx/ doc10327/06-19-CapAndTradeCosts.pdf; *The Economic Effects of Legislation to Reduce Greenhouse-Gas Emissions* (September 2009), www.cbo.gov/ftpdocs/105xx/doc 10573/09-17-Greenhouse-Gas.pdf; *The Use of Offsets to Reduce Greenhouse Gases*, Issue Brief (August 3, 2009), www.cbo.gov/ftpdocs/104xx/doc10497/08-03-Offsets .pdf; and *How Regulatory Standards Can Affect a Cap-andTrade Program for Greenhouse Gases* (September 16, 2009), www.cbo.gov/ftpdocs/105xx/doc10562/09-16-Capand Standard s.pdf.

Analyses from other sources include the following:

Sergey Paltsev and others, *The Cost of Climate Policy in the United States,* Report No. 173 (Cambridge, Mass.: Massachusetts Institute of Technology, Joint Program on the

Science and Policy of Global Change, April 2009), "Appendix C: Analysis of the Waxman-Markey American Clean Energy and Security Act of 2009 (H.R. 2454)," available at http://globalchange.mit.edu/files/document/ MITJPSPGC_Rpt173.pdf. To perform the analysis, the authors used the Integrated Global System Model, including the Emissions Prediction and Policy Analysis model (IGSM-EPPA).

Department of Energy, Energy Information Administration, Office of Integrated Analysis and Forecasting, *Energy Market and Economic Impacts of H.R. 2454, the American Clean Energy and Security Act of 2009*, SR/OIAF/2009-05 (August 2009), available at www.eia.doe.gov/oiaf/service rpt/hr2454/pdf/sroiaf%282009%2905.pdf. The analysis was conducted using the National Energy Modeling System (NEMS).

Environmental Protection Agency, Office of Atmospheric Programs, *EPA Analysis of the American Clean Energy and Security Act of 2009, H.R. 2454 in the 111th Congress* (June 23, 2009), available at www.epa.gov/climatechange/ economics/pdfs/ HR2454_Analysis.pdf and www.epa.gov/ climatechange/economics/ pdfs/HR2454_ Analysis_Appendix.pdf. The analysis was performed using the Inter- temporal General Equilibrium Model of the U.S. (IGEM) and the Applied Dynamic Analysis of the Global Economy (ADAGE).

CRA International, *Impact on the Economy of the American Clean Energy and Security Act of 2009 (H.R. 2454)* (prepared for the National Black Chamber of Commerce, May 2009), available at www.crai.com/uploadedFiles/ Publications/impact-on-the-economy-of-the-americanclean-energy-and-security-act-of-2009.pdf. The analysis was conducted using the Multi-Region National-North American Electricity and Environment Model (MRN-NEEM).

A few other organizations have reported estimates of the economic impact of H.R. 2454, but their estimates appear not to have incorporated key features of the bill, such as the full range of greenhouse gases covered and the potential for the banking of allowances and the extensive use of international offsets.

Estimates of the economic cost of this proposal vary because of differences in assumptions about a number of factors: future economic growth and emission trends, firms' and households' responses to the policy, the future cost and availability of various technologies, the climate policies pursued by other countries, and the availability and cost of credible international offsets. Moreover, no single model or scenario addresses the full complexity of the domestic and global economies or the full range of activities associated with greenhouse-gas emissions, and thus no framework provides a comprehensive treatment of all the potential effects of any climate policy. For example, few models include an explicit channel through which a climate policy would influence the pace of development of new technologies (although at least some effects of policies on technological development are included implicitly in the responses of consumers and producers of energy to changes in energy prices). In addition, none of the models incorporates the effect of climate change itself on economic activity or incomes. As a consequence, the results provide information about the costs of policies to reduce greenhouse-gas emissions but do not provide enough information to do cost-benefit analyses of such policies. Moreover, since most models use many similar assumptions, even the wide range of reported estimates does not illustrate the full degree of uncertainty regarding the costs of mitigating emissions.

All of the models used to estimate the effects of climate policies provide estimates of the annual prices of allowances and offsets necessary to achieve the specified levels of emissions. Most of those models also provide estimates of macroeconomic impacts, such as changes in inflation- adjusted gross domestic product and personal consumption expenditures. Very few, however, provide comprehensive estimates of changes in households' economic well-being— what economists refer to as welfare impacts or the economic burden.

## Changes in Energy Use and Emissions

The qualitative findings of the leading models are similar. In nearly all of the reported scenarios, changes in the demand for energy and reductions in overall energy use are only modest in the near term—that is, roughly through 2025. Instead, emission reductions over that period would stem primarily from shifting the mix of fuels used in electricity generation away from coal and toward natural gas, and from domestic and international offsets (particularly the sequestration of carbon in forests). The heavy use of offsets predicted by the models presumes that offsets could be provided with only modest administrative and other costs; if offsets proved not to be available at modest cost, allowance prices and the economic costs of the proposal would be much higher.

Over the longer term, the use of domestic and international offsets would be constrained by the limits specified in the legislation, and reductions would come increasingly from the energy sector. Technological developments would play a critical role in that process, and some of the variation among results reflects a lack of consensus about which technologies would be adopted. Energy conservation and most renewable energy sources are projected to play relatively limited roles over the entire period, mainly because most kinds of renewable energy provide power intermittently. Instead, a substantial increase in the use of nuclear power plays a dominant role under some assumptions, while significant increases in the use of biofuels or carbon capture and sequestration play a much more important role under other assumptions.

All of the reported estimates that take into account the potential to bank emissions find that during the early years of the program, covered entities would reduce emissions and purchase offsets to a greater extent than necessary to meet the cap's specified for those years. They would thus accumulate a substantial quantity of allowances that they could use in later years. Those allowances, combined with the extensive purchase of offsets in later years, would allow covered entities to maintain levels of emissions that were much larger than the levels specified by the annual caps for those later years.

## Allowance Prices

The left panel of Figure 1 illustrates the range of estimates of allowance prices required to induce the emission reductions specified by H.R. 2454 according to several prominent models; for comparability, all of the estimates are presented in constant 2009 dollars per metric ton of $CO_2$ equivalent.[18] The estimates shown in the figure all reflect the full range of greenhouse gases covered by the bill, the banking of allowances, and the extensive use of international offsets. However, the estimates incorporate varying assumptions about economic growth, policy implementation, households' and firms' responses, the development and cost of various types of technology over time, and the availability of offsets.

For 2020, estimates of the allowance price vary by nearly a factor of three, ranging from under $18 to almost $50. The allowance prices rise over time at rates that reflect modelers' assumptions about rates of return that firms would require to bank allowances; for 2050, allowance prices range from about $70 to nearly $160 per metric ton of $CO_2$ equivalent. (If offsets were assumed not to be available, projected allowance prices would be substantially higher.) CBO's estimate of allowance prices lies near the middle of the range shown here, but it rises slightly faster than in most other estimates because CBO uses a rate of return for the banking of allowances that is slightly higher than that used by most other modeling groups.

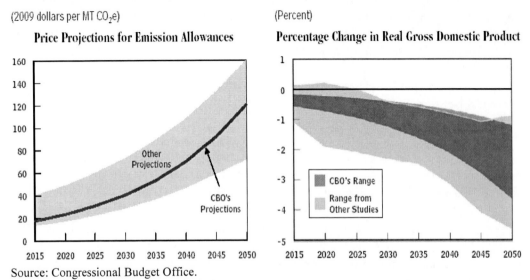

Source: Congressional Budget Office.

Notes: MT $CO_2$e = metric tons of carbon dioxide equivalent (or the amount of $CO_2$ that would cause an equivalent amount of warming).

The figures illustrate a range of estimates of the impacts of the emission reductions specified in the cap-and-trade portions of H.R. 2454, the American Clean Energy and Security Act of 2009. The estimates shown in the figure all reflect the full range of green- house gases covered by the bill, the banking of allowances, and the extensive use of international offsets. However, the estimates incorporate varying assumptions about economic growth, policy implementation, households' and firms' responses, the development and cost of various types of technology over time, and the availability of offsets.

The projections displayed in the figures were produced by the Energy Information Administration, the Environmental Protection Agency, CRA International, Massachusetts Institute of Technology, the Brookings Institution, and the Congressional Budget Office.

Figure 1. Projections under the American Clean Energy and Security Act of 2009

## Macroeconomic Impact

All of the models reporting macroeconomic impacts project that the emission reductions required by H.R. 2454 would slightly dampen the growth of GDP over the long term. (One model projects small increases in the early years of the program). Quantitative estimates of the losses in GDP and consumption vary among studies, depending in large part on differences in assumptions about the availability of offsets (reduced availability of offsets increases the emission reductions required in the energy sector and thus increases economic costs) and differences in assumptions about the sensitivity of energy use to changes in prices (reduced sensitivity increases the price increases required to reach emission targets and thus increases economic costs). On the basis of those estimates and its own analysis, CBO concluded that H.R. 2454 would slightly reduce real GDP—by roughly 0.25 percent to 0.75 percent in 2020 and by between 1.0 percent and 3.5 percent in 2050 (see the right panel of Figure 1).[19] By way of comparison, CBO projects that real GDP will be roughly two-and-a-half times as large in 2050 as it is today. Losses in consumption and overall well-being would probably be smaller than losses in GDP.

Unchecked increases in greenhouse-gas emissions would also tend to reduce output compared with a situation where climate change did not occur—especially later in this century as emissions accumulated in the atmosphere. Nonetheless, CBO concludes that the net effects on GDP of restricting emissions in the United States—combining the effects of diverting resources to reduce emissions and moderating losses in GDP by averting warming—are likely to be negative over the next few decades because most of the benefits from averting warming are expected to accrue in the second half of the 21st century and beyond.

## Impact on Employment

H.R. 2454 would cause a significant shift in the composition of employment over time. Production and employment would shift away from industries related to the production of carbon-based energy and energy-intensive goods and services and toward the production of alternative and lower-emission energy sources, goods that use energy more efficiently, and non-energy-intensive goods and services. Those shifts in employment would occur gradually over a long period, as the cap on emissions became progressively more stringent and the allowance price became progressively higher. The experience of the U.S. economy over the past half-century in adjusting to a sustained decline in manufacturing employment strongly suggests that the economy can absorb such long-term changes and maintain high levels of overall employment. As a result, CBO concludes that the cap would probably have only a small effect on total employment in the long run.

Nevertheless, the employment effects of H.R. 2454 could be substantial for some workers, families, and communities. Labor markets would take time to adjust to shifts in demand. Job losses would be concentrated in particular industries and in particular geographic regions. Some workers would probably end up working fewer hours or at lower wages than they did previously, and some might leave the labor force entirely. Involuntary job losses could significantly reduce the lifetime earnings of some affected workers. Several

134          Congressional Budget Office

provisions of H.R. 2454 would subsidize the development and deployment of technologies that reduced emissions or would subsidize production by specific industries and firms, tending to dampen the bill's effects on employment—especially in industries and areas where they are expected to be most severe.

## The Distribution of Costs

Few analysts have assessed how the costs and transfers associated with the cap-and-trade program established under H.R. 2454 would be distributed among groups of households. As a rough indication of those distributional impacts, CBO estimated the effect on households' purchasing power of the costs of complying with the policy (by reducing emissions and purchasing allowances and offsets) minus any associated compensation (such as freely allocated allowances, earnings from sales of allowances, and profits from producing offsets).[20] CBO concluded that under H.R. 2454 the loss of aggregate purchasing power would increase from about 0.1 percent in 2012 to 0.8 percent in 2050. Those losses would be distributed in ways that tend to benefit lower-income households. In 2020, households in the highest income quintile would see a loss in purchasing power of 0.1 percent of after-tax income, and households in the middle quintile would experience a loss equivalent to 0.6 percent. Households in the lowest quintile, by contrast, would see an average gain of 0.7 percent. By 2050, households in the highest quintile would see a loss in purchasing power of 0.7 percent of after-tax income, and households in the middle quintile would see a loss of 1.1 percent. Those in the lowest quintile would see a 2.1 percent increase.

---

This brief was prepared by Robert Shackleton of CBO's Macroeconomic Analysis Division. In addition to the reports mentioned in the text and footnotes of this brief, the following CBO publications focus on policy choices associated with climate change: *The Potential for Carbon Sequestration in the United States* (September 2007), *Evaluating the Role of Prices and R&D in Reducing Carbon Dioxide Emissions* (September 2006), and *Shifting the Cost Burden of a Carbon Cap-and-Trade Program* (July 2003). Those and other reports on climate policy are available at CBO's Web site, www.cbo.gov/ publications/collections/collections.cfm?collect=9.

Douglas W. Elmendorf
Director

---

## End Notes

[1] For a discussion of expected effects, see Congressional Budget Office, *Potential Impacts of Climate Change in the United States* (May 2009), available at www.cbo.gov/ftpdocs/101xx/doc10107/ 05-04-ClimateChange_ forWeb.pdf.

[2] For a general discussion of the economics of climate change, see Congressional Budget Office, *The Economics of Climate Change: A Primer* (April 2003), available at www.cbo.gov/ftpdocs/41xx/ doc4171/04-25-Climate

Change.pdf. For a general discussion of uncertainty and climate change, see Congressional Budget Office, *Uncertainty in Analyzing Climate Change: Policy Implications* (January 2005), available at www.cbo.gov/ftpdocs/60xx/doc 6061/01-24-ClimateChange.pdf.

[3] See Environmental Protection Agency, *Inventory of U.S. Greenhouse Gas Emissions and Sinks: 1990–2006* (April 15, 2008).

[4] For example, in 2000, the Energy Information Administration (EIA) overpredicted the amount of energy-related $CO_2$ that would be emitted in 2006 by 350 million tons, or by about 6 percent, largely because it overpredicted manufacturing output. EIA has subsequently reduced its projection of such emissions in 2020 by about 18 percent, resulting in lower estimates of future incremental costs for meeting any given target. (EIA's projections of greenhouse-gas emissions are particularly important for analysts of climate policy because nearly every estimate of the costs of mitigating those emissions takes the EIA reference case as its baseline. For this reason and others, the range of estimates among analysts does not fully represent the actual degree of uncertainty about mitigation costs.)

[5] For further discussion, see Congressional Budget Office, *Policy Options for Reducing $CO_2$ Emissions* (February 2008), available at www.cbo.gov/ftpdocs/89xx/doc8934/02-12-Carbon.pdf; and Congressional Budget Office, *How Regulatory Standards Can Affect a Cap-and-Trade Program for Greenhouse Gases,* Issue Brief (September 16, 2009), available at www.cbo.gov/ftpdocs/105xx/ doc10562/09-16-CapandStandards.pdf.

[6] The greater efficiency of taxes arises because the incremental cost of reducing emissions is relatively sensitive to the quantity of emission reductions (especially within a given year), while the incremental benefit is expected to be relatively insensitive to the quantity of reductions (and errors in estimating incremental benefits and costs are not expected to be correlated). The latter is true primarily because scientists are very unsure about the extent to which additional reductions would alter future climate outcomes. For further discussion, see Congressional Budget Office, *Potential Impacts of Climate Change in the United States*, pp. 14–17; and Congressional Budget Office, *Limiting Carbon Dioxide Emissions: Prices Versus Caps,* Issue Brief (March 15, 2005), available at www.cbo.gov/ftpdocs/61xx/doc6148/03-15-PriceVSQuantity .pdf.

[7] For a detailed description of cap-and-trade programs, see Congressional Budget Office, *Policy Options for Reducing $CO_2$ Emissions.*

[8] For a discussion of potential outcomes associated with alternative design features that are intended to provide firms with greater flexibility in shifting emission reductions over time, see Statement of Douglas W. Elmendorf, Director, Congressional Budget Office, before the House Committee on Ways and Means, *Flexibility in the Timing of Emission Reductions Under a Cap-and-Trade Program* (March 26, 2009), available at www.cbo.gov/ftpdocs/100xx/ doc10020/03-26-Cap-Trade_Testimony.pdf.

[9] A given quantity of reductions in greenhouse-gas emissions could be achieved at a lower cost if the cap covered more types of gases and more sources of emissions. (For example, although $CO_2$ from the burning of fossil fuels accounts for roughly 80 percent of domestic greenhouse-gas emissions, some cuts in emissions of other greenhouse gases, such as methane or nitrous oxide, could be achieved at a relatively low cost.) Policy proposals are often expressed in terms of reducing emissions by a certain percentage; and the greater the required percentage reduction, the greater the costs are likely to be. However, the percentage usually applies to covered emissions rather than to total emissions, and a given percentage target is likely to be more expensive to achieve if it covers more sources: For instance, a 90 percent reduction in $CO_2$ emissions from the burning of fossil fuels would not be as stringent as a 90 percent reduction in all greenhouse-gas emissions.

[10] For further discussion of offsets, see Congressional Budget Office, *The Use of Offsets to Reduce Greenhouse Gases,* Issue Brief (August 3, 2009), available at www.cbo.gov/ftpdocs/104xx/ doc10497/08-03-Offsets.pdf.

[11] One important exception involves $CO_2$ capture and storage. Capture would occur mainly at power plants, so if a policy imposed caps at the point of production (at the minemouth for coal, for example), it would also have to provide additional allowances as credit for capture and storage.

[12] See Congressional Budget Office, *The Economic Effects of Legislation to Reduce Greenhouse-Gas Emissions* (September 2009), available at www.cbo.gov/ftpdocs/105xx/doc10573/09-17- Greenhouse-Gas.pdf.

[13] See Congressional Budget Office, *Trade-Offs in Allocating Allowances for $CO_2$ Emissions,* Issue Brief (April 25, 2007), available at www.cbo.gov/ftpdocs/89xx/doc8946/04-25-Cap_Trade.pdf.

[14] In addition, higher energy prices would interact with the distortions of economic behavior imposed by the existing tax system, tending to magnify those distortions in some cases but to offset them in others.

[15] A useful recent such study is McKinsey & Company, *Pathways to a Low-Carbon Economy: Version 2 of the Global Greenhouse Gas Abatement Cost Curve* (McKinsey & Company, January 2009), available at https://solutions.mckinsey.com/ClimateDesk/ default.aspx.

[16] Almost none of those models provide estimates of the economic effects of climate change or the benefits of averting climate damage, so they usually cannot be used to compare costs and benefits of different policies.

[17] For a more detailed discussion of the studies reviewed in this section, see Congressional Research Service, *Climate Change: Costs and Benefits of the Cap-and-Trade Provisions of H.R. 2454*, CRS Report for Congress R40809 (September 14, 2009).

[18] Somewhat confusingly, estimates are often provided in different years' constant dollars and in different physical quantities as well, such as metric tons of carbon—MTC—instead of MT $CO_2$e. Because of inflation, $10 per

MT $CO_2$e in constant 2009 dollars is equivalent to roughly \$8 per MT $CO_2$e in constant 2000 dollars; and because carbon dioxide is composed only partly of carbon, it is equivalent to roughly \$37 per MTC in constant 2009 dollars.

[19] See Congressional Budget Office, *The Economic Effects of Legislation to Reduce Greenhouse-Gas Emissions.*

[20] See Congressional Budget Office, *The Economic Effects of Legislation to Reduce Greenhouse-Gas Emissions.*

In: Sources and Reduction of Greenhouse Gas Emissions ISBN: 978-1-61668-856-1
Editor: Steffen D. Saldana © 2010 Nova Science Publishers, Inc.

*Chapter 7*

# EMISSIONS OF GREENHOUSE GASES IN THE UNITED STATES 2008

## *United States Energy Information Administration*

### GREENHOUSE GAS EMISSIONS OVERVIEW

## Total Emissions

### *Summary*

- Total U.S. greenhouse gas emissions in 2008 were 2.2 percent below the 2007 total.
- The decline in total emissions—from 7,209.8 million metric tons carbon dioxide equivalent (MMTCO$_2$e) in 2007 to 7,052.6 MMTCO$_2$e in 2008—was largely the result of a 177.8-MMTCO$_2$e drop in carbon dioxide (CO$_2$) emissions. There were small percentage increases in emissions of other greenhouse gases, but their absolute contributions to the change in total emissions were relatively small: 14.8 MMTCO$_2$e growth for methane (CH$_4$), 0.4 MMTCO$_2$e growth for nitrous oxide (N$_2$O), and 5.3 MMTCO$_2$e growth for the man-made gases with high global warming potentials (high-GWP gases). As a result, the increases in emissions of these gases were more than offset by the drop in CO$_2$ emissions (Table 1).
- The decrease in U.S. CO$_2$ emissions in 2008 resulted primarily from three factors: higher energy prices— especially during the summer driving season—that led to a drop in petroleum consumption; economic contraction in three out of four quarters of the year that resulted in lower energy demand for the year as a whole in all sectors except the commercial sector; and lower demand for electricity along with lower carbon intensity of electricity supply.
- Methane emissions totaled 737.4 MMTCO$_2$e in 2008 (Figure 1), up by 14.8 MMTCO$_2$e (2 percent) from 2007. Most of the increase came from coal mining and from natural gas production and processing. Emissions from petroleum systems

decreased. Emissions from stationary combustion—primarily from wood combustion for residential heating—increased.
- Emissions of nitrous oxide ($N_2O$) increased by 0.4 MMTCO$_2$e (0.1 percent).
- Based on a partial estimate, U.S. emissions of highGWP gases totaled 175.6 MMTCO$_2$e in 2008—5.4 MMTCO$_2$e above the 2007 level. The increase resulted mainly from higher emissions levels for hydrofluorocarbons (HFCs, up by 5.0 MMTCO$_2$e).

## Energy-Related Carbon Dioxide Emissions by Fuel and End-Use Sector

*Summary*

- Energy-related $CO_2$ emissions dominate total U.S. greenhouse gas emissions (see Figure 1). The figures below show the shares of energy-related $CO_2$ emissions accounted for by major energy fuels and by energy end-use sectors.

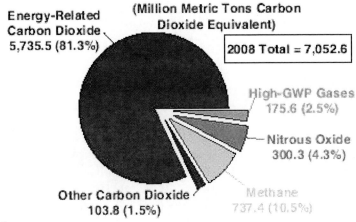

Source: EIA estimates.

Figure 1. U.S. Greenhouse Gas Emissions by Gas, 2008

### U.S. Anthropogenic Greenhouse Gas Emissions, 1990, 2007, and 2008

|  | 1990 | 2007 | 2008 |
|---|---|---|---|
| Estimated Emissions (Million Metric Tons CO$_2$e).. | 6,187.4 | 7,209.8 | 7,052.6 |
| Change from 1990 (Million Metric Tons CO$_2$e) |  | 1,022.4 | 865.1 |
| *(Percent)* |  | *16.5%* | *14.0%* |
| Average Annual Change from 1990 *(Percent)* |  | *0.9%* | *0.7%* |
| Change from 2007 (Million Metric Tons CO$_2$e) |  |  | -157.3 |
| *(Percent)* |  |  | *-2.2%* |

**Table 1. U.S. Emissions of Greenhouse Gases, Based on Global Warming Potential, 1990-2008 (Million Metric Tons Carbon Dioxide Equivalent)**

| Gas | 1990 | 1995 | 2000 | 2002 | 2003 | 2004 | 2005 | 2006 | 2007 | 2008 |
|---|---|---|---|---|---|---|---|---|---|---|
| Carbon Dioxide | 5,022.3 | 5,341.5 | 5,886.4 | 5,849.1 | 5,908.8 | 6,009.9 | 6,029.0 | 5,928.7 | 6,017.0 | 5,839.3 |
| Methane | 783.5 | 756.2 | 683.0 | 673.3 | 681.6 | 686.6 | 691.8 | 706.3 | 722.7 | 737.4 |
| Nitrous Oxide | 279.3 | 305.6 | 289.8 | 283.7 | 283.0 | 302.2 | 304.0 | 305.2 | 299.8 | 300.3 |
| High-GWP Gases[a] | 102.3 | 119.0 | 150.5 | 148.1 | 141.8 | 153.5 | 157.8 | 160.5 | 170.3 | 175.6 |
| **Total** | **6,187.4** | **6,522.3** | **7,009.8** | **6,954.2** | **7,015.2** | **7,152.1** | **7,182.6** | **7,100.8** | **7,209.8** | **7,052.6** |
| Difference from 2000 | — | — | — | -55.6 | 5.5 | 142.3 | 172.8 | 91.0 | 200.0 | 42.8 |
| Percent Difference from 2000 | — | — | — | -0.8 | 0.1 | 2.0 | 2.5 | 1.3 | 2.9 | 0.6 |

[a] Hydrofluorocarbons (HFCs), perfluorocarbons (PFCs), and sulfur hexafluoride ($SF_6$).

Notes: Data in this table are revised from the data contained in the previous EIA report, *Emissions of Greenhouse Gases in the United States 2007*, DOE/EIA-0573(2007) (Washington, DC, December 2008). Totals may not equal sum of components due to independent rounding.

Sources: **Emissions**: EIA estimates. **Global Warming Potentials**: Intergovernmental Panel on Climate Change, *Climate Change 2007: The Physical Science Basis: Errata* (Cambridge, UK: Cambridge University Press, 2008), web site http://ipcc-wg1.ucar.edu/wg1/Report/AR4WG1_Errata_2008-12-01.pdf.

- Petroleum is the largest fossil fuel source for energy-related $CO_2$ emissions, contributing 42 percent of the total (Figure 2).
- Coal is the second-largest fossil fuel contributor, at 37 percent. Although coal produces more $CO_2$ per unit of energy, petroleum consumption—in terms of British thermal units (Btu)—made up 44.6 percent of total fossil fuel energy consumption in 2008, as compared with coal's 26.8 percent.
- Natural gas, with a carbon intensity that is 55 percent of the carbon intensity for coal and 75 percent of the carbon intensity for petroleum, accounted for 28.5 percent of U.S. fossil energy use in 2008 but only 21 percent of total energy-related $CO_2$ emissions.
- In Figure 3 below, emissions are divided into three categories: emissions from the direct use of fossil fuels in homes (for example, natural gas for heating), commercial buildings, and industry; emissions from fuel use for transportation (principally, petroleum); and emissions from the conversion of primary energy to electricity in the electric power sector.
- The electric power sector is the largest source, accounting for 40.6 percent of all energy-related $CO_2$ emissions. The electric power sector consists of those entities whose primary business is the production of electricity (NAICS-22).
- The transportation sector is the second-largest source, at 33.1 percent of the total. Those emissions are principally from the combustion of motor gasoline, diesel fuel, and jet fuel.
- Direct fuel use in the residential and commercial sectors (mainly, for heating) and the use of fuels to produce process heat in the industrial sector account for 26.3 percent of total emissions.

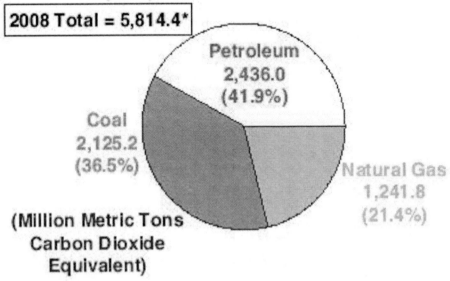

*Includes small amounts of $CO_2$ from non-biogenic municipal solid waste and geothermal energy (0.2 percent of total).
Source: EIA estimates.

Figure 2. U.S. Energy-Related Carbon Dioxide Emissions by Major Fuel, 2008

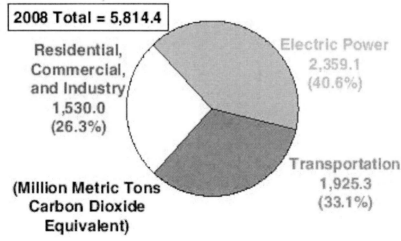

Source: EIA estimates.

Figure 3. U.S. Energy-Related Carbon Dioxide Emissions by End-Use Sector, 2008

## Decomposition of U.S. Greenhouse Gas Changes

### *Summary*

- From 2007 to 2008, the greenhouse gas intensity of the U.S. economy—measured as metric tons carbon dioxide equivalent (MTCO$_2$e) emitted per million dollars of real gross domestic product (GDP)—fell by 2.6 percent.
- Economic growth of 0.4 percent in 2008, coupled with a 2.2-percent decrease in total greenhouse gas emissions, accounted for the decrease (improvement) in U.S. greenhouse gas intensity from 2007 to 2008 (Table 2).
- Because energy-related CO$_2$ is such a large component of greenhouse gas emissions, it is helpful to analyze energy-related CO$_2$ emissions by using an equation known as the *Kaya identity*. The Kaya identity relates percent changes in energy-related CO$_2$ emissions to changes in the economy through the following approximation:

$$\%\Delta CO_2 \approx \%\Delta GDP + \%\Delta(Energy/GDP) + \%\Delta(CO_2/Energy),$$

where %Δ represents percentage change.

As indicated in Figure 4, energy intensity (Energy/GDP) has gone down in every year since 2000. The carbon intensity of the energy supply (CO$_2$/Energy) has gone down in some years and up in others. While GDP growth was positive in all years from 2000 through 2008, it has varied. In 2008, economic growth was low (0.4 percent), while there were decreases in both energy intensity (-2.5 percent) and carbon intensity (-0.8 percent), leading to a 2.9-percent decline in energy-related CO$_2$ emissions.

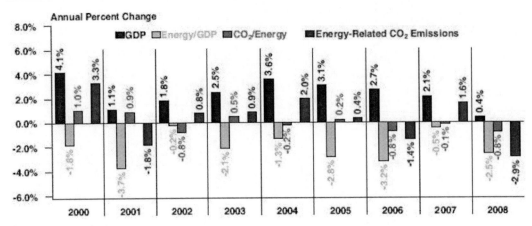

Source: EIA estimates.

Figure 4. Annual Changes in Kaya Identity Factors (GDP, Energy/GDP, and CO$_2$/Energy) and Energy-Related Carbon Dioxide Emissions, 2000-2008

## Greenhouse Gas Emissions in the U.S. Economy

The diagram on page 5 illustrates the flow of U.S. greenhouse gas emissions in 2008, from their sources to their end uses. The left side shows CO$_2$ quantities, by fuel sources, and quantities for other gases; the right side shows their distribution by sector. The center of the diagram indicates the split between CO$_2$ emissions from direct fuel combustion and electricity conversion. Adjustments indicated at the top of the diagram for U.S. territories and international bunker fuels correspond to greenhouse gas reporting requirements developed by the United Nations Framework Convention on Climate Change (UNFCCC).

**CO$_2$.** CO$_2$ emission sources include energy-related emissions (primarily from fossil fuel combustion) and emissions from industrial processes. The energy subtotal (5,814 MMTCO$_2$e) includes petroleum, coal, and natural gas consumption and smaller amounts from non-biogenic municipal solid waste and some forms of geothermal power generation. The energy subtotal also includes emissions from nonfuel uses of fossil fuels, mainly as inputs to other products. Industrial process emissions (104 MMTCO$_2$e) include cement manufacture, limestone and dolomite calcination, soda ash manufacture and consumption, carbon dioxide manufacture, and aluminum production. The sum of the energy subtotal and industrial processes equals unadjusted CO$_2$ emissions (5,918 MMTCO$_2$e). The energy component of unadjusted emissions can be divided into direct fuel use (3,381 MMTCO$_2$e) and fuel converted to electricity (2,433 MMTCO$_2$e).

**Non-CO$_2$ Gases.** Methane (737 MMTCO$_2$e) and nitrous oxide (300 MMTCO$_2$e) sources include emissions related to energy, agriculture, waste management, and industrial processes. Other, high-GWP gases (176 MMTCO$_2$e) include hydrofluorocarbons (HFCs), perfluorocarbons (PFCs), and sulfur hexafluoride (SF$_6$). These gases have a variety of uses in the U.S. economy, including refrigerants, insulators, solvents, and aerosols; as etching, cleaning, and firefighting agents; and as cover gases in various manufacturing processes.

## Table 2. U.S. Greenhouse Gas Intensity and Related Factors, 1990-2008

| | 1990 | 1995 | 2000 | 2002 | 2003 | 2004 | 2005 | 2006 | 2007 | 2008 |
|---|---|---|---|---|---|---|---|---|---|---|
| Gross Domestic Product | | | | | | | | | | |
| (Billion 2005 Dollars) | 8,033.9 | 9,093.7 | 11,226.0 | 11,553.0 | 11,840.7 | 12,263.8 | 12,638.4 | 12,976.2 | 13,254.1 | 13,312.2 |
| Greenhouse Gas Emissions | | | | | | | | | | |
| (MMTCO$_2$e) | 6,187.4 | 6,522.3 | 7,009.8 | 6,954.2 | 7,015.2 | 7,152.1 | 7,182.6 | 7,100.8 | 7,209.8 | 7,052.6 |
| Greenhouse Gas Intensity | | | | | | | | | | |
| (MTCO$_2$e per Million 2005 Dollars) | 770.2 | 717.2 | 624.4 | 601.9 | 592.5 | 583.2 | 568.3 | 547.2 | 544.0 | 529.8 |
| **Change from Previous ear (Percent)** | | | | | | | | | | |
| Energy-Related CO$_2$ Emissions | — | 1.0 | 3.3 | 0.8 | 0.8 | 2.0 | 0.4 | -1.3 | 1.6 | -2.9 |
| Gross DomesticProduct(GDP) | — | 2.5 | 4.1 | 1.8 | 2.5 | 3.6 | 3.1 | 2.7 | 2.1 | 0.4 |
| Energy/GDP | — | -0.3 | -1.8 | -0.2 | -2.1 | -1.3 | -2.8 | -3.2 | -0.5 | -2.5 |
| CO$_2$/Energy | — | -1.2 | 1.0 | -0.8 | 0.5 | -0.2 | 0.2 | -0.7 | -0.1 | -0.8 |

Note: Data in this table are revised from the data contained in the previous EIA report, *Emissions of Greenhouse Gases in the United States 2007, DOE/EIA-0573(2007)* (Washington, DC, December 2008).

Sources: **Emissions**: EIA estimates. **GDP**: U.S. Department of Commerce, Bureau of Economic Analysis, web site www.bea.gov (November 12,2009).

# 144 United States Energy Information Administration

**Adjustments.** In keeping with the UNFCCC, $CO_2$ emissions from U.S. Territories (48 $MMTCO_2e$) are added to the U.S. total, and $CO_2$ emissions from fuels used for international transport (both oceangoing vessels and airplanes) (127 $MMTCO_2e$) are subtracted to derive total U.S. greenhouse gas emissions (7,053 $MMTCO_2e$).

**Emissions by End-Use Sector.** $CO_2$ emissions by end-use sectors are based on EIA's estimates of energy consumption (direct fuel use and purchased electricity) by sector and on the attribution of industrial process emissions by sector. $CO_2$ emissions from purchased electricity are allocated to the end-use sectors based on their shares of total electricity sales. Non-$CO_2$ gases are allocated by direct emissions in those sectors plus emissions in the electric power sector that can be attributed to the end-use sectors based on electricity sales.

**Residential** emissions (1,244 $MMTCO_2e$) include energy-related $CO_2$ emissions (1,230 $MMTCO_2e$) and non-$CO_2$ emissions (14 $MMTCO_2e$). The non-$CO_2$ sources include methane and nitrous oxide emissions from direct fuel use. Non-$CO_2$ indirect emissions attributable to purchased electricity, including methane and nitrous oxide emissions from electric power generation and $SF_6$ emissions related to electricity transmission and distribution, are also included.

Emissions in the **commercial** sector (1,353 $MMTCO_2e$) include both energy-related $CO_2$ emissions (1,084 $MMTCO_2e$) and non-$CO_2$ emissions (269 $MMTCO_2e$). The non-$CO_2$ emissions include direct emissions from landfills, wastewater treatment plants, commercial refrigerants, and stationary combustion emissions of methane and nitrous oxide. Non-$CO_2$ indirect emissions attributable to purchased electricity, including methane and nitrous oxide emissions from electric power generation and $SF_6$ emissions related to electricity transmission and distribution, are also included.

**Industrial** emissions (2,510 $MMTCO_2e$) include $CO_2$ emissions (1,706 $MMTCO_2e$)—which can be broken down between combustion (1,602 $MMTCO_2e$) and process emissions (104 $MMTCO_2e$)—and non-$CO_2$ emissions (804 $MMTCO_2e$). The non-$CO_2$ direct emissions include emissions from agriculture (methane and nitrous oxide), coal mines (methane), petroleum and natural gas pipelines (methane), industrial process emissions (methane, nitrous oxide, HFCs, PFCs and $SF_6$), and direct stationary combustion emissions of methane and nitrous oxide. Non-$CO_2$ indirect emissions attributable to purchased electricity, including methane and nitrous oxide emissions from electric power generation and $SF_6$ emissions related to electricity transmission and distribution, are also included.

**Transportation** emissions (1,946 $MMTCO_2e$) include energy-related $CO_2$ emissions from mobile source combustion (1,819 $MMTCO_2e$); and non-$CO_2$ emissions (127 $MMTCO_2e$). The non-$CO_2$ emissions include methane and nitrous oxide emissions from mobile source combustion and HFC emissions from the use of refrigerants for mobile source air-conditioning units.

# Emissions of Greenhouse Gases in the United States 2008

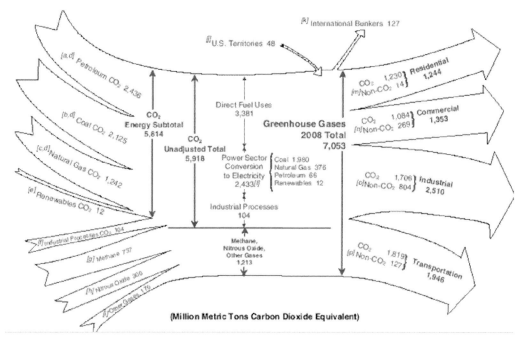

**Diagram Notes**

[a] emissions related to petroleum consumption (includes 84.0 MMTCO$_2$ of non-fuel-related emissions).

[b] CO$_2$ emissions related to coal consumption (includes 0.5 MMTCO$_2$ of non-fuel-related emissions).

[c] CO$_2$ emissions related to natural gas consumption (includes 18.1 MMTCO$_2$ of non-fuel-related emissions).

[d] Excludes carbon sequestered in nonfuel fossil products.

[e] CO$_2$ emissions from the plastics portion of municipal solid waste (11.2 MMTCO$_2$) combusted for electricity generation and very small amounts (0.4 MMTCO$_2$) of geothermal-related emissions.

[f] Includes mainly direct process emissions. Some combustion emissions are included from waste combustion outside the electric power sector and flaring of non-marketed natural gas.

[g] Includes methane emissions related to energy, agriculture, waste management, and industrial processes.

[h] Includes nitrous oxide emissions related to agriculture, energy, industrial processes, and waste management.

[i] Includes hydrofluorocarbons, perfluorocarbons, and sulfur hexafluoride.

[j] Includes only energy-related CO$_2$ emissions from fossil fuels. Emissions are allocated to end-use sectors in proportion to U.S. ratios. Therefore, the sector CO$_2$ values shown here do not match the values in the carbon dioxide chapter.

[k] Includes vessel bunkers and jet fuel consumed for international travel. Under the UNFCCC, these emissions are not included in country emission inventories. Emissions are subtracted from the transportation sector total.

[l] CO$_2$ emissions from electricity generation in the commercial and industrial sectors are included in those sectors.

[m] Non-CO$_2$: Direct stationary combustion emissions of methane and nitrous oxide plus indirect power sector emissions of methane, nitrous oxide, and other greenhouse gases.

[n] Non-CO$_2$: Direct stationary combustion emissions of methane and nitrous oxide plus indirect power sector emissions of methane, nitrous oxide, and other greenhouse gases. Additional direct emissions include emissions from landfills, wastewater treatment, and commercial refrigerants.

[o] Non-$CO_2$: Direct stationary combustion emissions of methane and nitrous oxide plus indirect power sector emissions of methane, nitrous oxide, and other greenhouse gases. In addition, all agricultural emissions are included in the industrial sector as well as direct process emissions of methane, nitrous oxide, and the other gases.

[p] Non-$CO_2$: Direct mobile combustion emissions of methane and nitrous oxide. Also, emissions related to transportation refrigerants are included.

Source: Estimates presented in this chapter. $CO_2$ emissions by end-use sector are based on EIA's estimates of energy consumption by sector and on industrial process emissions. $CO_2$ emissions from the electric power sector are allocated to the end-use sectors based on electricity sales to the sector. Non-$CO_2$ emissions by end-use sector are allocated by direct emissions in those sectors plus indirect emissions from the electric power sector allocated by electricity sales. Data are preliminary. Totals may not equal sum of components due to independent rounding.

## Distribution of Total U.S. Greenhouse Gas Emissions by End-Use Sector, 2008

| Greenhouse Gas and Source | Sector | | | | |
|---|---|---|---|---|---|
| | Residential | Commercial | Industrial | Transportation | Total |
| **Carbon Dioxide** | **Million Metric Tons Carbon Dioxide Equivalent** | | | | |
| Energy-Related (adjusted) | 1,230.3 | 1,084.1 | 1,602.3 | 1,818.8 | **5,735.5** |
| Industrial Processes | — | — | 103.8 | — | **103.8** |
| **Total CO$_2$** | **1,230.3** | **1,084.1** | **1,706.1** | **1,818.8** | **5,839.2** |
| **Methane** | | | | | |
| **Energy** | | | | | |
| Coal Mining | — | — | 82.0 | — | **82.0** |
| Natural Gas Systems | — | — | 178.9 | — | **178.9** |
| Petroleum Systems | — | — | 22.1 | — | **22.1** |
| Stationary Combustion | 4.5 | 1.0 | 1.8 | — | **7.3** |
| Stationary Combustion: Electricity | 0.3 | 0.3 | 0.2 | — | **0.8** |
| Mobile Sources | — | — | — | 4.6 | **4.6** |
| **Waste Management** | | | | | |
| Landfills | — | 184.3 | — | — | **184.3** |
| Domestic Wastewater Treatment | — | 17.6 | — | — | **17.6** |
| Industrial Wastewater Treatment. | — | — | 10.2 | — | **10.2** |
| **Industrial Processes** | — | — | 4.7 | — | **4.7** |
| **Agricultural Sources** | | | | | |
| Enteric Fermentation | — | — | 148.6 | — | **148.6** |
| Animal Waste | — | — | 64.5 | — | **64.5** |
| Rice Cultivation | — | — | 10.6 | — | **10.6** |

## Emissions of Greenhouse Gases in the United States 2008

**(Continued)**

| Greenhouse Gas and Source | Sector | | | | |
|---|---|---|---|---|---|
| | Residential | Commercial | Industrial | Transportation | Total |
| Crop Residue Burning | — | — | 1.3 | — | **1.3** |
| **Total Methane** | **4.8** | **203.2** | **524.8** | **4.6** | **737.4** |
| **Nitrous Oxide** | | | | | |
| **Agriculture** | | | | | |
| Nitrogen Fertilization of Soils | — | — | 165.0 | — | **165.0** |
| Solid Waste of Animals | — | — | 52.3 | — | **52.3** |
| Crop Residue Burning | — | — | 0.6 | — | **0.6** |
| **Energy Use** | | | | | |
| Mobile Combustion | — | — | — | 48.8 | **48.8** |
| Stationary Combustion | 0.9 | 0.3 | 4.1 | — | **5.4** |
| Stationary Combustion: Electricity | 3.6 | 3.4 | 2.7 | — | **9.7** |
| **Industrial Sources** | — | — | 15.1 | — | **15.1** |
| **Waste Management** | | | | | |
| Human Sewage in Wastewater | — | 3.0 | — | — | **3.0** |
| Waste Combustion | — | — | — | — | **0.0** |
| Waste Combustion: Electricity | 0.1 | 0.1 | 0.1 | — | **0.4** |
| **Total Nitrous Oxide** | **4.6** | **6.9** | **239.9** | **48.8** | **300.2** |
| **Hydrofluorocarbons (HFCs)** | | | | | |
| HFC-23 | — | — | 21.9 | — | **21.9** |
| HFC-32 | — | 1.2 | — | — | **1.2** |
| HFC-125 | — | 22.1 | — | — | **22.1** |
| HFC-134a | — | — | — | 73.6 | **73.6** |
| HFC-143a | — | 22.5 | — | — | **22.5** |
| HFC-236fa | — | 1.4 | — | — | **1.4** |
| **Total HFCs** | **0.0** | **47.2** | **21.9** | **73.6** | **142.7** |
| **Perfluorocarbons (PFCs)** | | | | | |
| $CF_4$ | — | — | 5.2 | — | **5.2** |
| $C_2F_6$ | — | — | 4.2 | — | **4.2** |
| $NF_3$, $C_3F_8$, and $C_4F_8$ | — | — | 0.7 | — | **0.7** |
| **Total PFCs** | **0.0** | **0.0** | **10.1** | **0.0** | **10.1** |
| **Other HFCs, PFCs/PFPEs** | — | 7.1 | — | — | **7.1** |

# 148 United States Energy Information Administration

**(Continued)**

| Greenhouse Gas and Source | Sector | | | | |
|---|---|---|---|---|---|
| | Residential | Commercial | Industrial | Transportation | Total |
| **Sulfur Hexafluoride (SF6)** | | | | | |
| $SF_6$: Utility | 4.5 | 4.3 | 3.3 | — | **12.1** |
| $SF_6$: Other | — | — | 3.7 | — | **3.7** |
| **Total $SF_6$** | **4.5** | **4.3** | **7.0** | **0.0** | **15.8** |
| **Total Non-$CO_2$** | **13.9** | **268.7** | **803.7** | **127.0** | **1,213.3** |
| **Total Emissions** | **1,244.1** | **1,352.8** | **2,509.8** | **1,945.8** | **7,052.6** |

**World Energy-Related Carbon Dioxide Emissions, 1990, 2006, and 2030**

| | 1990 | 2006 | 2030[*] |
|---|---|---|---|
| Estimated Emissions (Million Metric Tons) | 21,518 | 29,017 | 40,178 |
| Change from 1990 (Million Metric Tons) | | 7,499 | 18,660 |
| *(Percent)* | | *34.8%* | *86.7%* |
| Average Annual Change from 1990 *(Percent)* | | *2.0%* | *1.6%* |
| Change from 2006 (Million Metric Tons) | | | 11,161 |
| *(Percent)* | | | *38.5%* |

[*]EIA, International Energy Outlook 2009.

## U.S. Emissions in a Global Perspective

### *Summary*

- Based on the 2008 emissions inventory report, total U.S. energy-related $CO_2$ emissions in 2006 (including nonfuel uses of fossil fuels) were estimated at 5,894 MMT—about 20 percent of the 2006 world total for energy-related $CO_2$ emissions, estimated at 29,017 MMT (see Table 3 on page 8).
- $CO_2$ emissions related to energy use in the mature economies of countries that are members of the Organization for Economic Cooperation and Development (OECD)—including OECD North America, OECD Europe, Japan, and Australia/New Zealand—were estimated at 13,582 MMT in 2006, or 47 percent of the world total. With the remaining 53 percent of worldwide energy-related $CO_2$ emissions (15,435 MMT) estimated to have come from non- OECD countries, 2006 was the second year in which emissions from the non-OECD economies surpassed those from the OECD economies (Figure 5).
- In EIA's *International Energy Outlook 2009* (*IEO2009*) reference case, projections of energy use and emissions are sensitive to economic growth rates and energy prices. Projections for a range of alternative growth and price scenarios are presented in *IEO2009*.
- U.S. energy-related $CO_2$ emissions are projected to increase by an average of 0.2 percent per year from 2006 to 2030 in the *IEO2009* reference case, while emissions from the non-OECD economies grow by 2.2 percent per year. Both rates are lower

than previous projections as a result of the 2008-2009 global recession and newly enacted energy policies. Consequently, the U.S. share of world $CO_2$ emissions is projected to fall to 15.4 percent in 2030 (6,207 MMT out of a global total of 40,178 MMT) (Figure 6).
- China's share of global energy-related $CO_2$ emissions is projected to grow from 21 percent in 2006 to 29 percent in 2030, and China accounts for 51 percent of the projected increase in world emissions over the period. India accounts for the second-largest share of the projected increase, 7 percent.

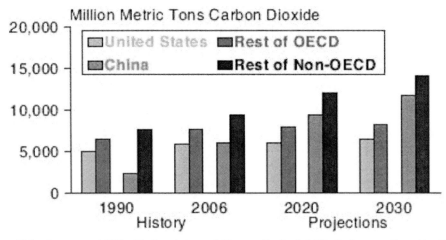

Sources: This chapter and EIA, *Updated Annual Energy Outlook 2009 Reference Case* (April 2009), web site www.eia.doe.gov/oiaf/servicerpt/stimulus.

Figure 5. World Carbon Dioxide Emissions by Region, 1990, 2006, 2020, and 2030

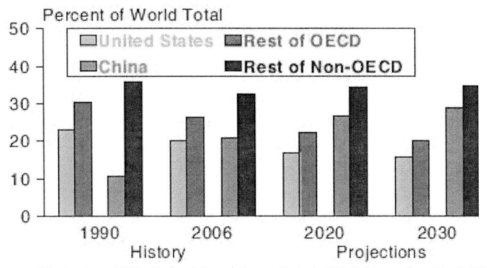

Sources: This chapter and EIA, *Updated Annual Energy Outlook 2009 Reference Case* (April 2009), web site www.eia. doe.gov/oiaf/servicerpt/stimulus.

Figure 6. Regional Shares of World Carbon Dioxide Emissions, 1990, 2006, 2020, and 2030

## U.S. Emissions in a Global Perspective

### Table 3. World Energy-Related Carbon Dioxide Emissions by Region, 1990-2030
### (Million Metric Tons Carbon Dioxide, Percent Share of World Emissions)

| Region/ Country | History[a] | | | Projections[a] | | | | | Average Annual Percent Change, 2006-2030[b] |
|---|---|---|---|---|---|---|---|---|---|
| | 1990 | 2005 | 2006 | 2010 | 2015 | 2020 | 2025 | 2030 | |
| **OECD** | | | | | | | | | |
| **OECD North America** | **5,793** | **7,006** | **6,936** | **6,739** | **6,889** | **7,046** | **7,262** | **7,495** | **0.3** |
| | **(26.9%)** | **(24.8%)** | **(23.9%)** | **(21.8%)** | **(20.9%)** | **(19.9%)** | **(19.2%)** | **(18.7%)** | **(5.0%)** |
| United States[c] | 5,020 | 5,974 | 5,894 | 5,746 | 5,830 | 5,905 | 6,047 | 6,207 | 0.2 |
| | (23.3%) | (21.1%) | (20.3%) | (18.6%) | (17.6%) | (16.7%) | (16.0%) | (15.4%) | (2.8%) |
| Canada | 471 | 629 | 611 | 622 | 645 | 675 | 705 | 731 | 0.7 |
| | (2.2%) | (2.2%) | (2.1%) | (2.0%) | (2.0%) | (1.9%) | (1.9%) | (1.8%) | (1.1%) |
| Mexico | 302 | 403 | 431 | 371 | 414 | 466 | 510 | 557 | 1.1 |
| | (1.4%) | (1.4%) | (1.5%) | (1.2%) | (1.3%) | (1.3%) | (1.3%) | (1.4%) | (1.1%) |
| **OECD Europe** | **4,149** | **4,424** | **4,429** | **4,335** | **4,368** | **4,450** | **4,489** | **4,519** | **0.1** |
| | **(19.3%)** | **(15.6%)** | **(15.3%)** | **(14.0%)** | **(13.2%)** | **(12.6%)** | **(11.9%)** | **(11.2%)** | **(0.8%)** |
| **OECD Asia** | **1,595** | **2,201** | **2,217** | **2,221** | **2,287** | **2,327** | **2,346** | **2,367** | **0.3** |
| | **(7.4%)** | **(7.8%)** | **(7.6%)** | **(7.2%)** | **(6.9%)** | **(6.6%)** | **(6.2%)** | **(5.9%)** | **(1.3%)** |
| Japan | 1,054 | 1,250 | 1,247 | 1,169 | 1,204 | 1,219 | 1,188 | 1,157 | -0.3 |
| | (4.9%) | (4.4%) | (4.3%) | (3.8%) | (3.6%) | (3.4%) | (3.1%) | (2.9%) | (-0.8%) |
| South Korea | 243 | 497 | 515 | 598 | 614 | 617 | 651 | 680 | 1.2 |
| | (1.1%) | (1.8%) | (1.8%) | (1.9%) | (1.9%) | (1.7%) | (1.7%) | (1.7%) | (1.5%) |
| Australia/ New Zealand | 298 | 454 | 455 | 454 | 469 | 491 | 507 | 530 | 0.6 |
| | (1.4%) | (1.6%) | (1.6%) | (1.5%) | (1.4%) | (1.4%) | (1.3%) | (1.3%) | (0.7%) |
| **Total OECD** | **11,537** | **13,631** | **13,582** | **13,295** | **13,544** | **13,823** | **14,097** | **14,381** | **0.2** |
| | **(53.6%)** | **(48.2%)** | **(46.8%)** | **(43.0%)** | **(41.0%)** | **(39.1%)** | **(37.3%)** | **(35.8%)** | **(7.2%)** |
| **Non-OECD** | | | | | | | | | |
| **Non-OECD Europe and Eurasia** | **4,246** | **2,889** | **2,886** | **3,069** | **3,234** | **3,323** | **3,362** | **3,422** | **0.7** |
| | **(19.7%)** | **(10.2%)** | **(9.9%)** | **(9.9%)** | **(9.8%)** | **(9.4%)** | **(8.9%)** | **(8.5%)** | **(4.8%)** |
| Russia | 2,393 | 1,699 | 1,704 | 1,803 | 1,894 | 1,945 | 1,950 | 1,978 | 0.6 |
| | (11.1%) | (6.0%) | (5.9%) | (5.8%) | (5.7%) | (5.5%) | (5.2%) | (4.9%) | (2.5%) |
| Other | 1,853 | 1,190 | 1,182 | 1,266 | 1,339 | 1,378 | 1,412 | 1,443 | 0.8 |
| | (8.6%) | (4.2%) | (4.1%) | (4.1%) | (4.1%) | (3.9%) | (3.7%) | (3.6%) | (2.3%) |
| **Non-OECD Asia** | **3,677** | **8,305** | **8,988** | **10,465** | **11,900** | **13,590** | **15,382** | **17,033** | **2.7** |
| | **(17.1%)** | **(29.4%)** | **(31.0%)** | **(33.9%)** | **(36.0%)** | **(38.4%)** | **(40.7%)** | **(42.4%)** | **(72.1%)** |
| China | 2,293 | 5,429 | 6,018 | 7,222 | 8,204 | 9,417 | 10,707 | 11,730 | 2.8 |
| | (10.7%) | (19.2%) | (20.7%) | (23.4%) | (24.8%) | (26.6%) | (28.3%) | (29.2%) | (51.2%) |
| India | 573 | 1,192 | 1,292 | 1,366 | 1,572 | 1,783 | 1,931 | 2,115 | 2.1 |
| | (2.7%) | (4.2%) | (4.5%) | (4.4%) | (4.8%) | (5.0%) | (5.1%) | (5.3%) | (7.4%) |
| Other Non-OECD Asia | 811 | 1,684 | 1,678 | 1,877 | 2,124 | 2,390 | 2,744 | 3,188 | 2.7 |

# Emissions of Greenhouse Gases in the United States 2008

**Table 3. (Continued)**

| Region/ Country | History[a] | | Projections[a] | | | | | | Average Annual Percent Change, 2006-2030[b] |
|---|---|---|---|---|---|---|---|---|---|
| | 1990 | 2005 | 2006 | 2010 | 2015 | 2020 | 2025 | 2030 | |
| | (3.8%) | (6.0%) | (5.8%) | (6.1%) | (6.4%) | (6.8%) | (7.3%) | (7.9%) | (13.5%) |
| **Middle East** | **704** | **1,393** | **1,456** | **1,686** | **1,830** | **1,939** | **2,088** | **2,279** | **1.9** |
| | (3.3%) | (4.9%) | (5.0%) | (5.5%) | (5.5%) | (5.5%) | (5.5%) | (5.7%) | (7.4%) |
| **Africa** | **659** | **985** | **982** | **1,086** | **1,161** | **1,239** | **1,325** | **1,409** | **1.5** |
| | (3.1%) | (3.5%) | (3.4%) | (3.5%) | (3.5%) | (3.5%) | (3.5%) | (3.5%) | (3.8%) |
| **Central and South America** | **695** | **1,093** | **1,123** | **1,311** | **1,368** | **1,437** | **1,547** | **1,654** | **1.6** |
| | (3.2%) | (3.9%) | (3.9%) | (4.2%) | (4.1%) | (4.1%) | (4.1%) | (4.1%) | (4.8%) |
| Brazil | 235 | 366 | 374 | 437 | 488 | 543 | 612 | 682 | 2.5 |
| | (1.1%) | (1.3%) | (1.3%) | (1.4%) | (1.5%) | (1.5%) | (1.6%) | (1.7%) | (2.8%) |
| Other Central/ South America | 460 | 727 | 749 | 874 | 881 | 894 | 935 | 972 | 1.1 |
| | (2.1%) | (2.6%) | (2.6%) | (2.8%) | (2.7%) | (2.5%) | (2.5%) | (2.4%) | (2.0%) |
| **Total Non-OECD** | **9,981** | **14,665** | **15,435** | **17,616** | **19,494** | **21,528** | **23,703** | **25,797** | **2.2** |
| | (46.4%) | (51.8%) | (53.2%) | (57.0%) | (59.0%) | (60.9%) | (62.7%) | (64.2%) | (92.8%) |
| **Total World** | **21,518** | **28,296** | **29,017** | **30,911** | **33,038** | **35,351** | **37,800** | **40,178** | **1.4** |

[a]Values adjusted for nonfuel sequestration.

[b]Values in parentheses indicate percentage share of total world absolute change from 2006 to 2030.

[c]Includes the 50 States and the District of Columbia.

Note: The U.S. numbers include carbon dioxide emissions attributable to geothermal energy and nonbiogenic materials in municipal solid waste.

Sources: **History:** Energy Information Administration (EIA), *International Energy Annual 2006* (June-December 2008), web site www.eia.doe. gov/iea/; and data presented in this chapter. **Projections:** EIA, *Annual Energy Outlook2009*, DOE/EIA-0383(2009) (Washington, DC, March 2009), Table 1, web site www.eia.doe.gov/oiaf/aeo; *Updated Annual Energy Outlook 2009 Reference Case* (Washington, DC, April 2009), web site www.eia.doe.gov/oiaf/servicerpt/stimulus; and *International Energy* Outlook 2009, DOE/EIA-0484(2009) (Washington, DC, May 2009), Table A10.

# Recent U.S. and International Developments in Global Climate Change

## United States

### Federal Actions

- The Consolidated Appropriations Act of 2008, which became Public Law 110-161 on December 26, 2007, directed the U.S. Environmental Protection Agency (EPA) to develop a mandatory reporting rule for greenhouse gases (GHGs). The Final Rule was signed by the Administrator of the EPA on September 22,2009. The Rule requires that emitters of GHGs from 31 different source categories report their emissions to the EPA. Approximately 80 to 85 percent of total U.S. GHG emissions from

152         United States Energy Information Administration

10,000 facilities are expected to be covered by the Rule. Reporters must begin to monitor their emissions on January 1, 2010; the first annual emissions reports willbe due in 2011.

- On April 2, 2007, the U.S. Supreme Court ruled that Section 202(a)(1) of the Clean Air Act (CAA) gives the EPA authority to regulate tailpipe emissions of GHGs. On April 17, 2009, the EPA Administrator signed a Proposed Endangerment and Cause or Contribute Findings for Greenhouse Gases. The proposal finds that the six key GHGs pose a threat to public health and welfare for current and future generations, and that GHG emissions from new motor vehicles and motor vehicle engines contribute to climate change. Finalization of the "Endangerment Finding" must occur before the EPA can implement its proposed standards for GHG emissions from stationary sources and from vehicles (see below). The Final Finding was sent to the White House Office of Management and Budget (OMB) on November 6, 2009.

- As one result of the Supreme Court's decision in 2007, the EPA drafted the Prevention of Significant Deterioration/Title V Greenhouse Gas Tailoring Rule. The draft rule, published in the *Federal Register* on October 27, 2009, limits the applicability of $CO_2$ emissions standards under the CAA to new and modified stationary sources that emit more than 25,000 $MTCO_2e$ annually, rather than applying the threshold of 250 tons per source for triggering the regulation of criteria pollutants specified in Title V of the CAA. At the 25,000 $MTCO_2e$ level, the EPA expects that 14,000 large industrial sources, which are responsible for nearly 70 percent of U.S. GHG emissions, will be required to obtain Title V operating permits. The threshold would cover power plants, refineries, and other large industrial operations but exempt small farms, restaurants, schools, and other small facilities.

- On September 15, 2009, the EPA and the U.S. Department of Transportation (DOT) jointly proposed new nationwide standards for corporate average fuel economy (CAFE) and GHG emissions standards for new light- and medium-duty vehicles. The proposal formalizes an agreement announced in May 2009 between the Administration and automobile industry stakeholders to accelerate the existing CAFE mandate and impose nationwide the tailpipe GHG standards sought by California. The proposed rule outlines fuel economy and GHG emissions standards for five model years 2012 through 2016 for cars sold in the United States. The Final Rule must be published by April2010 if it is to take effect on schedule for model year 2012.[1]

- The American Recovery and Reinvestment Act of 2009 ("The Stimulus Bill") was signed into law by President Obama on February 17, 2009. Under the Act, the U.S. Department of Energy (DOE) received $36.7 billion to fund renewable energy, carbon capture and storage, energy efficiency, and smart grid projects, among others. The projects are expected to provide reductions in both energy use and GHG emissions.

- On May 26, 2009, the EPA published a Notice of Proposed Rulemaking for the national Renewable Fuel Standard (RFS2), as revised by the Energy Independence and Security Act of 2007 (EISA). The revised statutory requirements establish new specific volume standards for cellulosic biofuel, biomass-based diesel, advanced biofuels, and total renewable fuel that must be used in transportation fuel each year. The revisions also include new definitions and criteria for both renewable fuels and the feedstocks used to produce them, including new GHG emission thresholds for renewable fuels. The EPA's proposed rulemaking includes guidelines on how life-cycle emissions from each type of renewable fuel would be calculated and compared against those of traditional fossil-based motor fuels. EPA's proposed method of life-cycle accounting includes GHG emissions from production and transport of the feedstock; land use change; production, distribution, and blending of the renewable fuel; and end use of the renewable fuel. The Final Rule establishing the standards for 2010 is expected to be published in December 2009.

## Congressional Initiatives

- On June 26,2009, the U. S. House of Representatives passed The American Clean Energy and Security Act of 2009 (ACESA). The legislation includes a Federal GHG emissions "cap-and-trade" program that would take effect in 2012. The declining emissions cap requires that total GHG emissions be 17 percent below 2005 levels by 2020 (5,056 $MMTCO_2e$) and 83 percent below 2005 levels by 2050 (1,035 $MMTCO_2e$). In addition to the cap-and-trade title, the legislation includes provisions for the use of domestic and international offsets to meet the requirements of the emissions cap in lieu of allowances; funding for carbon capture and sequestration projects; standards and programs designed to boost energy efficiency; and a combined renewable electricity and energy efficiency standard of 20 percent by 2020.
- In the Senate, two major pieces of a comprehensive energy and climate bill have passed out of their respective committees. In June, the Energy and Natural Resources Committee voted out an energy package that includes a renewable electricity and efficiency standard of 15 percent by 2021, along with provisions addressing energy project financing, expanded oil and natural gas leasing, and energy efficiency. In November, the Environment and Public Works Committee passed a GHG cap-and-trade bill that borrows much of its structure and content from the House-passed ACESA. Notably, the emissions cap was tightened to 20 percent below 2005levels by 2020. Several other committees are expected to weigh in before a final bill is crafted and brought to the Senate floor.

## Regional and State Efforts
Although ACESA includes a provision that, when enacted, would prohibit State and regional cap-and-trade programs from operating between 2012 and 2017, activity continues across the various regional GHG emissions reduction initiatives in the United States.

- On September 9,2009, the Regional Greenhouse Gas Initiative (RGGI) held its fifth auction of $CO_2$ emissions allowances. Over the three auctions held so far in 2009, approximately 90 million allowances have been sold, generating more than $432.8 million to be used by the participating States for energy efficiency and renewable energy programs. Auction 6 is scheduled for December 2009. Additionally, the RGGI participants signed a letter of intent on December 31, 2008, to develop a low-carbon fuel standard for transportation fuels. A draft program Memorandum of

Understanding is due by December 31, 2009. Connecticut, Delaware, Maine, Maryland, Massachusetts, New Hampshire, New Jersey, New York, Rhode Island, and Vermont are signatory States to the RGGI agreement.

- The Western Climate Initiative (WCI) continues to develop its proposed comprehensive regional market-based cap-and-trade program. The program seeks to reduce emissions across participating States to 15 percent below 2005 levels by 2020. On July 16,2009, the partnership released its "Essential Requirements for Mandatory Reporting," which will require facilities in participating States and provinces that emit more than 10,000 $MTCO_2e$ annually to report their GHG emissions. During 2009, the WCI Committees also released draft white papers and scoping documents related to offset limits, offset definition and eligibility, competitiveness, and early reduction allowances. The objective of the partnership remains the launch of its emissions trading system on January 1, 2012. Participating U.S. States include Arizona, California, Montana, New Mexico, Oregon, Utah, and Washington. Canadian provinces participating include British Columbia, Manitoba, Ontario, and Quebec.

- On June 8, 2009, members of the Midwestern Greenhouse Gas Reduction Accord (MGGRA) released their Advisory Group's draft final recommendations for a regional cap-and-trade program. The Advisory Group recommends emission reduction targets of 18 to 20 percent below 2005 levels by 2020 and 80 percent below 2005 levels by 2050. Sectors for which emission caps are proposed include electricity generation and imports, industrial combustion and process sources, transportation fuels, and residential, commercial, and industrial fuels not otherwise covered. Only entities emitting more than 25,000 $MTCO_2e$ annually would be capped. The Advisory Group's recommendations also include a proposal that one-third of allowances should be auctioned initially, with the remainder sold for a small fee. The program would transition to full auction over time. Member States include Iowa, Illinois, Kansas, Michigan, Minnesota, and Wisconsin, as well as the Canadian province of Manitoba. Observer States include Indiana, Ohio, and South Dakota, as well as the Canadian province of Ontario.

- On September 25, 2009, the State of California became the second government body in the United States (after Boulder, Colorado) to impose a fee on carbon emissions. The California Air Resources Board voted to apply the GHG emissions fee in order to fund implementation of the State's GHG cap-and-trade program, which was established in 2006 under Assembly Bill 32. The fee—which is expected to raise $63 million per year from approximately 350 large emitting entities in the State—will go into effect in fiscal year 2010-2011 and will be apportioned among entities on the basis of their annual GHG emissions, at a cost of approximately 15.5 cents per $MTCO_2e$.

# International: United Nations Framework Convention on Climate Change and the Kyoto Protocol

## United Nations Framework Convention on Climate Change and the Kyoto Protocol

### COP-14 and CMP-4

In December 2008, the Fourteenth Conference of the Parties to the United Nations Framework Convention on Climate Change (COP-14) and the Fourth Meeting of the Parties to the Kyoto Protocol (CMP-4) were held in Poznañ, Poland. Key areas of discussion included the following:

- Governments resolved in Poznañ to shift into "full negotiating mode" in hopes of delivering a comprehensive new climate change agreement in December 2009 in Copenhagen.
- Work programs for 2009 were established for the Parties for parallel negotiations under the Kyoto Protocol and the UNFCCC in 2009.
- The European Union called on developed countries as a group to reduce their emissions to 30 percent below 1990 levels by 2020 and on developing countries to reduce their emissions by 15 to 30 percent below business-as-usual levels.
- Ministers encouraged the Parties to view the global economic crisis as an opportunity to address climate change and, simultaneously, contribute to economic recovery, rather than as a hindrance to progress on climate change.

### Major Economies Forum

On March 28, 2009, President Obama launched the Major Economies Forum on Energy and Climate. The Forum is intended to facilitate dialog among the major economies, leading up to an agreement at the Fifteenth Conference of the Parties to the United Nations Framework Convention on Climate Change (COP-15) in Copenhagen, Denmark. Further, the Forum seeks to advance joint ventures to develop clean energy resources. Discussions have centered on adaptation, mitigation, measuring, reporting and verification, and technological cooperation. There are 17 major economies participating in the Forum: Australia, Brazil, Canada, China, the European Union, France, Germany, India, Indonesia, Italy, Japan, South Korea, Mexico, Russia, South Africa, the United Kingdom, and the United States.

### Hydrofluorocarbon Emissions

On September 15, 2009, the United States, Canada, and Mexico expressed their support for a proposal to include HFCs—a class of gases with high GWPs that are used primarily in refrigeration and air conditioning applications—under the Montreal Protocol. The proposal calls for reductions in the consumption and production of HFCs around the world, with developed nations taking the lead. The Montreal Protocol, which was signed in 1987, mandates the phaseout of production and use of ozone-depleting hydrochlorofluorocarbons (HCFCs), which have largely been replaced by the use of HFCs, resulting in increased GHG emissions worldwide. At the 21st Meeting of the Parties in Egypt in November, nations considered whether to alter the Montreal Protocol, but no agreement was reached. Further discussion of the issue is expected at COP-15.

## Units for Measuring Greenhouse Gases

Emissions data are reported here in metric units, as favored by the international scientific community. Metric tons are relatively intuitive for users of U.S. measurement units, because 1 metric ton is about 10 percent heavier than a short ton.

Throughout this chapter, emissions of carbon dioxide and other greenhouse gases are given in carbon dioxide equivalents. In the case of carbon dioxide, emissions denominated in the molecular weight of the gas or in carbon dioxide equivalents are the same. Carbon dioxide equivalent data can be converted to carbon equivalents by multiplying by 12/44.

Emissions of other greenhouse gases (such as methane) can also be measured in carbon dioxide equivalent units by multiplying their emissions (in metric tons) by their GWPs. Carbon dioxide equivalents are the amount of carbon dioxide by weight emitted into the atmosphere that would produce the same estimated radiative forcing as a given weight of another radiatively active gas.

Carbon dioxide equivalents are computed by multiplying the weight of the gas being measured (for example, methane) by its estimated GWP (which is 25 for methane). In 2008, the IPCC Working Group I released Errata to its Fourth Assessment Report, *Climate Change 2007: The Physical Science Basis*.[2] The Errata revise the reported GWPs for a small number of high-GWP gases. The GWPs published in the Errata to the Fourth Assessment Report (AR4) were used in the calculation of carbon dioxide equivalent emissions for this chapter. Table 4 on page 13 summarizes the GWP values from the Second, Third, and Fourth Assessment Reports.

## Units for Measuring Greenhouse Gases

### Table 4. Greenhouse Gases and 100-Year Net Global Warming Potentials

| Greenhouse Gas | Chemical Formula | Global Warming Potential | | |
|---|---|---|---|---|
| | | SAR[a] | TAR[b] | AR4[c] |
| **Carbon Dioxide** | $CO_2$ | 1 | 1 | 1 |
| **Methane** | $CH_4$ | 21 | 23 | 25 |
| **Nitrous Oxide** | $N_2O$ | 310 | 296 | 298 |
| **Hydrofluorocarbons** | | | | |
| HFC-23 (Trifluoromethane) | $CHF_3$ | 11,700 | 12,000 | 14,800 |
| HFC-32 (Difluoromethane) | $CH_2F_2$ | 650 | 550 | 675 |
| HFC-41 (Monofluoromethane) | $CH_3F$ | 150 | 97 | 92 |
| HFC-125 (Pentafluoroethane) | $CHF_2CF_3$ | 2,800 | 3,400 | 3,500 |
| HFC-134 (1,1,2,2-Tetrafluoroethane) | $CHF_2CHF_2$ | 1,000 | 1,100 | 1,100 |
| HFC-134a (1,1,1,2-Tetrafluoroethane | $CH_2FCF_3$ | 1,300 | 1,300 | 1,430 |
| HFC-143 (1,1,2-Trifluoroethane) | $CHF_2CH_2F$ | 300 | 330 | 353 |
| HFC-143a (1,1,1-Trifluoroethane) | $CF_3CH_3$ | 3,800 | 4,300 | 4,470 |
| HFC-152 (1,2-Difluoroethane) | $CH_2FCH_2F$ | — | 43 | 53 |

# Emissions of Greenhouse Gases in the United States 2008

**Table 4. (Continued)**

| Greenhouse Gas | Chemical Formula | Global Warming Potential | | |
|---|---|---|---|---|
| | | SAR[a] | TAR[b] | AR4[c] |
| HFC-152a (1,1-Difluoroethane) | $CH_3CHF_2$ | 140 | 120 | 124 |
| HFC-161 (Ethyl Fluoride) | $CH_3CH_2F$ | — | 12 | 12 |
| HFC-227ea (Heptafluoropropane) | $CF_3CHFCF_3$ | 2,900 | 3,500 | 3,220 |
| HFC-236cb (1,1,1,2,2,3-Hexafluoropropane) | $CH_2FCF_2CF_3$ | — | 1,300 | 1,340 |
| HFC-152a (1,1-Difluoroethane) | $CH_3CHF_2$ | 140 | 120 | 124 |
| HFC-236ea (1,1,1,2,3,3-Hexafluoropropane) | $CHF_2CHFCF_3$ | — | 1,200 | 1,370 |
| HFC-236fa(1,1,1,3,3,3-Hexafluoropropane) | $CF_3CH_2CF_3$ | 6,300 | 9,400 | 9,810 |
| HFC-245ca (1,1,2,2,3-Pentafluoropropane) | $CH_2FCF_2CHF_2$ | 560 | 640 | 693 |
| HFC-245fa (1,1,1 ,3,3-Pentafluoropropane) | $CHF_2CH_2CF_3$ | — | 950 | 1,030 |
| HFC-365mfc (Pentafluorobutane) | $CF_3CH_2CF_2CH_3$ | — | 890 | 794 |
| HFC-43-10mee (Decafluoropentane) | $CF_3CHFCHFCF_2CF_3$ | 1,300 | 1,500 | 1,640 |
| **Perfluorocarbons** | | | | |
| Perfluoromethane | $CF_4$ | 6,500 | 5,700 | 7,390 |
| Perfluoroethane | $C_2F_6$ | 9,200 | 11,900 | 12,200 |
| Perfluoropropane | $C_3F_8$ | 7,000 | 8,600 | 8,830 |
| Perfluorobutane (FC 3-1-10) | $C_4F_{10}$ | 7,000 | 8,600 | 8,860 |
| Perfluorocyclobutane | $c-C_4F_8$ | 8,700 | 10,000 | 10,300 |
| Perfluoropentane | $C_5F_{12}$ | 7,500 | 8,900 | 9,160 |
| Perfluorohexane (FC 5-1-14) | $C_6F_{14}$ | 7,400 | 9,000 | 9,300 |
| **Sulfur Hexafluoride** | $SF_6$ | 23,900 | 22,200 | 22,800 |
| **Nitrogen Triflouride** | $NF_3$ | — | 10,800 | 17,200 |

Sources:

[a]Intergovernmental Panel on Climate Change, *Climate Change 1995: The Science of Climate Change* (Cambridge, UK: Cambridge University Press, 1996). This document was part of the Second Assessment Report (SAR) by the Intergovernmental Panel on Climate Change.

b Intergovernmental Panel on Climate Change, *Climate Change 2001: The Scientific Basis* (Cambridge, UK: Cambridge University Press, 2001), web site www.ipcc.ch/ipccreports/tar. This document was part of the Third Assessment Report (TAR) by the Intergovernmental Panel on Climate Change.

[c]Intergovernmental Panel on Climate Change, *Climate Change 2007: The Physical Science Basis: Errata* (Cambridge, UK: Cambridge University Press, 2008), web site http://ipcc-wg1.ucar.edu/wg1/Report/AR4WG1_Errata_2008-12-01.pdf. This document describes errata in parts of the Fourth Assessment Report (AR4) by the Intergovernmental Panel on Climate Change.

## Methodology Updates for This Chapter

### *Carbon Dioxide*

For the first time, the nonfuel calculations in this chapter for carbon sequestered in petrochemical products include updates from EIA's 2006 Manufacturing Energy Consumption Survey.

### *Methane*

For the first time, emissions from the treatment of wastewater produced by ethanol production and by petroleum refining are included in this inventory for all years.

Methane emissions factors for stationary combustion have been revised to the values published in the 2006 IPCC guidelines, resulting in a significant decline in calculated emissions from this source category.

Emissions factors for methane from ethylene, ethylene dichloride, and methanol associated with chemical production and from sinter and coke in iron and steel production have been updated to match revised values in the 2006 IPCC guidelines.

Additional emissions factors, conversion factors, and constants applied to the calculation of agriculture- and livestock-related emissions have been updated on the basis of the most recent values published by the EPA or IPCC, as applicable.

### *Nitrous Oxide*

Emissions factors for direct and indirect emissions of $N_2O$ from nitrogen fertilization of agricultural soils and from runoff of fertilizer and manure applied to soils have been revised to the values provided in the 2006 IPCC guidelines. The indirect emission factor for $N_2O$ from domestic wastewater also has been updated, from the 1996 to the 2006 IPCC value.

According to the 2006 IPCC guidelines, the fraction of nitrogen from manure and synthetic fertilizers that is volatized no longer is subtracted from the nitrogen total before the direct emission factor is applied.

Because final data on fertilizer consumption in 2008 are not yet available, emissions of nitrous oxide from the use of synthetic nitrogen fertilizers in 2008 are based on a preliminary estimate of the change in fertilizer consumption from 2007.

### *High-GWP Gases*

For the first time, the emissions class "Other HFCs" has been calculated using AR4 GWP values. TAR GWP values were used in previous editions of this chapter.

### *Land Use*

Because of the complexity of the methodologies used to calculate emissions and sequestration from land use, land-use change, and forests, changes in this year's report are discussed in the chapter itself.

# CARBON DIOXIDE EMISSIONS

## Total Emissions

### *Summary*

- Total U.S. carbon dioxide emissions in 2008, compared with 2007 emissions (Figure 7), fell by 177.8 million metric tons (MMT), or 3.0 percent, to 5,839.3 MMT. The decrease—the largest over the 18-year period beginning with the 1990 baseline—puts 2008 emissions 47.1 MMT below the 2000 level.
- The important factors that contributed to the decrease in carbon dioxide emissions in 2008 included higher energy prices, especially during the summer driving season, slowing economic growth, and a decrease in the carbon intensity of energy supply.
- Energy-related carbon dioxide emissions account for 98 percent of U.S. carbon dioxide emissions (Table 5). The vast majority of carbon dioxide emissions come from fossil fuel combustion, with smaller amounts from the nonfuel use of energy inputs, and the total adjusted for emissions from U.S. Territories and international bunker fuels. Other sources include emissions from industrial processes, such as cement and limestone production.

**U.S. Anthropogenic Carbon Dioxide Emissions, 1990, 2007, and 2008**

|  | 1990 | 2007 | 2008 |
|---|---|---|---|
| Estimated Emissions (Million Metric Tons) | 5,022.3 | 6,017.0 | 5,839.3 |
| Change from 1990 (Million Metric Tons) |  | 994.7 | 817.0 |
| *(Percent)* |  | *19.8%* | *16.3%* |
| Average Annual Change from 1990 *(Percent)* |  | *1.1%* | *0.8%* |
| Change from 2007 (Million Metric Tons) |  |  | -177.8 |
| *(Percent)* |  |  | *-3.0%* |

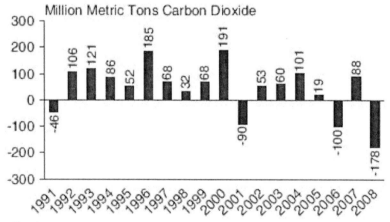

Source: EIA estimates.

Figure 7. Annual Change in U.S. Carbon Dioxide Emissions, 1990-2008

## Energy-Related Emissions

*Summary*

- Energy-related carbon dioxide emissions account for more than 80 percent of U.S. greenhouse gas emissions. EIA breaks energy use into four end-use sectors (Table 6), and emissions from the electric power sector are attributed to the end-use sectors. Growth in energy-related carbon dioxide emissions since 1990 has resulted largely from increases associated with electric power generation and transportation fuel use. All other energy-related carbon dioxide emissions (from direct fuel use in the residential, commercial, and industrial sectors) have been either flat or declining in recent years (Figure 8). In 2008, however, emissions from both electric power and transportation fuel use were down—by 2.1 percent and 4.7 percent, respectively.
- Reasons for the long-term growth in electric power and transportation sector emissions include: increased demand for electricity for computers and electronics in homes and offices; strong growth in demand for commercial lighting and cooling; substitution of new electricity-intensive technologies, such as electric arc furnaces for steelmaking, in the industrial sector; and increased demand for transportation services as a result of relatively low fuel prices and robust economic growth in the 1990s and early 2000s. Likewise, the recent declines in emissions from both the transportation and electric power sectors are tied to the economy, with people driving less and consuming less electricity in 2008 than in 2007.

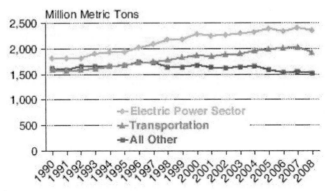

Source: EIA estimates.

Figure 8. U.S. Energy-Related Carbon Dioxide Emissions by Sector, 1990-2008

**U.S. Energy-Related Carbon Dioxide Emissions, 1990, 2007, and 2008**

|  | 1990 | 2007 | 2008 |
|---|---|---|---|
| Estimated Emissions (Million Metric Tons) | 5,020.1 | 5,986.4 | 5,814.4 |
| Change from 1990 (Million Metric Tons) |  | 966.3 | 794.4 |
| *(Percent)* |  | *19.2%* | *15.8%* |
| Average Annual Change from 1990 *(Percent)* |  | *1.0%* | *0.8%* |
| Change from 2007 (Million Metric Tons) |  |  | -171.9 |
| *(Percent)* |  |  | *-2.9%* |

**Table 5. U.S. Carbon Dioxide Emissions from Energy and Industry, 1990-2008 (Million Metric Tons Carbon Dioxide)**

| Fuel Type or Process | 1990 | 1995 | 2000 | 2002 | 2003 | 2004 | 2005 | 2006 | 2007 | 2008 |
|---|---|---|---|---|---|---|---|---|---|---|
| **Energy Consumption** | | | | | | | | | | |
| Petroleum | 2,185.9 | 2,208.4 | 2,461.3 | 2,469.9 | 2,516.7 | 2,605.4 | 2,625.7 | 2,594.9 | 2,588.6 | 2,436.0 |
| Coal | 1,803.4 | 1,899.9 | 2,138.1 | 2,077.2 | 2,115.6 | 2,140.3 | 2,161.0 | 2,129.9 | 2,154.5 | 2,125.2 |
| Natural Gas | 1,024.7 | 1,183.7 | 1,240.6 | 1,229.5 | 1,194.6 | 1,195.4 | 1,176.1 | 1,157.1 | 1,231.7 | 1,241.8 |
| Renewables[a] | 6.1 | 10.2 | 10.4 | 13.0 | 11.7 | 11.4 | 11.5 | 11.8 | 11.6 | 11.6 |
| **Energy Subtotal** | **5,020.1** | **5,302.3** | **5,850.4** | **5,789.6** | **5,838.6** | **5,952.5** | **5,974.3** | **5,893.7** | **5,986.4** | **5,814.4** |
| *Nonfuel Use Emissions[b]* | *97.1* | *104.9* | *109.8* | *103.7* | *101.7* | *110.8* | *104.7* | *108.9* | *106.7* | *100.2* |
| *Nonfuel Use Sequestration[c]* | *252.7* | *286.6* | *308.2* | *296.0* | *292.1* | *315.4* | *305.1* | *301.9* | *293.7* | *264.2* |
| Adjustments to Energy | -82.9 | -63.1 | -61.7 | -38.2 | -28.7 | -44.6 | -48.8 | -70.9 | -74.9 | -79.0 |
| **Adjusted Energy Subtotal** | **4,937.2** | **5,239.1** | **5,788.7** | **5,751.4** | **5,809.9** | **5,907.9** | **5,925.5** | **5,822.8** | **5,911.5** | **5,735.5** |
| Other Sources | 85.1 | 102.3 | 97.8 | 97.7 | 98.9 | 102.0 | 103.5 | 106.0 | 105.6 | 103.8 |
| **Total** | **5,022.3** | **5,341.5** | **5,886.4** | **5,849.1** | **5,908.8** | **6,009.9** | **6,029.0** | **5,928.7** | **6,017.0** | **5,839.3** |

[a]Includes emissions from electricity generation using nonbiogenic municipal solid waste and geothermal energy.

[b]Emissions from nonfuel uses are included in the energy subtotal above.

[c]The Btu value of carbon sequestered by nonfuel uses is subtracted from energy consumption before emissions are calculated.

Notes: Data in this table are revised from the data contained in the previous EIA report, *Emissions of Greenhouse Gases in the United States 2007,* DOE/EIA-0573(2007) (Washington, DC, December 2008). Totals may not equal sum of components due to independent rounding. Adjusted energy subtotal includes U.S. Territories but excludes international bunker fuels.

Source: EIA estimates.

162 United States Energy Information Administration

- Other U.S. energy-related carbon dioxide emissions have remained flat or declined, for reasons that include increased efficiencies in heating technologies, declining activity in older "smokestack" industries, and the growth of less energy-intensive industries, such as computers and electronics.

## Carbon Capture and Storage: A Potential Option for Reducing Future Emissions

The possibility of future constraints on greenhouse gas emissions has heightened interest in carbon capture and storage (CCS) technologies as an option to control $CO_2$ emissions. The U.S. Department of Energy (DOE) has received increased funding for the continued development of new CCS technologies,[3] and as the scale and scope of CCS projects grow, it will be important for EIA to track volumes of carbon stored, so that they can be subtracted appropriately in greenhouse gas inventory estimates.

The United States emits about 1.9 billion metric tons of $CO_2$ annually from coal-fired power plants—33 percent of total energy-related $CO_2$ emissions and 81 percent of $CO_2$ emissions from the U.S. electric power sector. Coal-fired power plants are the most likely source of $CO_2$ for storage; however, other sources are possible.

CCS involves three steps: capture of $CO_2$ from a fossil-fueled power plant or other industrial process; transport of the compressed gas via pipeline to a storage site; and injection and storage in a geologic formation.

**$CO_2$ Capture:** There are three types of $CO_2$ capture: post-combustion, pre-combustion, and oxy-combustion. Post-combustion capture is a well-known technology, which currently is used to a limited degree. It involves capture of $CO_2$ from flue gases after a fossil fuel has been burned. Pre-combustion capture involves gasifying the fossil fuel, instead of using direct combustion. The $CO_2$ can be captured readily from the gasification exhaust stream. For oxy-combustion capture, coal is burned in pure oxygen instead of air, so that the resulting exhaust contains only $CO_2$ and water vapor. Systems that use these technologies currently are being developed to capture at least 90 percent of emitted $CO_2$.[4]

**Pipeline Transportation:** Captured $CO_2$ emissions are transported most commonly as highly pressurized gas through pipeline networks to storage sites. Currently, more than 1,550 miles of pipeline transport some 48 MMT per year in the United States from natural and anthropogenic sources, mostly to oil fields in Texas and New Mexico for enhanced oil recovery (EOR).[5] As is done for natural gas pipelines, fugitive emissions from the transport of gaseous $CO_2$ will need to be accounted for in EIA's greenhouse gas inventories.[6]

**Geological Storage:** Three main types of geological formation—each with varying capacities—currently are viewed as possible reservoirs for the storage of captured $CO_2$: oil and gas reservoirs, saline formations, and unmineable coal seams (see figure below).

**Table 6. U.S. Energy-Related Carbon Dioxide Emissions by End-Use Sector, 1990-2008 (Million Metric Tons Carbon Dioxide)**

| Sector | 1990 | 1995 | 2000 | 2002 | 2003 | 2004 | 2005 | 2006 | 2007 | 2008 |
|---|---|---|---|---|---|---|---|---|---|---|
| Residential | 958.6 | 1,035.5 | 1,179.8 | 1,197.6 | 1,224.9 | 1,221.9 | 1,254.5 | 1,186.7 | 1,235.1 | 1,220.1 |
| Commercial | 785.1 | 845.1 | 1,013.1 | 1,018.0 | 1,026.1 | 1,043.3 | 1,059.6 | 1,034.9 | 1,070.3 | 1,075.1 |
| Industrial | 1,689.5 | 1,739.5 | 1,784.7 | 1,683.3 | 1,690.3 | 1,728.5 | 1,671.4 | 1,657.8 | 1,655.2 | 1,589.1 |
| Transportation | 1,586.9 | 1,682.2 | 1,872.7 | 1,890.7 | 1,897.4 | 1,958.9 | 1,988.7 | 2,014.3 | 2,025.7 | 1,930.1 |
| **Total** | **5,020.1** | **5,302.3** | **5,850.4** | **5,789.6** | **5,838.6** | **5,952.5** | **5,974.3** | **5,893.7** | **5,986.4** | **5,814.4** |
| Electricity Generation[a] | 1,814.6 | 1,947.9 | 2,293.5 | 2,270.5 | 2,298.8 | 2,331.3 | 2,396.8 | 2,343.5 | 2,409.1 | 2,359.1 |

[a]Electric power sector emissions are distributed across the end-use sectors. Emissions allocated to sectors are unadjusted. Adjustments are made to total emissions only.

Notes: Data in this table are revised from the data contained in the previous EIA report, *Emissions of Greenhouse Gases in the United States 2007*, DOE/EIA-0573(2007) (Washington, DC, December 2008). Totals may not equal sum of components due to independent rounding.

Source: EIA estimates.

## U.S. Potential for Short-Term and Long-Term Carbon Storage Projects

| Storage Sitea | Geologic Formation | Source of Carbon Capture | Expected Start of Operation | Expected Duration (Years) |
|---|---|---|---|---|
| **Short-Term Projects** | | | | |
| SaskPower Plant: Poplar Dome Storage (1) | Sandstone Formation | Pulverized Combustion Retrofit (Canada) | 2011 | 2 |
| Nugget Sandstone/ Riley Ridge (1) | Sandstone Formation | Natural Gas Processing Plant | 2015 | 2.5 |
| Williston Basin $CO_2$ Sequestration and EOR (2) | Depleted Oil Field | Post-Combustion Capture Facility | 2012 | 4 |
| Decatur Sequestration (3) | Mt. Simon Saline Formation | Ethanol Plant | 2010 | 3 |
| Michigan Basin (4) | Deep Saline Reservoir | Natural Gas Processing Plant | 2010 | 1 |
| TAME Ethanol Plant (4) | Sandstone | Ethanol Plant | 2012 | 4 |
| Cranfield Early Test (5) | Depleted Oil Field | Natural Source | 2010 | 1.5 |
| Permian Basin, Texas (6) | — | — | — | — |
| Paradox Basin, Utah (6) | Deep Saline Formation | Natural Sources | 2010 | 3.5 |
| San Juan Basin/Allison Unit, New Mexico (6) | Enhanced Coalbed Methane | Natural Sources | 2010 | 1 |
| Farnham Dome, Utah (6) | Deep Saline Reservoir | Natural Sources | 2010 | 2.5 |
| Entrada Sandstone (6) | Deep Saline Reservoir | Natural Gas Processing Plant | 2010 | 4 |
| Kimberlina, San Joaquin Basin (7) | Sandstone Formation | Oxyfuel Combustion | 2012 | 4 |
| AEP Alstom Mountaineer (8) | Deep Saline Reservoir | Coal-Fired Plant, Post-Combustion | 2009 | 2 |
| AEP Pleasant Prairie (8) | Deep Saline Reservoir | Coal-Fired Plant, Post-Combustion | 2010 | 1.5 |
| **Long-Term Projects** | | | | |
| Wallula Energy Resource Center (1) | Basalt Formation | IGCC/Pre-Combustion | 2020 | — |
| Plant Barry (5) | Depleted Oil Field | Coal-Fired Plant, Post-Combustion | 2011 | — |
| Great Plains Synfuels Plant (8) | Canada Oil Reservoir EOR | Synthetic Fuel Plant | 2000 | — |
| ExxonMobil LaBarge, Wyoming (8) | EOR | Natural Gas Processing Plant | 2010 | — |
| Basin Electric Power Cooperative, Antelope Valley Station (8) | EOR | Coal-Fired Plant, Post-Combustion | 2012 | — |

**(Continued)**

| Storage Site[a] | Geologic Formation | Source of Carbon Capture | Expected Start of Operation | Expected Duration (Years) |
|---|---|---|---|---|
| AEP Alstom Northeastern (8) | EOR | Coal-Fired Plant, Post-Combustion | 2012 | — |
| WA Parish Plant (8) | EOR | Coal-Fired Plant, Post-Combustion | 2012 | — |
| Duke Energy Corporation (8) | Deep Saline Aquifer | Coal Gasification Power Plant | 2015 | — |
| Hydrogen Energy California Project (8) | EOR | IGCC for Petroleum Coke | 2015 | — |
| FutureGen (8) | Deep Saline Formation | IGCC Plant | 2020 | — |

[a]Storage site affiliations: (1) Big Sky Carbon Sequestration Partnership; (2) Plains $CO_2$ Reduction Partnership (PCOR); (3) Midwest Geological Sequestration Consortium (MGSC); (4) Midwest Regional Carbon Seqestration Partnership (MRCSP); (5) Southeast Regional Carbon Sequestration Partnership (SECARB); (6) Southwest Regional Partnership on Carbon Sequestration (SWP); (7) West Coast Regional Carbon Sequestration Partnership (WESTCARB); (8) Existing, integrated, or independent project.

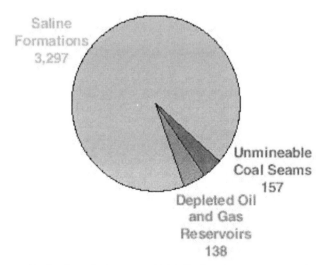

Source: National Energy Technology Laboratory, *2008 Carbon Sequestration Atlas of the United States and Canada*.

U.S. Carbon Dioxide Storage Potential (Billion Metric Tons)

*Oil and Gas Reservoirs:* Currently in the United States, about 48 MMT of $CO_2$ per year is injected into oil and gas fields for EOR.[7] $CO_2$ also may be pumped into oil and gas reservoirs strictly for storage: as a result of EOR operations, about 9 MMT of $CO_2$ is stored per year.[8] Storage capacity for $CO_2$ in depleted oil and gas fields in the United States and Canada currently is estimated at 138 billion metric tons.[9] Worldwide, $CO_2$ storage capacity in EOR projects and other depleted oil and gas fields is estimated at 675 to 1,200 billion metric tons.[10]

166       United States Energy Information Administration

*Saline Formations:* A second type of geologic formation that could be used to store $CO_2$ is saline formations, which have an estimated worldwide storage capacity of up to 20,000 billion metric tons.[11] These formations have the potential to trap $CO_2$ in pore spaces, and many large point sources of $CO_2$ emissions are relatively close to saline formations. The United States and Canada have an estimated combined storage capacity of 3,300 to 12,600 billion metric tons in saline formations.[12]

*Unmineable Coal Seams:* When $CO_2$ is injected into an unmineable coal seam, it displaces methane and remains sequestered in the bed. Although the method is relatively untested, and the resulting methane recovery would add cost to the CCS process, sales of the methane could provide some cost offsets.[13] Coal seam sequestration has an estimated storage capacity of 10 to 200 billion metric tons worldwide,[14] including an estimated 157 to 178 billion metric tons of capacity in the United States and Canada.[15]

The table on page 19 lists CCS projects that currently are either operating or actively being prepared for deployment. At present, there are few commercial-scale projects in operation that integrate carbon capture from a coal-fired power plant with transportation to a permanent storage site; however, a number of projects and locations have been proposed. Given the possibility of delays and project cancellations, it is unlikely that all the projects listed will become operational on the dates planned. On the other hand, other projects that are not included in the table may come to fruition.

## Residential Sector

### *Summary*

- Residential sector carbon dioxide emissions originate primarily from:
  - Direct fuel consumption (principally, natural gas) for heating and cooking
  - Electricity for cooling (and heating), appliances, lighting, and increasingly for televisions, computers, and other household electronic devices (Table 7).
- Energy consumed for heating in homes and businesses has a large influence on the annual fluctuations in energy-related carbon dioxide emissions.
  - The 5.6-percent increase in heating degree-days in 2008 was one of the few upward pressures on emissions in 2008 (Figure 9).
  - Although annual changes in cooling degree-days have a smaller impact on energy demand, the 8.7-percent decrease in 2008 offset some of the upward pressure from the increase in heating degree-days.
- In the longer run, residential emissions are affected by population growth, income, and other factors. From 1990 to 2008:
  - Residential sector carbon dioxide emissions grew by an average of 1.3 percent per year.
  - U.S. population grew by an average of 1.1 percent per year.
  - Per-capita income (measured in constant dollars) grew by an average of 1.7 percent per year.

**Residential Sector Carbon Dioxide Emissions, 1990, 2007, and 2008**

|  | 1990 | 2007 | 2008 |
|---|---|---|---|
| Estimated Emissions (Million Metric Tons) | 958.6 | 1,235.1 | 1,220.1 |
| Change from 1990 (Million Metric Tons) |  | 276.5 | 261.5 |
| *(Percent)* |  | *28.8%* | *27.3%* |
| Average Annual Change from 1990 *(Percent)* |  | *1.5%* | *1.3%* |
| Change from 2007 (Million Metric Tons) |  |  | -15.0 |
| *(Percent)* |  |  | *-1.2%* |

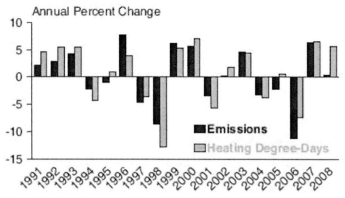

Source: EIA estimates.

Figure 9. Annual Changes in U.S. Heating Degree-Days and Residential Sector $CO_2$ Emissions from Direct Fuel Combustion, 1990-2008

- Energy efficiency improvements for homes and appliances have offset much of the growth in the number and size of housing units. As a result, direct fuel emissions from petroleum, coal, and natural gas consumed in the residential sector in 2008 were only 1.5 percent higher than in 1990.

## Commercial Sector

### *Summary*

- Commercial sector emissions (Table 8) are largely the result of energy use for lighting, heating, and cooling in commercial structures, such as office buildings, shopping malls, schools, hospitals, and restaurants.
- The commercial sector was the only sector that showed positive growth in emissions in 2008.
- Lighting accounts for a larger component of energy demand in the commercial sector (approximately 18 percent of total demand in 2007) than in the residential sector (approximately 11 percent of the total).
- Commercial sector emissions are affected less by weather than are residential sector emissions: heating and cooling accounted for approximately 38 percent of energy

demand in the residential sector in 2007 but only about 21 percent in the commercial sector.[16]
- In the longer run, trends in emissions from the commercial sector parallel economic trends. Commercial sector emissions grew at an average annual rate of 1.8 percent from 1990 to 2008—slightly more than the growth in real income per capita (Figure 10).
- Emissions from direct fuel consumption in the commercial sector declined from 1990 to 2008, while the sector's electricity-related emissions increased by an average of 2.4 percent per year.

**Commercial Sector Carbon Dioxide Emissions, 1990, 2007, and 2008**

|  | 1990 | 2007 | 2008 |
|---|---|---|---|
| Estimated Emissions (Million Metric Tons) | 785.1 | 1,070.3 | 1,075.1 |
| Change from 1990 (Million Metric Tons) |  | 285.3 | 290.0 |
| *(Percent)* |  | *36.3%* | *36.9%* |
| Average Annual Change from 1990 *(Percent)* |  | *1.8%* | *1.8%* |
| Change from 2007 (Million Metric Tons) |  |  | 4.8 |
| *(Percent)* |  |  | *0.4%* |

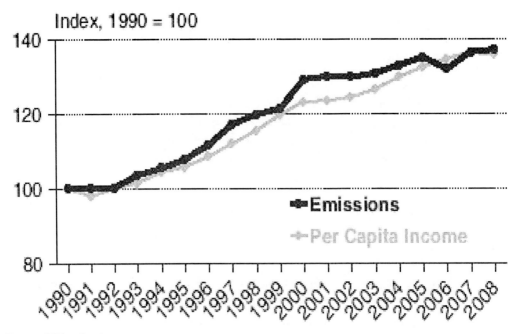

Source: EIA estimates.

Figure 10. U.S. Commercial Sector $CO_2$ Emissions and Per Capita Income, 1990-2008

**Table 7. U.S. Carbon Dioxide Emissions from Residential Sector Energy Consumption, 1990-2008**
**(Million Metric Tons Carbon Dioxide)**

| Fuel | 1990 | 1995 | 2000 | 2002 | 2003 | 2004 | 2005 | 2006 | 2007 | 2008 |
|---|---|---|---|---|---|---|---|---|---|---|
| **Petroleum** | | | | | | | | | | |
| Liquefied Petroleum Gas | 23.0 | 25.5 | 35.6 | 34.2 | 35.6 | 33.5 | 32.6 | 28.7 | 30.4 | 28.6 |
| Distillate Fuel | 71.6 | 66.2 | 66.2 | 62.9 | 66.2 | 67.6 | 62.5 | 52.1 | 53.1 | 50.1 |
| Kerosene | 4.6 | 5.4 | 6.8 | 4.3 | 5.1 | 6.1 | 6.1 | 4.8 | 3.2 | 1.4 |
| **Petroleum Subtotal** | **99.2** | **97.1** | **108.6** | **101.5** | **106.8** | **107.2** | **101.1** | **85.5** | **86.7** | **80.1** |
| Coal | 3.0 | 1.7 | 1.1 | 1.2 | 1.2 | 1.1 | 0.8 | 0.6 | 0.7 | 0.7 |
| Natural Gas | 238.3 | 262.9 | 270.8 | 266.1 | 277.5 | 264.5 | 262.7 | 237.5 | 256.8 | 265.0 |
| Electricity[a] | 618.2 | 673.9 | 799.3 | 828.9 | 839.5 | 849.1 | 889.9 | 863.0 | 890.9 | 874.4 |
| **Total** | **958.6** | **1,035.5** | **1,179.8** | **1,197.6** | **1,224.9** | **1,221.9** | **1,254.5** | **1,186.7** | **1,235.1** | **1,220.1** |

[a]Share of total electric power sector carbon dioxide emissions weighted by sales to the residential sector.

Notes: Data in this table are revised from the data contained in the previous EIA report, *Emissions of Greenhouse Gases in the United States 2007*, DOE/EIA-0573(2007) (Washington, DC, December 2008). Totals may not equal sum of components due to independent rounding.

Source: EIA estimates.

**Table 8. U.S. Carbon Dioxide Emissions from Commercial Sector Energy Consumption, 1990-2008**
**(Million Metric Tons Carbon Dioxide)**

| Fuel | 1990 | 1995 | 2000 | 2002 | 2003 | 2004 | 2005 | 2006 | 2007 | 2008 |
|---|---|---|---|---|---|---|---|---|---|---|
| **Petroleum** | | | | | | | | | | |
| Motor Gasoline | 7.9 | 1.3 | 3.1 | 3.2 | 4.2 | 3.4 | 3.2 | 3.4 | 4.2 | 4.0 |
| Liquefied Petroleum Gas | 4.1 | 4.5 | 6.3 | 6.0 | 6.3 | 5.9 | 5.8 | 5.1 | 5.4 | 5.0 |
| Distillate Fuel | 39.2 | 35.0 | 35.9 | 32.5 | 35.2 | 34.4 | 32.7 | 29.4 | 28.1 | 26.5 |
| Residual Fuel | 18.1 | 11.1 | 7.2 | 6.3 | 8.8 | 9.7 | 9.1 | 5.9 | 5.9 | 5.6 |
| Kerosene | 0.9 | 1.6 | 2.1 | 1.2 | 1.3 | 1.5 | 1.6 | 1.1 | 0.7 | 0.3 |
| **Petroleum Subtotal[a]** | **70.1** | **53.6** | **54.7** | **49.2** | **55.8** | **54.9** | **52.4** | **44.8** | **44.3** | **41.4** |
| Coal | 11.8 | 11.1 | 8.8 | 8.6 | 7.8 | 9.8 | 9.2 | 6.2 | 6.7 | 6.4 |
| Natural Gas | 142.3 | 164.3 | 172.5 | 171.1 | 173.7 | 170.0 | 163.2 | 154.0 | 164.2 | 169.9 |
| Electricity[b] | 560.8 | 616.1 | 777.2 | 789.2 | 788.7 | 808.6 | 834.8 | 830.0 | 855.2 | 857.3 |
| **Total** | **785.1** | **845.1** | **1,013.1** | **1,018.0** | **1,026.1** | **1,043.3** | **1,059.6** | **1,034.9** | **1,070.3** | **1,075.1** |

[a]Includes small amounts of petroleum coke.

[b]Share of total electric power sector carbon dioxide emissions weighted by sales to the commercial sector.

Notes: Data in this table are revised from the data contained in the previous EIA report, *Emissions of Greenhouse Gases in the United States 2007*, DOE/EIA-0573(2007) (Washington, DC, December 2008). Totals may not equal sum of components due to independent rounding.

Source: EIA estimates.

# Industrial Sector

## *Summary*

- Unlike commercial sector emissions, trends in U.S. industrial sector emissions (Table 9) have not followed aggregate economic growth trends but have been tied to trends in energy-intensive industries. In 2008, industrial carbon dioxide emissions fell by 4.0 percent from their 2007 level and were 5.9 percent (100.4 MMT) below their 1990 level. Decreases in industrial sector carbon dioxide emissions have resulted largely from a structural shift away from energy-intensive manufacturing in the U.S. economy.
- Coke plants consumed 22.1 million short tons of coal in 2008, down from 38.9 million short tons in 1990. Other industrial coal consumption declined from 76.3 million short tons in 1990 to 54.5 million short tons in 2008, as reflected by the drop in emissions from coal shown in Figure 11.

**Industrial Sector Carbon Dioxide Emissions, 1990, 2007, and 2008**

|  | 1990 | 2007 | 2008 |
|---|---|---|---|
| Estimated Emissions (Million Metric Tons) | 1,689.5 | 1,655.2 | 1,589.1 |
| Change from 1990 (Million Metric Tons) |  | -37.4 | -100.4 |
| *(Percent)* |  | *-2.0%* | *-5.9%* |
| Average Annual Change from 1990 *(Percent)* |  | *-0.1%* | *-0.3%* |
| Change from 2007 (Million Metric Tons) |  |  | -66.1 |
| *(Percent)* |  |  | *-4.0%* |

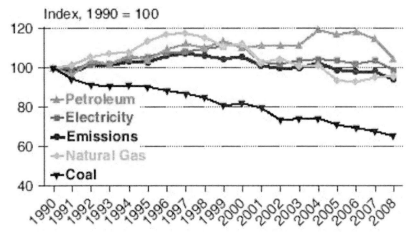

Source: EIA estimates.

Figure 11. U.S. Industrial Sector $CO_2$ Emissions and Major Industrial Fuel Use, 1990-2008

- The share of manufacturing activity represented by less energy-intensive industries, such as computer chip and electronic component manufacturing, has increased, while the share represented by the more energy-intensive industries has fallen.

- By fuel, only total petroleum and net imports of coke in 2008 were above 1990 levels for the industrial sector. As mentioned above, coal use has fallen since 1990, and natural gas use, which rose in the 1990s, has fallen since 2000.

## Transportation Sector

*Summary*

- Transportation sector carbon dioxide emissions in 2008 were 95.6 MMT lower than in 2007 but still 343.2 MMT higher than in 1990 (Table 10).
- The transportation sector has led all U.S. end-use sectors in emissions of carbon dioxide since 1999; however, with higher fuel prices and slower economic growth in 2008, emissions from the transportation sector fell by 4.7 percent from their 2007 level.
- Petroleum combustion is the largest source of carbon dioxide emissions in the transportation sector.
- Increases in ethanol fuel consumption in recent years have mitigated the growth in transportation sector emissions. Reported emissions from energy inputs to ethanol production plants are counted in the industrial sector).
- Transportation sector emissions from gasoline and diesel fuel combustion since 1990 generally have paralleled total vehicle miles traveled (Figure 12).

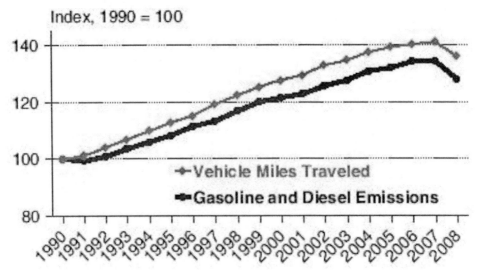

Source: EIA estimates.

Figure 12. U.S. Vehicle Miles Traveled and $CO_2$ Emissions from Gasoline and Diesel Transportation Fuel Use, 1990-2008

**Table 9. U.S. Carbon Dioxide Emissions from Industrial Sector Energy Consumption, 1990-2008 (Million Metric Tons Carbon Dioxide)**

| Fuel | 1990 | 1995 | 2000 | 2002 | 2003 | 2004 | 2005 | 2006 | 2007 | 2008 |
|---|---|---|---|---|---|---|---|---|---|---|
| **Petroleum** | | | | | | | | | | |
| MotorGasoline | 13.2 | 14.1 | 10.6 | 21.7 | 22.6 | 25.9 | 24.7 | 26.0 | 21.0 | 20.1 |
| Liquefied Petroleum Gas | 40.6 | 47.3 | 59.1 | 56.9 | 52.4 | 57.7 | 53.7 | 58.0 | 57.0 | 54.2 |
| Distillate Fuel | 83.9 | 82.4 | 87.4 | 87.7 | 82.6 | 88.3 | 91.8 | 91.7 | 91.9 | 87.3 |
| Residual Fuel | 30.6 | 24.5 | 17.0 | 13.2 | 15.5 | 17.5 | 19.9 | 16.4 | 12.7 | 11.9 |
| Asphalt and Road Oil | 0.0 | 0.0 | 0.0 | 0.0 | 0.0 | 0.0 | 0.0 | 0.0 | 0.0 | 0.0 |
| Lubricants | 6.9 | 6.6 | 7.0 | 6.4 | 5.9 | 6.0 | 5.9 | 5.8 | 6.0 | 5.6 |
| Kerosene | 0.9 | 1.1 | 1.1 | 1.0 | 1.7 | 2.0 | 2.8 | 2.1 | 1.0 | 0.4 |
| Petroleum Coke | 63.8 | 66.9 | 74.1 | 76.2 | 76.0 | 82.1 | 79.7 | 82.2 | 80.0 | 76.2 |
| Other Petroleum | 127.3 | 113.6 | 116.8 | 127.3 | 139.7 | 141.6 | 140.6 | 150.2 | 147.6 | 129.7 |
| **Petroleum Subtotal** | **367.2** | **356.5** | **373.1** | **390.4** | **396.5** | **421.1** | **419.2** | **432.5** | **417.2** | **385.3** |
| **Coal** | **256.8** | **231.5** | **210.0** | **188.1** | **190.2** | **190.7** | **182.0** | **178.3** | **173.7** | **167.5** |
| **Coal Coke Net Imports** | **0.5** | **7.0** | **7.5** | **6.9** | **5.8** | **15.7** | **5.0** | **6.9** | **2.9** | **4.7** |
| **Natural Gas** | **432.5** | **490.0** | **480.7** | **449.1** | **431.7** | **432.0** | **397.9** | **394.3** | **403.6** | **409.0** |
| **Electricity[a]** | **632.5** | **654.6** | **713.4** | **648.8** | **666.1** | **668.9** | **667.2** | **645.8** | **657.7** | **622.6** |
| **Total[b]** | **1,689.5** | **1,739.5** | **1,784.7** | **1,683.3** | **1,690.3** | **1,728.5** | **1,671.4** | **1,657.8** | **1,655.2** | **1,589.1** |

[a]Share of total electric power sector carbon dioxide emissions weighted by sales to the industrial sector.

[b]Includes emissions from nonfuel uses of fossil fuels. See Table 12 for details by fuel category.

Notes: Data in this table are revised from the data contained in the previous EIA report, *Emissions of Greenhouse Gases in the United States 2007*, DOE/EIA-0573(2007) (Washington, DC, December 2008). Totals may not equal sum of components due to independent rounding.

Source: EIA estimates.

**Transportation Sector Carbon Dioxide Emissions, 1990, 2007, and 2008**

|  | 1990 | 2007 | 2008 |
|---|---|---|---|
| Estimated Emissions (Million Metric Tons) | 1,586.9 | 2,025.7 | 1,930.1 |
| Change from 1990 (Million Metric Tons) |  | 438.8 | 343.2 |
| *(Percent)* |  | *27.7%* | *21.6%* |
| Average Annual Change from 1990 *(Percent)* |  | *1.4%* | *1.1%* |
| Change from 2007 (Million Metric Tons) |  |  | -95.6 |
| *(Percent)* |  |  | *-4.7%* |

## Electric Power Sector

### *Summary*

- The electric power sector transforms primary energy fuels into electricity. The sector consists of companies whose primary business is the generation of electricity.
- Carbon dioxide emissions from electric power generation declined by 2.1 percent in 2008 (Figure 13 and Table 11). The drop resulted from a decrease of 38.7 billion kilowatthours (1.0 percent) in the sector's total electricity generation and a 1.1-percent reduction in the carbon intensity of the electricity
- The lower overall carbon intensity of power generation in 2008 was the result of a 50-percent increase (17.6 billion kilowatthours) in generation from wind resources.
- Other non-carbon sources combined accounted for an additional 1 billion kilowatthours of increased generation, despite a slight decline in generation from nuclear power.
- Electricity generation from all fossil fuels fell by 57.4 billion kilowatthours from 2007 to 2008.

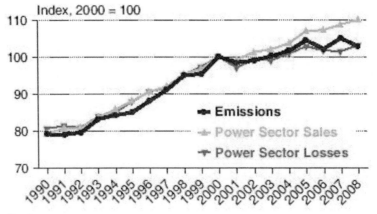

Source: EIA estimates.

Figure 13. U.S. Electric Power Sector Energy Sales and Losses and $CO_2$ Emissions from Primary Fuel Combustion, 1990-2008

## Table 10. U.S. Carbon Dioxide Emissions from Transportation Sector Energy Consumption, 1990-2008
### (Million Metric Tons Carbon Dioxide)

| Fuel | 1990 | 1995 | 2000 | 2002 | 2003 | 2004 | 2005 | 2006 | 2007 | 2008 |
|---|---|---|---|---|---|---|---|---|---|---|
| **Petroleum** | | | | | | | | | | |
| Motor Gasoline | 966.2 | 1,029.8 | 1,122.0 | 1,156.1 | 1,159.9 | 1,181.3 | 1,184.2 | 1,186.9 | 1,187.4 | 1,134.9 |
| Liquefied Petroleum Gas | 1.4 | 1.1 | 0.7 | 0.8 | 1.0 | 1.1 | 1.7 | 1.6 | 1.3 | 1.2 |
| Jet Fuel | 222.6 | 222.1 | 253.8 | 236.8 | 231.5 | 239.8 | 246.3 | 239.5 | 238.0 | 226.3 |
| Distillate Fuel | 267.8 | 306.9 | 377.8 | 394.5 | 414.5 | 433.9 | 444.4 | 469.2 | 472.3 | 445.7 |
| Residual Fuel | 80.1 | 71.7 | 69.9 | 53.3 | 45.0 | 58.3 | 66.0 | 71.4 | 78.3 | 74.1 |
| Lubricants[a] | 6.5 | 6.2 | 6.7 | 6.0 | 5.6 | 5.6 | 5.6 | 5.5 | 5.6 | 5.2 |
| Aviation Gasoline | 3.1 | 2.7 | 2.5 | 2.3 | 2.1 | 2.2 | 2.4 | 2.3 | 2.2 | 2.0 |
| **Petroleum Subtotal** | **1,547.7** | **1,640.5** | **1,833.4** | **1,849.8** | **1,859.5** | **1,922.2** | **1,950.7** | **1,976.4** | **1,985.1** | **1,889.4** |
| **Coal** | **0.0** | **0.0** | **0.0** | **0.0** | **0.0** | **0.0** | **0.0** | **0.0** | **0.0** | **0.0** |
| **Natural Gas** | **36.1** | **38.4** | **35.7** | **37.2** | **33.4** | **32.0** | **33.1** | **33.2** | **35.4** | **35.9** |
| **Electricity[b]** | **3.2** | **3.2** | **3.6** | **3.6** | **4.5** | **4.7** | **4.9** | **4.7** | **5.2** | **4.9** |
| **Total** | **1,586.9** | **1,682.2** | **1,872.7** | **1,890.7** | **1,897.4** | **1,958.9** | **1,988.7** | **2,014.3** | **2,025.7** | **1,930.1** |

[a]Includes emissions from nonfuel uses of fossil fuels. See Table 12 for details by fuel category.

[b]Share of total electric power sector carbon dioxide emissions weighted by sales to the transportation sector.

Notes: Data in this table are revised from the data contained in the previous EIA report, *Emissions of Greenhouse Gases in the United States 2007*, DOE/EIA-0573(2007) (Washington, DC, December 2008). Totals may not equal sum of components due to independent rounding.

Source: EIA estimates.

**Table 11. U.S. Carbon Dioxide Emissions from Electric Power Sector Energy Consumption, 1990-2008**
**(Million Metric Tons Carbon Dioxide)**

| Fuel | 1990 | 1995 | 2000 | 2002 | 2003 | 2004 | 2005 | 2006 | 2007 | 2008 |
|---|---|---|---|---|---|---|---|---|---|---|
| **Petroleum** | | | | | | | | | | |
| Residual Fuel Oil | 91.6 | 44.6 | 68.6 | 51.9 | 68.5 | 69.3 | 69.1 | 28.4 | 31.3 | 18.8 |
| DistillateFuel Oil | 7.1 | 7.9 | 12.8 | 9.3 | 11.8 | 8.1 | 8.4 | 5.4 | 6.5 | 5.2 |
| Petroleum Coke | 3.1 | 8.2 | 10.1 | 17.9 | 17.8 | 22.7 | 24.9 | 21.8 | 17.5 | 15.8 |
| **Petroleum Subtotal** | **101.8** | **60.7** | **91.5** | **79.1** | **98.1** | **100.1** | **102.3** | **55.6** | **55.3** | **39.7** |
| **Coal** | **1,531.2** | **1,648.7** | **1,910.8** | **1,872.4** | **1,910.7** | **1,922.9** | **1,963.9** | **1,937.8** | **1,970.6** | **1,945.9** |
| **Natural Gas** | **175.5** | **228.2** | **280.9** | **306.0** | **278.3** | **296.8** | **319.1** | **338.2** | **371.7** | **362.0** |
| **Municipal Solid Waste** | **5.7** | **9.9** | **10.0** | **12.6** | **11.3** | **11.1** | **11.1** | **11.4** | **11.2** | **11.2** |
| **Geothermal** | **0.4** | **0.3** | **0.4** | **0.4** | **0.4** | **0.4** | **0.4** | **0.4** | **0.4** | **0.4** |
| **Total** | **1,814.6** | **1,947.9** | **2,293.5** | **2,270.5** | **2,298.8** | **2,331.3** | **2,396.8** | **2,343.5** | **2,409.1** | **2,359.1** |

Notes: Data in this table are revised from the data contained in the previous EIA report, *Emissions of Greenhouse Gases in the United States 2007*, DOE/EIA-0573(2007) (Washington, DC, December2008). Emissions for total fuel consumption are allocated to end-use sectors in proportion to electricity sales. Totals may not equal sum of components due to independent rounding.

Source: EIA estimates.

**Electric Power Sector Carbon Dioxide Emissions, 1990, 2007, and 2008**

|  | 1990 | 2007 | 2008 |
|---|---|---|---|
| Estimated Emissions (Million Metric Tons) | 1,814.6 | 2,409.1 | 2,359.1 |
| Change from 1990 (Million Metric Tons) |  | 594.4 | 544.5 |
| *(Percent)* |  | *32.8%* | *30.0%* |
| Average Annual Change from 1990 *(Percent)* |  | *1.7%* | *1.5%* |
| Change from 2007 (Million Metric Tons) |  |  | -50.0 |
| *(Percent)* |  |  | *-2.1%* |

# Nonfuel Uses of Energy Inputs

## *Summary*

- Nonfuel uses of fossil fuels (for purposes other than their energy value) create carbon dioxide emissions and also sequester carbon in nonfuel products.
- In 2008, carbon dioxide emissions from nonfuel uses of energy inputs totaled 100.2 MMT—6.1 percent below the 2007 total (Table 12).
- Carbon sequestration from nonfuel uses of energy inputs in 2008 included 264.2 MMTCO$_2$e that was embedded in plastics and other nonfuel products rather than emitted to the atmosphere (see Table 13 on page 26).
- The 2008 sequestration total was 10.1 percent below the 2007 total.

**Carbon Dioxide Emissions from Nonfuel Uses of Energy Inputs, 1990, 2007, and 2008**

|  | 1990 | 2007 | 2008 |
|---|---|---|---|
| Estimated Emissions (Million Metric Tons) | 97.1 | 106.7 | 100.2 |
| Change from 1990 (Million Metric Tons) |  | 9.5 | 3.0 |
| *(Percent)* |  | *9.8%* | *3.1%* |
| Average Annual Change from 1990 *(Percent)* |  | *0.6%* | *0.2%* |
| Change from 2007 (Million Metric Tons) |  |  | -6.5 |
| *(Percent)* |  |  | *-6.1%* |

**Carbon Sequestration from Nonfuel Uses of Energy Inputs, 1990, 2007, and 2008**

|  | 1990 | 2007 | 2008 |
|---|---|---|---|
| Estimated Sequestration (Million Metric Tons CO$_2$e) | 252.7 | 293.7 | 264.2 |
| Change from 1990 (Million Metric Tons CO$_2$e) |  | 41.0 | 11.5 |
| *(Percent)* |  | *16.2%* | *4.5%* |
| Average Annual Change from 1990 *(Percent)* |  | *0.9%* | *0.2%* |
| Change from 2007 (Million Metric Tons CO$_2$e) |  |  | -29.5 |
| *(Percent)* |  |  | *-10.1%* |

# 178     United States Energy Information Administration

## Table 12. U.S. Carbon Dioxide Emissions from Nonfuel Use of Energy Fuels, 1990-2008
### (Million Metric Tons Carbon Dioxide)

| End Useand Type | 1990 | 1995 | 2000 | 2002 | 2003 | 2004 | 2005 | 2006 | 2007 | 2008 |
|---|---|---|---|---|---|---|---|---|---|---|
| **Industrial** | | | | | | | | | | |
| Petroleum | | | | | | | | | | |
| Liquefied Petroleum Gases | 14.8 | 19.5 | 20.4 | 19.9 | 19.0 | 19.4 | 18.3 | 18.7 | 18.9 | 17.7 |
| Distillate Fuel Oil | 0.3 | 0.3 | 0.4 | 0.4 | 0.5 | 0.5 | 0.6 | 0.6 | 0.6 | 0.6 |
| Residual FuelOil | 1.9 | 2.1 | 2.0 | 1.7 | 1.9 | 2.1 | 2.3 | 2.5 | 2.5 | 2.5 |
| Asphalt and Road Oil | 0.0 | 0.0 | 0.0 | 0.0 | 0.0 | 0.0 | 0.0 | 0.0 | 0.0 | 0.0 |
| Lubricants | 6.9 | 6.6 | 7.0 | 6.4 | 5.9 | 6.0 | 5.9 | 5.8 | 6.0 | 5.6 |
| Other *(Subtotal)* | *51.6* | *52.0* | *54.2* | *53.1* | *52.9* | *61.0* | *56.1* | *59.8* | *57.1* | *52.4* |
| Pentanes Plus | 1.1 | 4.1 | 3.2 | 2.3 | 2.3 | 2.3 | 2.0 | 1.4 | 1.8 | 1.6 |
| Petrochemical Feed | 33.6 | 36.0 | 36.8 | 33.5 | 36.5 | 41.8 | 38.4 | 40.1 | 37.1 | 32.2 |
| Petroleum Coke | 9.1 | 6.8 | 7.2 | 9.8 | 8.2 | 13.2 | 11.1 | 13.1 | 12.5 | 12.4 |
| Special Naphtha | 7.8 | 5.2 | 7.1 | 7.5 | 5.9 | 3.7 | 4.6 | 5.1 | 5.7 | 6.2 |
| Waxes and Miscellaneous | 0.0 | 0.0 | 0.0 | 0.0 | 0.0 | 0.0 | 0.0 | 0.0 | 0.0 | 0.0 |
| *Petroleum Subtotal* | *75.5* | *80.5* | *84.1* | *81.5* | *80.2* | *89.0* | *83.2* | *87.3* | *85.2* | *78.8* |
| Coal | 0.5 | 0.7 | 0.6 | 0.5 | 0.5 | 0.5 | 0.5 | 0.5 | 0.5 | 0.5 |
| Natural Gas | 14.7 | 17.5 | 18.4 | 15.7 | 15.5 | 15.6 | 15.4 | 15.7 | 15.4 | 15.7 |
| **Industrial Subtotal** | **90.6** | **98.6** | **103.1** | **97.7** | **96.1** | **105.2** | **99.1** | **103.5** | **101.0** | **94.9** |
| **Transportation** | | | | | | | | | | |
| Lubricants | 6.5 | 6.2 | 6.7 | 6.0 | 5.6 | 5.6 | 5.6 | 5.5 | 5.6 | 5.2 |
| **Total** | **97.1** | **104.9** | **109.8** | **103.7** | **101.7** | **110.8** | **104.7** | **108.9** | **106.7** | **100.2** |

Notes: Emissions from nonfuel use of energy fuels are included in the energy consumption tables in this chapter. Data in this table are revised from unpublished data used to produce the previous EIA report, *Emissions of Greenhouse Gases in the United States 2007*, DOE/EIA-0573(2007) (Washington, DC, December 2008). Totals may not equal sum of components due to independent rounding.

Sources: EIA estimates.


## Nonfuel Uses of Energy Inputs

### Table 13. U.S. Carbon Sequestered by Nonfuel Use of Energy Fuels, 1990-2008
### (Million Metric Tons Carbon Dioxide Equivalent)

| End Use and Type | 1990 | 1995 | 2000 | 2002 | 2003 | 2004 | 2005 | 2006 | 2007 | 2008 |
|---|---|---|---|---|---|---|---|---|---|---|
| **Industrial** | | | | | | | | | | |
| Petroleum | | | | | | | | | | |
| Liquefied Petroleum Gases | 59.2 | 78.2 | 81.7 | 79.6 | 76.0 | 77.6 | 73.2 | 74.6 | 75.8 | 70.9 |
| Distillate Fuel | 0.3 | 0.3 | 0.4 | 0.4 | 0.5 | 0.5 | 0.6 | 0.6 | 0.6 | 0.6 |
| Residual Fuel | 1.9 | 2.1 | 2.0 | 1.7 | 1.9 | 2.1 | 2.3 | 2.5 | 2.5 | 2.5 |

# Emissions of Greenhouse Gases in the United States 2008 — 179

## Table 13. (Continued)

| End Use and Type | 1990 | 1995 | 2000 | 2002 | 2003 | 2004 | 2005 | 2006 | 2007 | 2008 |
|---|---|---|---|---|---|---|---|---|---|---|
| Asphalt and Road Oil | 88.5 | 89.1 | 96.4 | 93.7 | 92.2 | 98.6 | 100.0 | 95.4 | 90.5 | 76.5 |
| Lubricants | 6.9 | 6.6 | 7.0 | 6.4 | 5.9 | 6.0 | 5.9 | 5.8 | 6.0 | 5.6 |
| Other *(Subtotal)* | *63.0* | *76.3* | *81.6* | *76.6* | *79.9* | *88.9* | *83.0* | *81.0* | *76.4* | *68.2* |
| Pentanes Plus | 4.4 | 16.2 | 12.7 | 9.2 | 9.0 | 9.1 | 8.0 | 5.7 | 7.4 | 6.3 |
| Petrochemical Feed | 46.0 | 50.0 | 57.7 | 55.1 | 59.2 | 69.1 | 64.2 | 63.2 | 57.5 | 49.9 |
| Petroleum Coke | 9.1 | 6.8 | 7.2 | 9.8 | 8.2 | 13.2 | 11.1 | 13.1 | 12.5 | 12.4 |
| Special Naphtha | 0.0 | 0.0 | 0.0 | 0.0 | 0.0 | 0.0 | 0.0 | 0.0 | 0.0 | 0.0 |
| Waxes and Miscellaneous | 12.6 | 10.2 | 11.2 | 12.3 | 11.6 | 10.7 | 10.7 | 12.0 | 11.5 | 12.0 |
| *Petroleum Subtotal* | *228.9* | *259.3* | *276.4* | *268.3* | *264.6* | *287.0* | *276.2* | *273.0* | *264.3* | *236.7* |
| Coal | 1.4 | 2.1 | 1.8 | 1.5 | 1.5 | 1.5 | 1.5 | 1.4 | 1.4 | 1.4 |
| Natural Gas | 15.9 | 19.0 | 23.3 | 20.2 | 20.4 | 21.3 | 21.8 | 22.0 | 22.4 | 20.9 |
| **Industrial Subtotal** | **246.2** | **280.4** | **301.6** | **290.0** | **286.5** | **309.8** | **299.4** | **296.5** | **288.1** | **259.0** |
| **Transportation** | | | | | | | | | | |
| Lubricants | 6.5 | 6.2 | 6.7 | 6.0 | 5.6 | 5.6 | 5.6 | 5.5 | 5.6 | 5.2 |
| **Total** | **252.7** | **286.6** | **308.2** | **296.0** | **292.1** | **315.4** | **305.1** | **301.9** | **293.7** | **264.2** |

Notes: Emissions from nonfuel use of energy fuels are included in the energy consumption tables in this chapter. Data in this table are revised from the data contained in the previous EIA report, Emissions of Greenhouse Gases in the United States 2007, DOE/EIA-0573(2007) (Washington, DC, December 2008). Totals may not equal sum of components due to independent rounding.

Sources: EIA estimates.

## Adjustments to Energy Consumption

### *Summary*

- EIA's greenhouse gas emissions inventory includes two "adjustments to energy consumption" (Table 14). First, the energy consumption and carbon dioxide emissions data in this chapter correspond to EIA's coverage of energy consumption, which includes the 50 States and the District of Columbia, but under the UNFCCC the United States is also responsible for emissions emanating from its Territories; therefore, their emissions are added to the U.S. total. Second, because the UNFCC definition of energy consumption excludes international bunker fuels, emissions from international bunker fuels are subtracted from the U.S. total. Similarly, because the UNFCC excludes emissions from military bunker fuels from national totals, they are subtracted from the U.S. total.
- The net adjustment in emissions has been negative in every year from 1990 to 2008, because emissions from international and military bunker fuels have always exceeded emissions from U.S. Territories. The net negative adjustment for 2008 was 79.0 MMT.

## Carbon Dioxide Emissions from U.S. Territories,* 1990, 2007, and 2008

| | 1990 | 2007 | 2008 |
|---|---|---|---|
| Estimated Emissions (Million Metric Tons) | 31.6 | 54.8 | 48.4 |
| Change from 1990 (Million Metric Tons) | | 23.1 | 16.8 |
| *(Percent)* | | *73.2%* | *53.0%* |
| Average Annual Change from 1990 *(Percent)* | | *3.3%* | *2.4%* |
| Change from 2007 (Million Metric Tons) | | | -6.4 |
| *(Percent)* | | | *-11.6%* |

*Added to total U.S. emissions.

## Carbon Dioxide Emissions from International Bunker Fuels,* 1990, 2007, and 2008

| | 1990 | 2007 | 2008 |
|---|---|---|---|
| Estimated Emissions (Million Metric Tons) | 114.5 | 129.7 | 127.4 |
| Change from 1990 (Million Metric Tons) | | 15.2 | 12.9 |
| *(Percent)* | | *13.2%* | *11.3%* |
| Average Annual Change from 1990 *(Percent)* | | *0.7%* | *0.6%* |
| Change from 2007 (Million Metric Tons) | | | -2.3 |
| *(Percent)* | | | *-1.8%* |

*Subtracted from total U.S. emissions.

## Table 14. U.S. Carbon Dioxide Emissions: Adjustments for U.S. Territories and International Bunker Fuels, 1990-2008 (Million Metric Tons Carbon Dioxide)

| Fuel | 1990 | 1995 | 2000 | 2002 | 2003 | 2004 | 2005 | 2006 | 2007 | 2008 |
|---|---|---|---|---|---|---|---|---|---|---|
| **Emissions from U.S. Territories** | | | | | | | | | | |
| Puerto Rico | 20.2 | 24.3 | 27.7 | 35.2 | 37.5 | 38.2 | 39.9 | 41.4 | 38.5 | 36.1 |
| U.S. Virgin Islands | 7.5 | 8.6 | 9.8 | 13.7 | 15.3 | 18.4 | 13.8 | 12.8 | 12.4 | 8.8 |
| American Samoa | 0.6 | 0.6 | 0.6 | 0.6 | 0.6 | 0.6 | 0.6 | 0.6 | 0.6 | 0.6 |
| Guam | 1.8 | 3.6 | 2.9 | 2.1 | 2.4 | 2.0 | 2.2 | 1.9 | 1.6 | 1.1 |
| U.S. Pacific Islands | 0.3 | 0.3 | 0.3 | 0.3 | 0.3 | 0.3 | 0.3 | 0.3 | 0.3 | 0.3 |
| Wake Island | 1.2 | 1.3 | 1.3 | 1.3 | 1.3 | 1.3 | 1.3 | 1.3 | 1.4 | 1.4 |
| **Subtotal[a]** | **31.6** | **38.6** | **42.6** | **53.3** | **57.4** | **60.7** | **58.1** | **58.4** | **54.8** | **48.4** |
| **Emissions from Bunker Fuels** | | | | | | | | | | |
| Marine Bunkers *(Subtotal)* | *62.7* | *47.0* | *37.9* | *24.4* | *20.0* | *29.6* | *29.8* | *50.3* | *51.5* | *56.1* |
| Distillate Fuel | 6.3 | 5.8 | 2.9 | 1.6 | 1.5 | 1.7 | 2.4 | 3.1 | 3.6 | 4.5 |
| Residual Fuel | 56.4 | 41.2 | 35.0 | 22.8 | 18.5 | 27.9 | 27.4 | 47.2 | 47.9 | 51.6 |
| Aviation Bun-kers *(Subtotal)* | *38.4* | *45.8* | *58.5* | *59.0* | *56.9* | *65.7* | *67.9* | *71.0* | *69.7* | *62.9* |
| U.S. Carriers | 18.7 | 21.3 | 26.2 | 23.9 | 23.4 | 26.7 | 28.6 | 28.8 | 29.7 | 29.6 |
| Foreign Carriers | 19.7 | 24.5 | 32.3 | 35.1 | 33.5 | 39.0 | 39.3 | 42.2 | 40.1 | 33.3 |
| Military Bunkers *(Subtotal)* | *13.4* | *8.9* | *7.9* | *8.1* | *9.2* | *10.1* | *9.2* | *8.0* | *8.4* | *8.4* |
| **Subtotal[b]** | **114.5** | **101.7** | **104.3** | **91.5** | **86.1** | **105.4** | **106.9** | **129.3** | **129.7** | **127.4** |
| **Net Adjustment** | **-82.9** | **-63.1** | **-61.7** | **-38.2** | **-28.7** | **-44.6** | **-48.8** | **-70.9** | **-74.9** | **-79.0** |

[a]Added to total U.S. emissions. [b]Subtracted from total U.S. emissions.

Note: Totals may not equal sum of components due to independent rounding.

Source: EIA estimates.

## Other Sources

### *Summary*

- "Other emissions sources" in total accounted for 1.8 percent (103.8 MMT) of all U.S. carbon dioxide emissions in 2008 (Figure 14).
- The largest source of U.S. carbon dioxide emissions other than fossil fuel consumption is cement manufacture (Table 15), where most emissions result from the production of clinker (consisting of calcium carbonate sintered with silica in a cement kiln to produce calcium silicate).
- Limestone consumption, especially for lime manufacture, is the source of 15 to 20 MMT of carbon dioxide emissions per year.
- In addition, "other sources" include: soda ash manufacture and consumption; carbon dioxide manufacture; aluminum manufacture; flaring of natural gas at the wellhead; carbon dioxide scrubbed from natural gas; and waste combustion in the commercial and industrial sectors.

**Carbon Dioxide Emissions from Other Sources, 1990, 2007, and 2008**

|  | 1990 | 2007 | 2008 |
|---|---|---|---|
| Estimated Emissions (Million Metric Tons) | 85.1 | 105.6 | 103.8 |
| Change from 1990 (Million Metric Tons) |  | 20.5 | 18.7 |
| *(Percent)* |  | *24.1%* | *22.0%* |
| Average Annual Change from 1990 *(Percent)* |  | *1.3%* | *1.1%* |
| Change from 2007 (Million Metric Tons) |  |  | -1.8 |
| *(Percent)* |  |  | *-1.7%* |

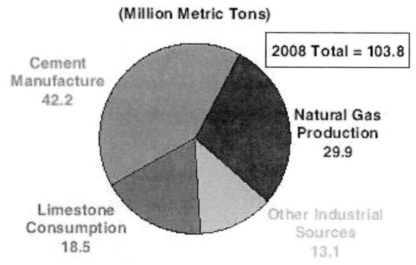

Source: EIA estimates.

Figure 14. U.S. Carbon Dioxide Emissions from Other Sources, 2008

## Table 15. U.S. Carbon Dioxide Emissions from Other Sources, 1990-2008
## (Million Metric Tons Carbon Dioxide)

| Source | 1990 | 1995 | 2000 | 2002 | 2003 | 2004 | 2005 | 2006 | 2007 | 2008 |
|---|---|---|---|---|---|---|---|---|---|---|
| **Cement Manufacture** | *33.3* | *36.9* | *41.3* | *43.0* | *43.2* | *45.7* | *46.1* | *46.7* | *45.4* | *42.2* |
| Clinker Production | 32.6 | 36.1 | 40.4 | 42.0 | 42.2 | 44.7 | 45.1 | 45.7 | 44.4 | 41.3 |
| Masonry Cement | 0.1 | 0.1 | 0.1 | 0.1 | 0.1 | 0.1 | 0.1 | 0.1 | 0.1 | 0.1 |
| Cement Kiln Dust | 0.7 | 0.7 | 0.8 | 0.8 | 0.8 | 0.9 | 0.9 | 0.9 | 0.9 | 0.8 |
| **Limestone Consumption** | *15.9* | *17.8* | *18.6* | *17.0* | *18.0* | *18.9* | *18.8* | *19.6* | *18.9* | *18.5* |
| Lime Manufacture | 12.4 | 14.5 | 15.4 | 14.1 | 15.1 | 15.7 | 15.7 | 16.5 | 15.9 | 15.5 |
| Iron Smelting | 1.7 | 1.2 | 1.1 | 0.9 | 0.9 | 1.0 | 0.8 | 0.9 | 0.8 | 0.8 |
| Steelmaking | 0.3 | 0.5 | 0.5 | 0.5 | 0.4 | 0.4 | 0.3 | 0.4 | 0.3 | 0.3 |
| Copper Refining | 0.1 | 0.2 | 0.1 | 0.1 | 0.1 | 0.1 | 0.1 | 0.1 | 0.1 | 0.1 |
| Glass Manufacture | 0.1 | 0.3 | 0.2 | 0.1 | 0.2 | 0.2 | 0.2 | 0.2 | 0.2 | 0.2 |
| Flue Gas Desulfurization | 0.7 | 0.9 | 1.2 | 1.3 | 1.3 | 1.4 | 1.5 | 1.5 | 1.5 | 1.5 |
| Dolomite Manufacture | 0.5 | 0.2 | 0.1 | 0.1 | 0.1 | 0.1 | 0.1 | 0.2 | 0.1 | 0.1 |
| **Natural Gas Production** | *23.1* | *33.9* | *23.8* | *24.4* | *24.5* | *24.3* | *25.3* | *26.6* | *28.5* | *29.9* |
| Carbon Dioxide in Natural Gas | 14.0 | 16.7 | 18.3 | 18.4 | 18.6 | 18.4 | 18.1 | 18.7 | 19.3 | 20.8 |
| Natural Gas Flaring | 9.1 | 17.2 | 5.5 | 6.0 | 5.9 | 5.8 | 7.2 | 7.8 | 9.1 | 9.1 |
| **Other** | *12.7* | *13.8* | *14.1* | *13.3* | *13.2* | *13.1* | *13.2* | *13.0* | *12.9* | *13.1* |
| Soda Ash Manufacture | 3.4 | 3.8 | 3.6 | 3.5 | 3.6 | 3.8 | 3.9 | 3.9 | 4.0 | 4.1 |
| Soda Ash Consumption | 0.5 | 0.8 | 0.6 | 0.4 | 0.6 | 0.6 | 0.6 | 0.6 | 0.6 | 0.5 |
| Carbon Dioxide Manufacture | 0.9 | 1.0 | 1.3 | 1.4 | 1.5 | 1.5 | 1.6 | 1.6 | 1.7 | 1.8 |
| Aluminum Manufacture | 5.9 | 4.9 | 5.4 | 4.0 | 4.0 | 3.7 | 3.6 | 3.3 | 3.7 | 3.9 |
| Shale Oil Production | 0.2 | * | * | * | * | * | * | * | * | * |
| Waste Combustion | 1.9 | 3.2 | 3.2 | 4.0 | 3.6 | 3.5 | 3.6 | 3.6 | 2.8 | 2.8 |
| **Total** | **85.1** | **102.3** | **97.8** | **97.7** | **98.9** | **102.0** | **103.5** | **106.0** | **105.6** | **103.8** |

*Less than 0.05 million metric tons.

Notes: Data in this table are revised from the data contained in the previous EIA report, *Emissions of Greenhouse Gases in the United States 2007*, DOE/EIA-0573(2007) (Washington, DC, December 2008). Totals may not equal sum of components due to independent rounding.

Source: EIA estimates.

# METHANE EMISSIONS

## Total Emissions

### *Summary*

- The major sources of U.S. methane emissions are energy production, distribution, and use; agriculture; and waste management (Figure 15).
- U.S. methane emissions in 2008 totaled 737.4 MMTCO$_2$e, 2.0 percent higher than the 2007 total of 722.7 MMTCO$_2$e (Table 16).
- Methane emissions declined steadily from 1990 to 2001, as emissions from coal mining and landfills fell.
- From 2002 to 2008, methane emissions rose as a result of moderate increases in emissions related to energy, agriculture, and waste management that more than offset a decline in industrial emissions of methane over the same period.
- The energy sector—including coal mining, natural gas systems, petroleum systems, and stationary and mobile combustion—is the largest source of U.S. methane emissions, accounting for 295.7 MMTCO$_2$e in 2008.
- Agricultural emissions (primarily from livestock management) and emissions from waste management (primarily landfills) also are large sources of U.S. methane emissions, contributing 225.0 and 212.1 MMTCO$_2$e, respectively, in 2008.

**Total U.S. Anthropogenic Methane Emissions, 1990, 2007, and 2008**

|  | 1990 | 2007 | 2008 |
|---|---|---|---|
| Estimated Emissions (Million Metric Tons CO$_2$e) | 783.5 | 722.7 | 737.4 |
| Change from 1990 (Million Metric Tons CO$_2$e) |  | -60.9 | -46.1 |
| *(Percent)* |  | -7.8% | -5.9% |
| Average Annual Change from 1990 *(Percent)* |  | -0.5% | -0.3% |
| Change from 2007 (Million Metric Tons CO$_2$e) |  |  | 14.8 |
| *(Percent)* |  |  | 2.0% |

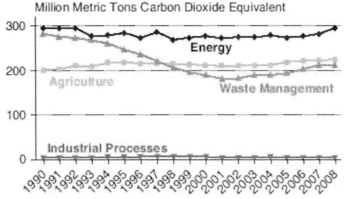

Source: EIA estimates.

Figure 15. U.S. Methane Emissions by Source, 1990-2008

# Table 16. U.S. Methane Emissions from Anthropogenic Sources, 1990-2008 (Million Metric Tons Carbon Dioxide Equivalent)

| Source | 1990 | 1995 | 2000 | 2002 | 2003 | 2004 | 2005 | 2006 | 2007 | 2008 |
|---|---|---|---|---|---|---|---|---|---|---|
| Energy Sources | 294.4 | 284.8 | 277.6 | 275.6 | 274.5 | 278.9 | 274.2 | 276.7 | 282.8 | 295.7 |
| Agricultural Sources | 201.5 | 219.1 | 211.1 | 210.4 | 212.0 | 212.5 | 218.8 | 221.8 | 222.6 | 225.0 |
| Waste Management | 283.0 | 246.8 | 188.6 | 182.0 | 189.8 | 189.6 | 193.7 | 202.6 | 212.1 | 212.1 |
| Industrial Processes | 4.6 | 5.6 | 5.7 | 5.3 | 5.2 | 5.6 | 5.1 | 5.2 | 5.2 | 4.7 |
| **Total** | **783.5** | **756.2** | **683.0** | **673.3** | **681.6** | **686.6** | **691.8** | **706.3** | **722.7** | **737.4** |

Notes: Data in this table are revised from the data contained in the previous EIA report, *Emissions of Greenhouse Gases in the United States 2007*, DOE/EIA-0573(2007) (Washington, DC, December 2008). Totals may not equal sum of components due to independent rounding.

Sources: Published and unpublished data used to produce *Emissions of Greenhouse Gases in the United States 2007*. Emissions calculations based on Intergovernmental Panel on Climate Change, *IPCC Guidelines for National Greenhouse Gas Inventories* (2006 and revised 1996 guidelines), web site www.ipcc-nggip.iges.or.jp/public/gl/invs6.html; and U.S. Environmental Protection Agency, *Inventory of U.S. Greenhouse Gas Emissions and Sinks: 1990-2007*, EPA 430-R-09-004 (Washington, DC, April 2009), web site www.epa.gov/climatechange/emissions/usinventoryreport.html.

## Energy Sources

### Summary

- Natural gas systems and coal mines are the major sources of methane emissions in the energy sector (Figure 16 and Table 17).
- U.S. methane emissions from natural gas systems grew from 1990 to 2008, largely because of increases in natural gas consumption.
- Emissions from coal mines declined from 1990 to 2002 and remained nearly steady through 2007. In 2008, emissions from ventilation of underground mines jumped by 24.6 percent, leading to a 15.3- percent increase in total mining emissions over their 2007 level. Much of the 2008 increase can be attributed to the larger number of gassy mines in operation throughout the year and to the fact that, as mining proceeds into deeper seams, more methane emissions tend to be produced per ton of coal mined.
- With domestic oil production dropping by 30 percent from 1990 to 2008, methane emissions from petroleum systems also declined.
- Residential wood consumption accounted for nearly 45 percent of U.S. methane emissions from stationary combustion in 2008.
- Methane emissions from passenger cars fell by 51.3 percent from 1990 to 2008, as the use of catalytic converters increased. A 9.2-percent drop in annual miles traveled by passenger cars from 2002 to 2008 also contributed to the decrease in emissions.

**Table 17. U.S. Methane Emissions from Energy Sources, 1990-2008 (Million Metric Tons Carbon Dioxide Equivalent)**

| Source | 1990 | 1995 | 2000 | 2002 | 2003 | 2004 | 2005 | 2006 | 2007 | 2008 |
|---|---|---|---|---|---|---|---|---|---|---|
| **Natural Gas Systems** | *140.4* | *149.9* | *164.5* | *167.8* | *166.8* | *168.6* | *168.1* | *170.4* | *176.5* | *178.9* |
| Production | 36.8 | 39.3 | 43.5 | 46.2 | 46.7 | 47.4 | 48.2 | 49.2 | 50.9 | 52.2 |
| Processing | 16.2 | 18.0 | 17.9 | 16.9 | 15.7 | 16.2 | 15.9 | 15.7 | 16.6 | 17.5 |
| Transmission and Storage | 52.6 | 53.7 | 60.4 | 60.6 | 59.6 | 58.5 | 58.3 | 56.2 | 60.1 | 60.3 |
| Distribution | 34.9 | 38.9 | 42.8 | 44.1 | 44.8 | 46.4 | 45.7 | 49.3 | 48.9 | 48.9 |
| **Coal Mining** | *106.4* | *91.4* | *74.1* | *69.7* | *69.8* | *73.3* | *70.3* | *71.4* | *71.1* | *82.0* |
| Surface | 11.6 | 12.2 | 13.5 | 14.2 | 13.8 | 14.3 | 14.7 | 15.5 | 15.3 | 15.7 |
| Underground | 94.8 | 79.1 | 60.6 | 55.6 | 56.0 | 59.0 | 55.6 | 55.9 | 55.8 | 66.4 |
| **Petroleum Systems** | *32.4* | *29.2* | *25.8* | *25.5* | *25.2* | *24.1* | *23.1* | *22.8* | *22.6* | *22.1* |
| Refineries | 0.6 | 0.6 | 0.7 | 0.7 | 0.7 | 0.7 | 0.7 | 0.7 | 0.7 | 0.6 |
| Exploration and Production | 31.6 | 28.4 | 25.0 | 24.7 | 24.4 | 23.3 | 22.3 | 21.9 | 21.8 | 21.3 |
| Crude Oil Transportation | 0.2 | 0.1 | 0.1 | 0.1 | 0.1 | 0.1 | 0.1 | 0.1 | 0.1 | 0.1 |
| **Stationary Combustion** | *9.1* | *8.8* | *8.1* | *7.5* | *7.7* | *7.9* | *7.9* | *7.4* | *7.8* | *8.1* |
| **Mobile Sources** | *6.1* | *5.5* | *5.1* | *5.1* | *4.9* | *5.0* | *4.9* | *4.8* | *4.8* | *4.6* |
| **Total** | **294.4** | **284.8** | **277.6** | **275.6** | **274.5** | **278.9** | **274.2** | **276.7** | **282.8** | **295.7** |

Notes: Data in this table are revised from the data contained in the previous EIA report, *Emissions of Greenhouse Gases in the United States 2007*, DOE/EIA-0573(2007) (Washington, DC, December 2008). Totals may not equal sum of components due to independent rounding.

Source: EIA estimates.

**Methane Emissions from Energy Sources, 1990, 2007, and 2008**

|  | 1990 | 2007 | 2008 |
|---|---|---|---|
| Estimated Emissions (Million Metric Tons $CO_2e$) | 294.4 | 282.8 | 295.7 |
| Change from 1990 (Million Metric Tons $CO_2e$) |  | -11.6 | 1.2 |
| *(Percent)* |  | -4.0% | 0.4% |
| Average Annual Change from 1990 *(Percent)* |  | -0.2% | 0.0% |
| Change from 2007 (Million Metric Tons $CO_2e$) |  |  | 12.9 |
| *(Percent)* |  |  | 4.6% |

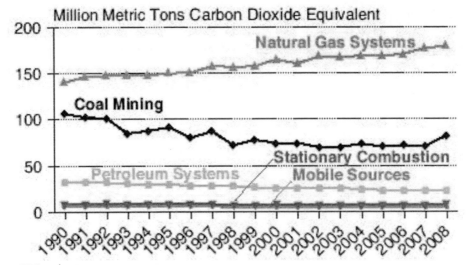

Source: EIA estimates.

Figure 16. U.S. Methane Emissions from Energy Sources, 1990-2008

## Agriculture

*Summary*

- Livestock management—including emissions from enteric fermentation (66 percent) and management of animal wastes (29 percent)—accounts for most of the U.S. methane emissions from agricultural activities (Table 18).
- Since 1990, there has been a shift in livestock management to larger facilities that manage waste in liquid systems, increasing the amount of methane generated from livestock waste. Increases in the U.S. swine population since 1990 have also contributed to the rise in methane emissions. Emissions of methane from animal waste rose by 2.0 percent from 2007 to 2008. Swine accounted for 45.9 percent (29.7 MMTCO$_2$e) and dairy cattle 36.2 percent (23.3 MMTCO$_2$e) of the 2008 total.
- Enteric fermentation (food digestion) in ruminant animals also produces methane emissions, and digestion by cattle accounts for 95 percent of U.S. methane emissions from this source. With little change in the cattle population since 1990, the level of

emissions from enteric fermentation has been relatively stable. Small declines in cattle and goat populations in 2008 were offset by increases in horse, goat and swine poulations, which caused methane emissions from enteric fermentation to increase slightly (by 0.5 MMTCO2e, or 0.3 percent) from their 2007 level.
- Despite lower crop yields, U.S. rice production rose by 7 percent in 2008 as a result of increases in area harvested in Missouri, Texas, Arkansas, Louisiana, and Mississippi. Consequently, emissions from this source increased by 7.1 percent from 2007 to 2008.
- Although emissions from crop residue burning grew by 2.0 percent in 2008, residue burning remains the smallest contributor to methane emissions from agriculture, representing less than 1 percent of total U.S. methane emissions from agriculture (Figure 17).

**Methane Emissions from Agricultural Sources, 1990, 2007, and 2008**

|  | 1990 | 2007 | 2008 |
|---|---|---|---|
| Estimated Emissions (Million Metric Tons $CO_2e$) | 201.5 | 222.6 | 225.0 |
| Change from 1990 (Million Metric Tons $CO_2e$) |  | 21.1 | 23.5 |
| *(Percent)* |  | *10.5%* | *11.7%* |
| Average Annual Change from 1990 *(Percent)* |  | *0.6%* | *0.6%* |
| Change from 2007 (Million Metric Tons $CO_2e$) |  |  | 2.5 |
| *(Percent)* |  |  | *1.1%* |

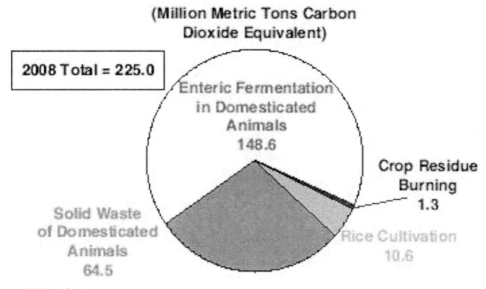

Source: EIA estimates.

Figure 17. U.S. Methane Emissions from Agriculture by Source, 2008

# 188 United States Energy Information Administration

**Table 18. U.S. Methane Emissions from Agricultural Sources, 1990-2008**
**(Million Metric Tons Carbon Dioxide Equivalent)**

| Source | 1990 | 1995 | 2000 | 2002 | 2003 | 2004 | 2005 | 2006 | 2007 | 2008 |
|---|---|---|---|---|---|---|---|---|---|---|
| Enteric Fermentation in Domesticated Animals | 145.4 | 155.1 | 143.9 | 142.6 | 144.0 | 142.9 | 145.4 | 147.4 | 148.1 | 148.6 |
| Solid Waste of Domesticated Animals | 45.0 | 51.9 | 54.9 | 55.6 | 56.2 | 56.4 | 60.6 | 63.3 | 63.3 | 64.5 |
| Rice Cultivation | 10.1 | 11.1 | 11.1 | 11.1 | 10.7 | 11.8 | 11.5 | 9.9 | 9.9 | 10.6 |
| Crop Residue Burning | 1.0 | 1.0 | 1.2 | 1.1 | 1.2 | 1.3 | 1.3 | 1.2 | 1.3 | 1.3 |
| **Total** | **201.5** | **219.1** | **211.1** | **210.4** | **212.0** | **212.5** | **218.8** | **221.8** | **222.6** | **225.0** |

Notes: Data in this table are revised from the data contained in the previous EIA report, *Emissions of Greenhouse Gases in the United States 2007*, DOE/EIA-0573(2007) (Washington, DC, December 2008). Totals may not equal sum of components due to independent rounding.

Source: EIA estimates.

## Waste Management

- Methane emissions from waste management are dominated by the decomposition of solid waste in municipal and industrial landfills (Table 19).
- Emissions from landfills declined substantially from 1990 to 2001 as a result of increases in recycling and in the recovery of landfill methane for energy; since 2001, increases in the total amount of waste deposited in landfills have resulted in increasing methane emissions (Figure 18).
- Rapid growth in methane recovery during the 1990s can be traced in part to the Federal Section 29 tax credit for alternative energy sources, which provided a subsidy of approximately 1 cent per kilowatthour for electricity generated from landfill gas before June 1998.
- The U.S. EPA's New Source Performance Standards and Emission Guidelines, which require large landfills to collect and burn landfill gas, have also played an important role in the growth of methane recovery.
- The American Recovery and Reinvestment Act of 2009, signed into law on February 17, 2009, included a 2-year extension (through December 31, 2012) of the production tax credit (PTC) for renewable energy, including waste-to-energy and landfill gas combustion.
- Wastewater treatment, including both domestic wastewater (two-thirds) and industrial wastewater (one-third), is responsible for 13.1 percent (27.8 MMTCO$_2$e) of methane emissions from waste management.
- Emissions from the treatment of wastewater from ethanol production and from petroleum refineries are included for the first time in the 2008 inventory, increasing the estimates of methane emissions from industrial wastewater in previous years by 8.0 to 11.4 percent. In 2008, emissions from wastewater at petroleum refineries

accounted for 7.3 percent (0.74 MMTCO₂e) of total emissions from industrial wastewater, and emissions from ethanol production accounted for 2.9 percent (0.30 MMTCO₂e).

**Methane Emissions from Waste Management, 1990, 2007, and 2008**

|  | 1990 | 2007 | 2008 |
|---|---|---|---|
| Estimated Emissions (Million Metric Tons CO₂e) | 283.0 | 212.1 | 212.1 |
| Change from 1990 (Million Metric Tons CO₂e) |  | -70.9 | -71.0 |
| *(Percent)* |  | *-25.1%* | *-25.1%* |
| Average Annual Change from 1990 *(Percent)* |  | *-1.7%* | *-1.6%* |
| Change from 2007 (Million Metric Tons CO₂e) |  |  | 0.0 |
| *(Percent)* |  |  | *0.0%* |

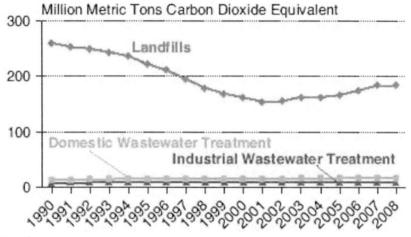

Source: EIA estimates.

Figure 18. U.S. Methane Emissions from Waste Management by Source, 1990-2008

**Table 19. U.S. Methane Emissions from Waste Management, 1990-2008**
**(Million Metric Tons Carbon Dioxide Equivalent)**

| Source | 1990 | 1995 | 2000 | 2002 | 2003 | 2004 | 2005 | 2006 | 2007 | 2008 |
|---|---|---|---|---|---|---|---|---|---|---|
| Landfills | 260.2 | 221.9 | 162.3 | 155.6 | 163.3 | 162.7 | 166.8 | 175.2 | 184.4 | 184.3 |
| Domestic Wastewater Treatment | 14.4 | 15.4 | 16.3 | 16.6 | 16.8 | 17.0 | 17.1 | 17.3 | 17.4 | 17.6 |
| Industrial WasteWater Treatment | 8.4 | 9.5 | 10.0 | 9.8 | 9.7 | 9.9 | 9.8 | 10.1 | 10.2 | 10.2 |
| Total | 283.0 | 246.8 | 188.6 | 182.0 | 189.8 | 189.6 | 193.7 | 202.6 | 212.1 | 212.1 |

Notes: Data in this table are revised from the data contained in the previous EIA report, *Emissions of Greenhouse Gases in the United States 2007*, DOE/EIA-0573(2007) (Washington, DC, December 2008). Totals may not equal sum of components due to independent rounding.
Source: EIA estimates.

## Industrial Processes

### *Summary*

- Methane emissions are generated by industrial processes in the production of iron and steel and in chemical production (Figure 19 and Table 20).
- Methane emissions from industrial processes declined by a net 0.6 MMTCO$_2$e (10.7 percent) from 2007 to 2008, as a result of declines in both chemical production and iron and steel production.
- Since 1990, methane emissions from industrial processes have increased by 0.1 MMTCO$_2$e (2.4 percent). A 32.4-percent decline (0.4 MMTCO$_2$e) in emissions from iron and steel production since 1990 has been offset by an increase of 0.5 MMTCO$_2$e (13.9 percent) in emissions from chemical production.
- Estimates of industrial emissions of methane in the 2008 inventory are approximately 50 to 65 percent higher for all years, as a result of applying the IPCC's revised emissions factors for methane from ethylene, ethylene dichloride, and methanol production in the chemical industry and for methane from sinter and coke in iron and steel production.

**Methane Emissions from Industrial Processes, 1990, 2007, and 2008**

|  | 1990 | 2007 | 2008 |
|---|---|---|---|
| Estimated Emissions (Million Metric Tons CO$_2$e) | 4.6 | 5.2 | 4.7 |
| Change from 1990 (Million Metric Tons CO$_2$e) |  | 0.7 | 0.1 |
| *(Percent)* |  | *14.7%* | *2.4%* |
| Average Annual Change from 1990 *(Percent)* |  | *0.8%* | *0.1%* |
| Change from 2007 (Million Metric Tons CO$_2$e) |  |  | -0.6 |
| *(Percent)* |  |  | *-10.7%* |

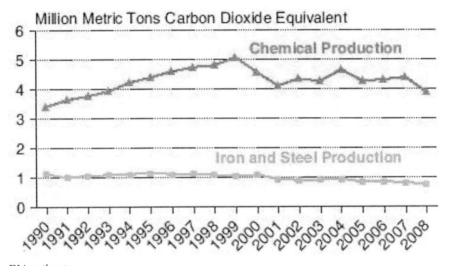

Source: EIA estimates.

Figure 19. U.S. Methane Emissions from Industrial Processes by Source, 1990-2008

Emissions of Greenhouse Gases in the United States 2008

**Table 20. U.S. Methane Emissions from Industrial Processes, 1990-2008
(Million Metric Tons Carbon Dioxide Equivalent)**

| Source | 1990 | 1995 | 2000 | 2002 | 2003 | 2004 | 2005 | 2006 | 2007 | 2008 |
|---|---|---|---|---|---|---|---|---|---|---|
| **Chemical Production** | | | | | | | | | | |
| Ethylene | 2.5 | 3.2 | 3.4 | 3.2 | 3.1 | 3.5 | 3.3 | 3.4 | 3.5 | 3.1 |
| Ethylene Dichloride | * | * | * | * | * | * | * | * | * | * |
| Styrene | 0.4 | 0.5 | 0.5 | 0.5 | 0.5 | 0.5 | 0.5 | 0.4 | 0.5 | 0.4 |
| Methanol | 0.2 | 0.3 | 0.3 | 0.3 | 0.2 | 0.3 | 0.1 | 0.1 | 0.1 | 0.1 |
| Carbon Black | 0.4 | 0.4 | 0.4 | 0.4 | 0.4 | 0.4 | 0.4 | 0.4 | 0.4 | 0.4 |
| **Subtotal** | **3.4** | **4.4** | **4.6** | **4.4** | **4.3** | **4.7** | **4.3** | **4.3** | **4.4** | **3.9** |
| **Iron and Steel Production** | | | | | | | | | | |
| Coke[a] | * | * | * | * | * | * | * | * | * | * |
| Sinter | * | * | * | * | * | * | * | * | * | * |
| Pig Iron | 1.1 | 1.1 | 1.1 | 0.9 | 0.9 | 1.0 | 0.8 | 0.9 | 0.8 | 0.8 |
| **Subtotal** | **1.1** | **1.2** | **1.1** | **0.9** | **0.9** | **1.0** | **0.9** | **0.9** | **0.8** | **0.8** |
| **Total** | **4.6** | **5.6** | **5.7** | **5.3** | **5.2** | **5.6** | **5.1** | **5.2** | **5.2** | **4.7** |

[a]Based on total U.S. production of metallurgical coke, including for uses other than iron and steel manufacture.

*Less than 0.05 million metric tons.

Notes: Data in this table are revised from the data contained in the previous EIA report, *Emissions of Greenhouse Gases in the United States 2007*, DOE/EIA-0573(2007) (Washington, DC, December 2008). Totals may not equal sum of components due to independent rounding.

Source: EIA estimates.

# NITROUS OXIDE EMISSIONS

## Total Emissions

### *Summary*

- U.S. nitrous oxide emissions in 2008 were 0.1 percent(0.4 MMTCO$_2$e) above their 2007 total (Table 21).

- Sources of U.S. nitrous oxide emissions include agriculture, energy use, industrial processes, and waste management. The largest source is agriculture, and the majority of agricultural emissions result from nitrogen fertilization of agricultural soils (75.7 percent) and the management of animal waste (24.0 percent).

- Annual U.S. nitrous oxide emissions rose from 1990 to 1994, then fell from 1994 to 2003 (Figure 20). They rose sharply from 2003 to 2008, largely as a result of increased use of synthetic fertilizers, which grew by more than 30 percent from 2005 to 2008.

- Since 2005, when the Renewable Fuels Standard was signed into law, U.S. ethanol ethanol production has more than doubled (a 130-percent increase from 2005 to 2008). Nearly all U.S. ethanol production is from corn, and with corn production rising by 8.9 percent since 2005, the percentage used for ethanol production also has risen, from 16.1 percent to 21.1 percent of total U.S. corn production. As the demand for corn has increased, use of synthetic fertilizer (a nitrous oxide emitter that is used most heavily in corn production) has risen by 8.2 percent.

**Total Anthropogenic Nitrous Oxide Emissions, 1990, 2007, and 2008**

|  | 1990 | 2007 | 2008 |
|---|---|---|---|
| Estimated Emissions (Million Metric Tons $CO_2e$) | 279.3 | 299.8 | 300.3 |
| Change from 1990 (Million Metric Tons $CO_2e$) |  | 20.5 | 20.9 |
| *(Percent)* |  | *7.3%* | *7.5%* |
| Average Annual Change from 1990 *(Percent)* |  | *0.4%* | *0.4%* |
| Change from 2007 (Million Metric Tons $CO_2e$) |  |  | 0.4 |
| *(Percent)* |  |  | *0.1%* |

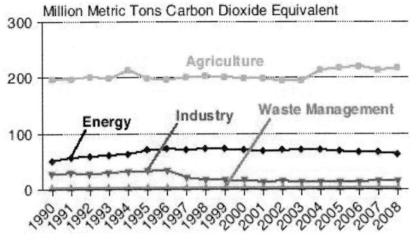

Source: EIA estimates.

Figure 20. U.S. Nitrous Oxide Emissions by Source, 1990-2008

**Table 21. U.S. Nitrous Oxide Emissions from Anthropogenic Sources, 1990-2008
(Million Metric Tons Carbon Dioxide Equivalent)**

| Source | 1990 | 1995 | 2000 | 2002 | 2003 | 2004 | 2005 | 2006 | 2007 | 2008 |
|---|---|---|---|---|---|---|---|---|---|---|
| **Agriculture** | | | | | | | | | | |
| Nitrogen Fertilization of Soils | 142.8 | 143.4 | 146.2 | 142.6 | 143.3 | 162.3 | 165.6 | 168.1 | 161.o | 165.0 |
| Solid Waste of Domesticated Animals | 53.2 | 55.4 | 52.4 | 51.7 | 51.3 | 51.0 | 51.7 | 52.2 | 52.3 | 52.3 |
| Crop Residue Burning | 0.4 | 0.4 | 0.5 | 0.5 | 0.5 | 0.6 | 0.6 | 0.6 | 0.5 | 0.6 |
| **Subtotal** | **196.4** | **199.2** | **199.1** | **194.8** | **195.1** | **213.9** | **217.9** | **220.8** | **213.9** | **217.9** |

**Table 21. (Continued)**

| Source | 1990 | 1995 | 2000 | 2002 | 2003 | 2004 | 2005 | 2006 | 2007 | 2008 |
|---|---|---|---|---|---|---|---|---|---|---|
| **Energy Use** | | | | | | | | | | |
| Mobile Combustion | 37.7 | 55.9 | 55.2 | 55.4 | 55.7 | 55.7 | 53.1 | 51.6 | 51.2 | 48.8 |
| Stationary Combustion | 13.9 | 14.5 | 15.7 | 15.1 | 15.3 | 15.6 | 15.7 | 15.4 | 15.6 | 15.1 |
| **Subtotal** | **51.6** | **70.4** | **71.0** | **70.4** | **71.0** | **71.4** | **68.8** | **67.1** | **66.8** | **63.9** |
| **Industrial Sources** | **28.8** | **33.1** | **16.7** | **15.3** | **13.7** | **13.7** | **14.0** | **14.0** | **15.9** | **15.1** |
| **Waste Management** | | | | | | | | | | |
| Human Sewage in Wastewater | 2.3 | 2.5 | 2.8 | 2.9 | 2.8 | 2.9 | 2.9 | 3.0 | 3.0 | 3.0 |
| Waste Combustion | 0.3 | 0.3 | 0.3 | 0.3 | 0.3 | 0.3 | 0.4 | 0.3 | 0.3 | 0.4 |
| **Subtotal** | **2.6** | **2.8** | **3.1** | **3.2** | **3.2** | **3.2** | **3.3** | **3.3** | **3.4** | **3.4** |
| **Total** | **279.3** | **305.6** | **289.8** | **283.7** | **283.0** | **302.2** | **304.0** | **305.2** | **299.8** | **300.3** |

Notes: Data in this table are revised from the data contained in the previous EIA report, *Emissions of Greenhouse Gases in the United States 2007*, DOE/EIA-0573(2007) (Washington, DC, December 2008). Totals may not equal sum of components due to independent rounding.

Sources: Estimates presented in this chapter. Emissions calculations based on Intergovernmental Panel on Climate Change, *IPCC Guidelines for National Greenhouse Gas Inventories* (2006 and revised 1996 guidelines), web site www.ipcc-nggip.iges.or.jp/public/gl/invs6.html; and U.S. Environmental Protection Agency, *Inventoryof U.S. Greenhouse Gas Emissions and Sinks: 1990-2007*, EPA 430-R-09-004 (Washington, DC, April 2009), web site www.epa.gov/climatechange/e missions/usinventoryreport.html.


# Agriculture

## *Summary*

- Agricultural sources, at 217.9 MMTCO$_2$e, account for 73 percent of U.S. nitrous oxide emissions. Nitrous oxide emissions from agricultural sources increased by 1.9 percent (4.0 MMTCO$_2$e) from 2007 to 2008 (Table 22).

- Three-quarters (165.0 MMTCO$_2$e) of U.S. agricultural emissions of nitrous oxide in 2008 is attributable to nitrogen fertilization of soils (Figure 21), including 145.2 MMTCO$_2$e from direct emissions and 19.8 MMTCO$_2$e from indirect emissions.

- Microbial denitrification of solid waste from domestic animals in the United States, primarily cattle, emitted 52.3 MMTCO$_2$e of nitrous oxide in 2008. The amount released is a function of animal size and manure production, the amount of nitrogen in the waste, and the method of managing the waste.

- Agricultural emissions of nitrous oxide are 23 to 28 percent lower in all years in the 2008 inventory than in the 2007 inventory, following the IPCC's downward revisions of direct and indirect emissions factors for nitrogen from the fertilization of agricultural soils.

**Nitrous Oxide Emissions from Agriculture, 1990, 2007, and 2008**

|  | 1990 | 2007 | 2008 |
|---|---|---|---|
| Estimated Emissions (Million Metric Tons $CO_2$e) | 196.4 | 213.9 | 217.9 |
| Change from 1990 (Million Metric Tons $CO_2$e) |  | 17.5 | 21.5 |
| *(Percent)* |  | *8.9%* | *10.9%* |
| Average Annual Change from 1990 *(Percent)* |  | *0.5%* | *0.6%* |
| Change from 2007 (Million Metric Tons $CO_2$e) |  |  | 4.0 |
| *(Percent)* |  |  | *1.9%* |

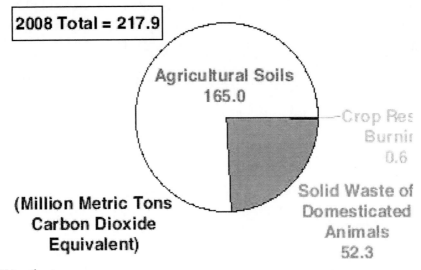

Source: EIA estimates.

Figure 21. U.S. Nitrous Oxide Emissions from Agriculture by Source, 2008

**Table 22. U.S. Nitrous Oxide Emissions from Agricultural Sources, 1990-2008**
**(Million Metric Tons Carbon Dioxide Equivalent)**

| Source | 1990 | 1995 | 2000 | 2002 | 2003 | 2004 | 2005 | 2006 | 2007 | 2008 |
|---|---|---|---|---|---|---|---|---|---|---|
| **Agricultural Soils** | | | | | | | | | | |
| *Direct Emissions* | | | | | | | | | | |
| Biological Fixation in Crops | 50.6 | 53.4 | 57.9 | 55.6 | 53.4 | 60.3 | 60.1 | 60.4 | 52.7 | 56.5 |
| Synthetic Nitrogen Fertilizers | 47.5 | 45.9 | 40.8 | 40.8 | 43.0 | 48.6 | 51.8 | 53.6 | 54.8 | 53.5 |
| Crop Residues | 22.0 | 21.9 | 27.0 | 25.7 | 25.6 | 30.2 | 29.3 | 28.9 | 28.1 | 29.9 |
| Other | 5.0 | 5.1 | 5.2 | 5.2 | 5.2 | 5.2 | 5.2 | 5.3 | 5.2 | 5.3 |
| *Total Direct Emissions* | *125.1* | *126.3* | *130.9* | *127.3* | *127.2* | *144.2* | *146.4* | *148.2* | *140.8* | *145.2* |
| *Indirect Emissions* | | | | | | | | | | |
| Soil Leaching | 11.1 | 10.7 | 9.6 | 9.6 | 10.0 | 11.3 | 12.0 | 12.4 | 12.7 | 12.4 |

## Table 22. (Continued)

| Source | 1990 | 1995 | 2000 | 2002 | 2003 | 2004 | 2005 | 2006 | 2007 | 2008 |
|---|---|---|---|---|---|---|---|---|---|---|
| Atmospheric Deposition | 6.6 | 6.4 | 5.7 | 5.7 | 6.0 | 6.8 | 7.2 | 7.4 | 7.6 | 7.7 |
| *Total Indirect Emissions* | *17.7* | *17.1* | *15.3* | *15.3* | *16.1* | *18.1* | *19.2* | *19.9* | *20.3* | *19.8* |
| **Solid Waste of Domesticated Animals** | | | | | | | | | | |
| Cattle | 49.0 | 51.1 | 48.2 | 47.7 | 47.4 | 47.0 | 47.5 | 47.9 | 47.9 | 47.8 |
| Swine | 1.3 | 1.4 | 1.4 | 1.4 | 1.4 | 1.4 | 1.4 | 1.4 | 1.5 | 1.6 |
| Poultry | 0.9 | 1.2 | 1.3 | 1.3 | 1.3 | 1.3 | 1.4 | 1.4 | 1.4 | 1.4 |
| Horses | 0.7 | 0.7 | 0.7 | 0.5 | 0.5 | 0.5 | 0.5 | 0.6 | 0.6 | 0.6 |
| Sheep | 1.1 | 0.8 | 0.6 | 0.6 | 0.5 | 0.5 | 0.5 | 0.5 | 0.5 | 0.5 |
| Goats | 0.3 | 0.3 | 0.2 | 0.2 | 0.2 | 0.2 | 0.4 | 0.4 | 0.4 | 0.4 |
| *Total Solid Waste* | *53.2* | *55.4* | *52.4* | *51.7* | *51.3* | *51.0* | *51.7* | *52.2* | *52.3* | *52.3* |
| *Crop Residue Burning* | *0.4* | *0.4* | *0.5* | *0.5* | *0.5* | *0.6* | *0.6* | *0.6* | *0.6* | *0.6* |
| **Total Agricultural Sources** | 196.4 | 199.2 | 199.1 | 194.8 | 195.1 | 213.9 | 217.9 | 220.8 | 213.9 | 217.9 |

Notes: Data in this table are revised from the data contained in the previous EIA report, *Emissions of Greenhouse Gases in the United States 2007*, DOE/EIA-0573(2007) (Washington, DC, December 2008). Totals may not equal sum of components due to independent rounding.

Source: EIA estimates.

# Energy Use

## *Summary*

- Emissions from energy sources made up about 21 percent of total U.S. nitrous oxide emissions in 2008. Nitrous oxide is a byproduct of fuel combustion in mobile and stationary sources (Figure 22).
- More than three-quarters of U.S. nitrous oxide emissions from energy use can be traced to mobile sources—motor vehicles, primarily passenger cars and light trucks (Table 23). Emissions from mobile sources dropped by 4.8 percent (2.4 MMTCO$_2$e) from 2007 to 2008, primarily because of a 5.2-percent decrease in emissions from passenger cars and light trucks. Vehicle miles traveled by passenger vehicles were 3.3 percent lower in 2008 than in 2007, as a result of higher gasoline prices and economic uncertainty.
- Nitrous oxide emissions from stationary combustion sources result predominantly from the burning of coal at electric power plants (9.2 MMTCO$_2$e, or 61.2 percent of all nitrous oxide from stationary combustion).

**U.S. Nitrous Oxide Emissions from Energy Use, 1990, 2007, and 2008**

|  | 1990 | 2007 | 2008 |
|---|---|---|---|
| Estimated Emissions (Million Metric Tons $CO_2e$) | 51.6 | 66.8 | 63.9 |
| Change from 1990 (Million Metric Tons $CO_2e$) |  | 15.2 | 12.3 |
| *(Percent)* |  | *29.4%* | *23.8%* |
| Average Annual Change from 1990 *(Percent)* |  | *1.5%* | *1.2%* |
| Change from 2007 (Million Metric Tons $CO_2e$) |  |  | -2.9 |
| *(Percent)* |  |  | *-4.3%* |

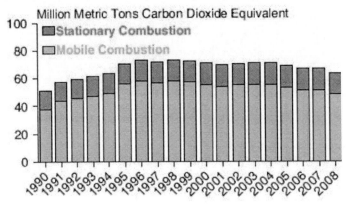

Source: EIA estimates.

Figure 22. U.S. Nitrous Oxide Emissions from Energy Use by Source, 1990-2008

**Table 23. U.S. Nitrous Oxide Emissions from Energy Use, 1990-2008**
**(Million Metric Tons Carbon Dioxide Equivalent)**

| Item | 1990 | 1995 | 2000 | 2002 | 2003 | 2004 | 2005 | 2006 | 2007 | 2008 |
|---|---|---|---|---|---|---|---|---|---|---|
| **Mobile Combustion** | | | | | | | | | | |
| Motor Vehicles | *34.1* | *52.3* | *51.3* | *51.8* | *52.3* | *52.0* | *49.2* | *47.7* | *47.2* | *44.9* |
| Passenger Cars | 21.7 | 31.4 | 29.3 | 28.7 | 28.3 | 27.2 | 25.5 | 24.1 | 23.5 | 21.9 |
| Light-DutyTrucks | 10.5 | 18.6 | 19.4 | 20.4 | 21.3 | 21.9 | 20.9 | 20.7 | 20.9 | 20.1 |
| Other Motor Vehicles | 1.8 | 2.2 | 2.6 | 2.7 | 2.7 | 2.8 | 2.8 | 2.8 | 2.8 | 2.8 |
| Other Mobile Sources | *3.6* | *3.6* | *3.9* | *3.6* | *3.4* | *3.8* | *3.9* | *4.0* | *4.0* | *3.9* |
| **Total** | 37.7 | 55.9 | 55.2 | 55.4 | 55.7 | 55.7 | 53.1 | 51.6 | 51.2 | 48.8 |
| **Stationary Combustion** | | | | | | | | | | |
| Residential and Commercial | 1.5 | 1.4 | 1.3 | 1.2 | 1.3 | 1.3 | 1.3 | 1.1 | 1.2 | 1.2 |

## Table 23. (Continued)

| Item | 1990 | 1995 | 2000 | 2002 | 2003 | 2004 | 2005 | 2006 | 2007 | 2008 |
|---|---|---|---|---|---|---|---|---|---|---|
| Industrial | 4.6 | 4.9 | 4.8 | 4.5 | 4.4 | 4.7 | 4.5 | 4.6 | 4.5 | 4.1 |
| Electric Power | 7.8 | 8.3 | 9.6 | 9.4 | 9.6 | 9.7 | 9.9 | 9.7 | 9.9 | 9.7 |
| **Total** | **13.9** | **14.5** | **15.7** | **15.1** | **15.3** | **15.6** | **15.7** | **15. 4** | **15.6** | **15.1** |
| **Total from Energy Use** | **51.6** | **70.4** | **71.0** | **70.4** | **71.0** | **71.4** | **68.8** | **67. 1** | **66.8** | **63.9** |

Notes: Data in this table are revised from the data contained in the previous EIA report, *Emissions of Greenhouse Gases in the United States 2007*, DOE/EIA-0573(2007) (Washington, DC, December 2008). Totals may not equal sum of components due to independent rounding.

Source: EIA estimates

## Industrial Sources

### *Summary*

- U.S. industrial sources emitted 15.1 MMTCO$_2$e of nitrous oxide in 2008, a decrease of 4.6 percent from 2007 (Table 24).
- The two industrial sources of nitrous oxide emissions are production of adipic acid and production of nitric acid.
- Nitric acid, a primary ingredient in fertilizers, usually is manufactured by oxidizing ammonia with a platinum catalyst. The oxidation process releases nitrous oxide emissions.
- Adipic acid is a fine white powder used primarily in the manufacture of nylon fibers and plastics. The three companies operating the U.S. plants manufacture adipic acid by oxidizing a ketone-alcohol mixture with nitric acid. The chemical reaction results in nitrous oxide emissions.
- A large decline in nitrous oxide emissions from industrial processes since 1996 (Figure 23) is a result of emissions control technology at three of the four adipic acid plants operating in the United States.

### U.S. Nitrous Oxide Emissions from Industrial Sources, 1990, 2007, and 2008

| | 1990 | 2007 | 2008 |
|---|---|---|---|
| Estimated Emissions (Million Metric Tons CO$_2$e) | 28.8 | 15.9 | 15.1 |
| Change from 1990 (Million Metric Tons CO$_2$e) | | -12.9 | -13.6 |
| *(Percent)* | | *-44.8%* | *-47.3%* |
| Average Annual Change from 1990 *(Percent)* | | *-3.4%* | *-3.5%* |
| Change from 2007 (Million Metric Tons CO$_2$e) | | | -0.7 |
| *(Percent)* | | | *-4.6%* |

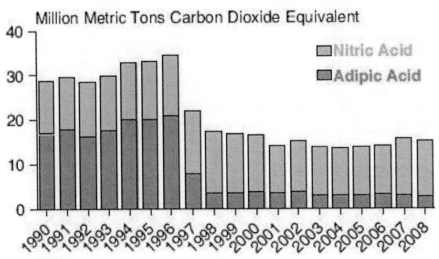

Source: EIA estimates.

Figure 23. U.S. Nitrous Oxide Emissions from Industry by Source, 1990-2008

**Table 24. U.S. Nitrous Oxide Emissions from Industrial Sources, 1990-2008**
**(Million Metric Tons Carbon Dioxide Equivalent)**

| Source | 1990 | 1995 | 2000 | 2002 | 2003 | 2004 | 2005 | 2006 | 2007 | 2008 |
|---|---|---|---|---|---|---|---|---|---|---|
| **Adipic Acid** | | | | | | | | | | |
| Controlled Sources | 1.0 | 1.1 | 1.6 | 1.6 | 1.6 | 1.7 | 1.6 | 1.7 | 1.6 | 1.5 |
| Uncontrolled Sources | 16.0 | 18.9 | 2.1 | 2.3 | 1.4 | 1.4 | 1.4 | 1.5 | 1.4 | 1.3 |
| **Subtotal** | 17.0 | 20.0 | 3.7 | 3.9 | 3.0 | 3.1 | 3.0 | 3.2 | 3.0 | 2.8 |
| **NitricAcid** | 11.8 | 13.1 | 12.9 | 11.4 | 10.7 | 10.6 | 11.0 | 10.8 | 12.8 | 12.3 |
| **Total Known Industrial Sources** | 28.8 | 33.1 | 16.7 | 15.3 | 13.7 | 13.7 | 14.0 | 14.0 | 15.9 | 15.1 |

Note: Data in this table are revised from the data contained in the previous EIA report, *Emissions of Greenhouse Gases in the United States 2007*, DOE/EIA-0573(2007) (Washington, DC, December 2008). Totals may not equal sum of components due to independent rounding.
Source: EIA estimates.

## Waste Management

### *Summary*

- In 2008, treatment of residential and commercial wastewater produced 89.5 percent (3.0 MMTCO$_2$e) of all nitrous oxide emissions from waste management. An additional 0.4 MMTCO$_2$e was emitted from the combustion of municipal solid waste (Figure 24 and Table 25).

- Estimates of nitrous oxide emissions from wastewater are directly related to population size and per-capita intake of protein.
- Nitrous oxide is emitted from wastewater that contains nitrogen-based organic materials, such as those found in human or animal waste. Factors that influence the amount of nitrous oxide generated from wastewater include temperature, acidity, biochemical oxygen demand, and nitrogen concentration.
- The emissions factor for nitrous oxide from sewage, which has been updated for the 2008 inventory in accordance with the IPCC's 2006 inventory guidelines, is 50 percent lower than in previous reports, resulting in a corresponding decline in reported nitrous oxide emissions from wastewater.

**U.S. Anthropogenic Nitrous Oxide Emissions from Waste Management, 1990, 2007, and 2008**

|  | 1990 | 2007 | 2008 |
|---|---|---|---|
| Estimated Emissions (Million Metric Tons CO$_2$e) | 2.6 | 3.4 | 3.4 |
| Change from 1990 (Million Metric Tons CO$_2$e) |  | 0.7 | 0.8 |
| *(Percent)* |  | *28.1%* | *29.5%* |
| Average Annual Change from 1990 *(Percent)* |  | *1.5%* | *1.4%* |
| Change from 2007 (Million Metric Tons CO$_2$e) |  |  | * |
| *(Percent)* |  |  | *1.1%* |

*Less than 0.05 million metric tons.

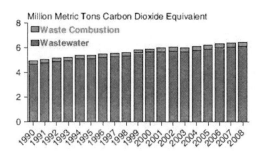

Source: EIA estimates.

Figure 24. U.S. Nitrous Oxide Emissions from Waste Management by Source, 1990-2008

**Table 25. U.S. Nitrous Oxide Emissions from Waste Management, 1990-2008 (Million Metric Tons Carbon Dioxide Equivalent)**

| Source | 1990 | 1995 | 2000 | 2002 | 2003 | 2004 | 2005 | 2006 | 2007 | 2008 |
|---|---|---|---|---|---|---|---|---|---|---|
| Domestic and Commercial Wastewater | 2.3 | 2.5 | 2.8 | 2.9 | 2.8 | 2.9 | 2.9 | 3.0 | 3.0 | 3.0 |
| Waste Combustion. | 0.3 | 0.3 | 0.3 | 0.3 | 0.3 | 0.3 | 0.4 | 0.3 | 0.3 | 0.4 |
| **Total** | **2.6** | **2.8** | **3.1** | **3.2** | **3.2** | **3.2** | **3.3** | **3.3** | **3.4** | **3.4** |

Note: Data in this table are revised from the data contained in the previous EIA report, *Emissions of Greenhouse Gases in the United States 2007*, DOE/EIA-0573(2007) (Washington, DC, December 2008). Totals may not equal sum of components due to independent rounding.

Source: EIA estimates.

# HIGH-GWP GASES

## Total Emissions

*Summary*

- Greenhouse gases with high global warming potential (high-GWP gases) are hydrofluorocarbons (HFCs), perfluorocarbons (PFCs), and sulfur hexafluoride ($SF_6$), which together represented 2.5 percent of U.S. greenhouse gas emissions in 2008.
- Emissions estimates for the high-GWP gases are provided to EIA by the EPA's Office of Air and Radiation. The estimates are derived from the EPA Vintaging Model.
- For this year's EIA inventory, 2007 values for PFCs and $SF_6$ are used as placeholders. The updated values will be available when the U.S. inventory is submitted to the UNFCCC in April 2010.
- Emissions of high-GWP gases have increased steadily since 1990 (Figure 25 and Table 26), largely because HFCs are being used to replace chlorofluorocarbons (CFCs), hydrochlorofluorocarbons (HCFCs), and other ozone-depleting substances that are being phased out under the terms of the Montreal Protocol, which entered into force on January 1, 1989.
- PFC emissions have declined since 1990 as a result of production declines in the U.S. aluminum industry as well as industry efforts to lower emissions per unit of output.
- Emissions of "other" HFCs, which are aggregated to protect confidential data, have been updated for 2008, showing a 4.4-percent increase from 2007.

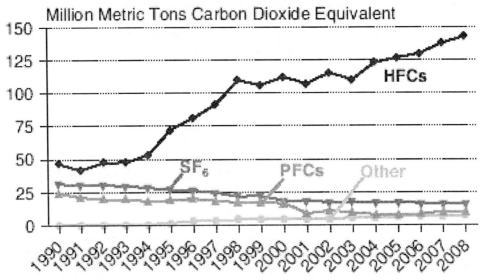

Source: U.S. EPA estimates.

Figure 25. U.S. Anthropogenic Emissions of High-GWP Gases, 1990-2008

# Emissions of Greenhouse Gases in the United States 2008

## U.S. Anthropogenic Emissions of High-GWP Gases, 1990, 2007, and 2008

|  | 1990 | 2007 | 2008 |
|---|---|---|---|
| Estimated Emissions (Million Metric Tons $CO_2e$) | 102.3 | 170.3 | 175.6 |
| Change from 1990 (Million Metric Tons $CO_2e$) |  | 68.0 | 73.3 |
| *(Percent)* |  | *66.4%* | *71.7%* |
| Average Annual Change from 1990 *(Percent)* |  | *3.0%* | *3.0%* |
| Change from 2007 (Million Metric Tons $CO_2e$) |  |  | 5.4 |
| *(Percent)* |  |  | *3.1%* |

**Table 26. U.S. Emissions of Hydrofluorocarbons, Perfluorocarbons, and Sulfur Hexafluoride, 1990-2008 (Million Metric Tons Carbon Dioxide Equivalent)**

| Gas | 1990 | 1995 | 2000 | 2002 | 2003 | 2004 | 2005 | 2006 | 2007 | 2008 |
|---|---|---|---|---|---|---|---|---|---|---|
| Hydro-fluoro-carbons | 46.3 | 72.0 | 111.8 | 115.1 | 110.0 | 122.8 | 126.6 | 129.8 | 137.7 | 142.7 |
| Perfluoro-carbons | 24.4 | 18.6 | 16.3 | 11.0 | 9.1 | 8.0 | 8.0 | 8.1 | 10.1 | 10.1 |
| Other HFCs, PFCs/ PFPEs | 0.3 | 1.6 | 4.1 | 4.7 | 5.4 | 5.8 | 6.1 | 6.4 | 6.8 | 7.1 |
| Sulfur Hexa-fluoride | 31.3 | 26.8 | 18.3 | 17.3 | 17.4 | 16.9 | 17.1 | 16.2 | 15.8 | 15.8 |
| **Total Emissions** | **102.3** | **119.0** | **150.5** | **148.1** | **141.8** | **153.5** | **157.8** | **160.5** | **170.3** | **175.6** |

Notes: Other HFCs, PFCs/PFPEs include HFC-152a, HFC-227ea, HFC-245fa, HFC-4310mee, and a variety of PFCs and perfluoropolyethers (PFPEs). They are grouped together to protect confidential data. Totals may not equal sum of components due to independent rounding.

Source: U.S. Environmental Protection Agency, Office of Air and Radiation, web site www.epa.gov/globalwarming/ (preliminary estimates, November 2009 for hydrofluorocarbons and other HFCs, PFCs/PFPEs; November 2008 for perfluorocarbons and sulfur hexafluoride).

## Hydrofluorocarbons

*Summary*

- HFCs are compounds that contain carbon, hydrogen, and fluorine. Although they do not destroy stratospheric ozone, they are powerful greenhouse gases.
- HFCs are used as solvents, residential and commercial refrigerants, firefighting agents, and propellants for aerosols.
- Emissions of substitutes for ozone-depleting substances, including HFC-32, HFC-125, HFC-134a, and HFC-236fa, have grown from trace amounts in 1990 to nearly 143 MMTCO$_2$e in 2008 (Table 27).

- Nearly 90 percent of the growth in HFC emissions since 1990 can be attributed to the use of HFCs as replacements for ozone-depleting substances. The market is expanding, with HFCs used in fire protection applications to replace Halon 1301 and Halon 1211.
- Since 2000, HFC-134a—used as a replacement for CFCs in air conditioners for passenger vehicles, trains, and buses—has accounted for the largest share of HFC emissions (Figure 26).
- Under the Clean Air Act, manufacture and import of HCFC-22, except for use as a feedstock and in equipment manufacture before 2010, are scheduled to be phased out by January 1, 2010. Manufacturers of HCFC-22 are using cost-effective methods to make voluntary reductions in the amount of HFC-23 that is created as a byproduct of HCFC-22 manufacture; however, HCFC-22 production remains a large and steady source of U.S. emissions of HFC-23.

**U.S. Anthropogenic Emissions of HFCs, 1990, 2007, and 2008**

|  | 1990 | 2007 | 2008 | 1990 |
|---|---|---|---|---|
| Estimated Emissions (Million Metric Tons $CO_2e$) |  | 46.3 | 137.7 | 142.7 |
| Change from 1990 (Million Metric Tons $CO_2e$) |  |  | 91.4 | 96.4 |
| *(Percent)* |  |  | *197.4%* | *208.3%* |
| Average Annual Change from 1990 *(Percent)* |  |  | *6.6%* | *6.5%* |
| Change from 2007 (Million Metric Tons $CO_2e$) |  |  |  | 5.0 |
| *(Percent)* |  |  |  | *3.7%* |

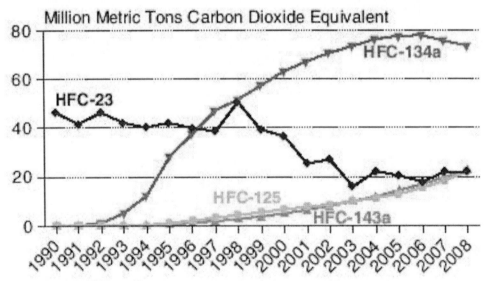

Source: U.S. EPA estimates.

Figure 26. U.S. Anthropogenic Emissions of HFCs, 1990-2008

Emissions of Greenhouse Gases in the United States 2008 203

**Table 27. U.S. Emissions of Hydrofluorocarbons, 1990-2008**
**(Million Metric Tons Carbon Dioxide Equivalent)**

| Gas | 1990 | 1995 | 2000 | 2002 | 2003 | 2004 | 2005 | 2006 | 2007 | 2008 |
|---|---|---|---|---|---|---|---|---|---|---|
| HFC-23 | 46.3 | 42.1 | 36.5 | 26.9 | 15.8 | 22.1 | 20.3 | 17.9 | 21.9 | 21.9 |
| HFC-32 | 0.0 | 0.0 | 0.0 | 0.1 | 0.2 | 0.3 | 0.4 | 0.6 | 0.9 | 1.2 |
| HFC-125 | 0.0 | 1.0 | 6.6 | 8.5 | 9.8 | 11.3 | 12.9 | 15.4 | 18.4 | 22.1 |
| HFC-134a | 0.0 | 27.9 | 62.9 | 70.5 | 73.3 | 76.2 | 77.5 | 77.7 | 75.5 | 73.6 |
| HFC-143a | 0.0 | 0.6 | 4.9 | 8.0 | 9.8 | 11.9 | 14.3 | 16.9 | 19.7 | 22.5 |
| HFC-36fa | 0.0 | 0.4 | 0.8 | 1.0 | 1.1 | 1.2 | 1.2 | 1.3 | 1.3 | 1.4 |
| **Total HFCs** | **46.3** | **72.0** | **111.8** | **115.1** | **110.0** | **122.8** | **126.6** | **129.8** | **137.7** | **142.7** |

Note: Totals may not equal sum of components due to independent rounding.

Source: U.S. Environmental Protection Agency, Office of Air and Radiation, web site www.epa.gov/globalwarming/ (preliminary estimates, November 2009).

## Perfluorocarbons

### *Summary*

- The two principal sources of PFC emissions are domestic aluminum production and semiconductor manufacture, which yield perfluoromethane ($CF_4$) and perfluoroethane ($C_2F_6$) (Figure 27 and Table 28).
- While PFC emissions from aluminum production have declined markedly since 1990, the decline has been offset in part by increased emissions from semiconductor manufacturing.
- Emissions from process inefficiencies during aluminum production (known as "anode effects") have been greatly reduced; in addition, high costs for alumina and energy have led to production cutbacks.
- Perfluoroethane is used as an etchant and cleaning agent in semiconductor manufacturing. The portion of the gas that does not react with the materials is emitted to the atmosphere.

### U.S. Anthropogenic Emissions of PFCs, 1990, 2007, and 2008

| | 1990 | 2007 | 2008 |
|---|---|---|---|
| Estimated Emissions (Million Metric Tons $CO_2e$) | 24.4 | 10.1 | NA |
| Change from 1990 (Million Metric Tons $CO_2e$) | | -14.4 | NA |
| *(Percent)* | | *-58.8%* | *NA* |
| Average Annual Change from 1990 *(Percent)* | | *-5.1%* | *NA* |
| Change from 2007 (Million Metric Tons $CO_2e$) | | | NA |
| *(Percent)* | | | *NA* |

NA = 2008 data not yet available.

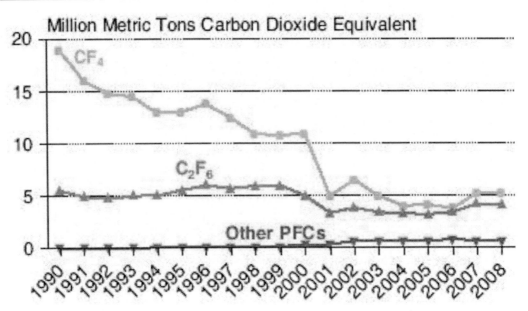

Source: U.S. EPA estimates.

Figure 27. U.S. Anthropogenic Emissions of PFCs, 1990-2008

**Table 28. U.S. Emissions of Perfluorocarbons, 1990-2008
(Million Metric Tons Carbon Dioxide Equivalent)**

| Gas | 1990 | 1995 | 2000 | 2002 | 2003 | 2004 | 2005 | 2006 | 2007 | 2008 |
|---|---|---|---|---|---|---|---|---|---|---|
| $CF_4$ | 18.9 | 13.0 | 10.9 | 6.5 | 4.9 | 4.0 | 4.1 | 3.8 | 5.2 | 5.2 |
| $C_2F_6$ | 5.5 | 5.6 | 5.0 | 3.8 | 3.4 | 3.3 | 3.2 | 3.4 | 4.2 | 4.2 |
| $NF_3$ | * | 0.1 | 0.2 | 0.5 | 0.5 | 0.5 | 0.4 | 0.7 | 0.5 | 0.5 |
| $C_3F_8$ | * | * | 0.2 | 0.1 | 0.1 | 0.1 | * | * | 0.1 | 0.1 |
| $C_4F_8$ | * | * | * | 0.1 | 0.1 | 0.1 | 0.1 | 0.1 | 0.1 | 0.1 |
| **Total PFCs** | **24.4** | **18.6** | **16.3** | **11.0** | **9.1** | **8.0** | **8.0** | **8.1** | **10.1** | **10.1** |

*Less than 0.05 million metric tons carbon dioxide equivalent.

Note: Totals may not equal sum of components due to independent rounding.

Source: U.S. Environmental Protection Agency, Office of Air and Radiation, web site www.epa.gov/globalwarming/ (estimates, November 2008); 2007 values are used as proxies for 2008.

## Sulfur Hexafluoride

### Summary

- $SF_6$, an excellent dielectric gas for high-voltage applications, is used primarily in electrical applications— as an insulator and arc interrupter for circuit breakers, switch gear, and other equipment in electricity transmission and distribution systems.
- Industry efforts to reduce emissions of $SF_6$ from electrical power systems have led to a decline in emissions since 1990 (Figure 28 and Table 29).

- $SF_6$ is also used in magnesium metal casting, as a cover gas during magnesium production, and as an atmospheric tracer for experimental purposes.
- Other, minor applications of $SF_6$ include leak detection and the manufacture of loudspeakers and lasers.

**U.S. Anthropogenic Emissions of $SF_6$, 1990, 2007, and 2008**

|  | 1990 | 2007 | 2008 |
|---|---|---|---|
| Estimated Emissions (Million Metric Tons CO₂e) | 31.3 | 15.8 | NA |
| Change from 1990 (Million Metric Tons CO₂e) |  | -15.5 | NA |
| *(Percent)* |  | *-49.5%* | *NA* |
| Average Annual Change from 1990 *(Percent)* |  | *-3.9%* | *NA* |
| Change from 2007 (Million Metric Tons CO₂e) |  |  | NA |
| *(Percent)* |  |  | *NA* |

NA = 2008 data not yet available.

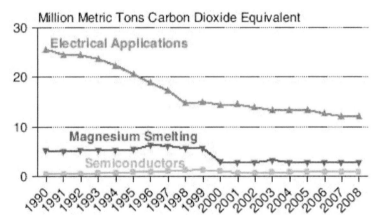

Source: U.S. EPA estimates.

Figure 28. U.S. Anthropogenic Emissions of $SF_6$ by Source, 1990-2008

**Table 29. U.S. Emissions of Sulfur Hexafluoride by Source, 1990-2008**
**(Million Metric Tons Carbon Dioxide Equivalent)**

| Source | 1990 | 1995 | 2000 | 2002 | 2003 | 2004 | 2005 | 2006 | 2007 | 2008 |
|---|---|---|---|---|---|---|---|---|---|---|
| Electrical Applications | 25.6 | 20.6 | 14.4 | 13.9 | 13.3 | 13.4 | 13.4 | 12.6 | 12.1 | 12.1 |
| Magnesium Smelting | 5.2 | 5.4 | 2.9 | 2.8 | 3.3 | 2.8 | 2.8 | 2.7 | 2.8 | 2.8 |
| Semiconductors | 0.5 | 0.9 | 1.0 | 0.6 | 0.8 | 0.8 | 0.9 | 0.9 | 0.9 | 0.9 |
| Total $SF_6$ | 31.3 | 26.8 | 18.3 | 17.3 | 17.4 | 16.9 | 17.1 | 16.2 | 15.8 | 15.8 |

Note: Totals may not equal sum of components due to independent rounding.
Source: U.S. Environmental Protection Agency, Office of Air and Radiation, web site www.epa.gov/globalwarming/ (estimates, November2009); 2007 values are used as proxies for 2008.

# LAND USE

## Overview

### Summary

- In 2007, land use, land-use change, and forests accounted for net carbon sequestration of 1,062.6 MMTCO$_2$e (Table 30), representing 17.7 percent of total U.S. CO$_2$ emissions.
- Net carbon sequestration from land use, land-use change, and forestry activities in 2007 was about 26 percent greater than in 1990 (Figure 29). The increase resulted primarily from a higher average annual rate of net carbon accumulation in forest carbon stocks.
- Sequestration from land use, land-use change, and forestry peaked in 2004 at 1,294.6 MMTCO$_2$e. By 2006 it had fallen to 1,050.5 MMTCO$_2$e, and in 2007 it rose slightly to 1,062.6 MMTCO$_2$e.
- Because forest land is the predominant category, its fluctuations drive the total. It is the category with the largest change since 1990 in absolute terms (249 MMTCO$_2$e).
- The largest percentage changes from 1990 to 2007 were seen for urban trees, which increased by 61 percent, and for grasslands remaining grasslands, which decreased by 91 percent.
- Among the categories of estimated greenhouse gas emissions, land-use change has a relatively high degree of uncertainty. As discussed on the following pages, the estimated values are highly dependent on the estimation methods used. Thus, countries that submit their emissions estimates to the UNFCCC generally provide two estimates: without land-use change (gross emissions) and with land-use change (net emissions).

**U.S. Carbon Sequestration from Land Use, Land-Use Change and Forestry, 1990, 2006, and 2007**

|  | 1990 | 2006 | 2007 |
|---|---|---|---|
| Estimated Sequestration (Million Metric Tons CO$_2$e) | 841.4 | 1,050.5 | 1,062.6 |
| Change from 1990 (Million Metric Tons CO$_2$e) |  | 209.1 | 221.1 |
| *(Percent)* |  | *24.9%* | *26.3%* |
| Average Annual Change from 1990 *(Percent)* |  | *1.4%* | *1.4%* |
| Change from 2006 (Million Metric Tons CO$_2$e) |  |  | 12.0 |
| *(Percent)* |  |  | *1.1%* |

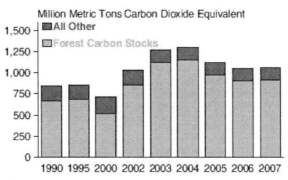

Source: U.S. EPA estimates.

Figure 29. U.S. Carbon Sequestration from Land Use, Land-Use Change, and Forestry, 1990-2007

**Table 30. Net U.S. Carbon Dioxide Sequestration from Land Use, Land-Use Change, and Forestry, 1990-2007 (Million Metric Tons Carbon Dioxide Equivalent)**

| Component | 1990 | 1995 | 2000 | 2002 | 2003 | 2004 | 2005 | 2006 | 2007 |
|---|---|---|---|---|---|---|---|---|---|
| Forest Land Remaining Forest Land[a] | 661.1 | 686.6 | 512.6 | 852.1 | 1,123.4 | 1,150.2 | 975.7 | 900.3 | 910.1 |
| Cropland Remaining Cropland[b] | 29.4 | 22.9 | 30.2 | 11.5 | 17.7 | 18.1 | 18.3 | 19.1 | 19.7 |
| Land Converted to Cropland | -2.2 | -2.9 | -2.4 | -4.8 | -5.9 | -5.9 | -5.9 | -5.9 | -5.9 |
| Grassland Remaining Grassland | 46.7 | 36.4 | 51.4 | 43.1 | 4.5 | 4.5 | 4.6 | 4.6 | 4.7 |
| Land Converted to Grassland | 22.3 | 22.5 | 32.0 | 28.3 | 26.7 | 26.7 | 26.7 | 26.7 | 26.7 |
| Settlements Remaining Settlements: | | | | | | | | | |
| Urban Trees[c] | 60.6 | 71.5 | 82.4 | 86.8 | 88.9 | 91.1 | 93.3 | 95.5 | 97.6 |
| Other: | | | | | | | | | |
| Landfilled Yard Trimmings and Food Scraps | 23.5 | 13.9 | 11.3 | 11.8 | 10.1 | 9.9 | 10.2 | 10.4 | 9.8 |
| Total Net Flux | 841.4 | 851.0 | 717.5 | 1,028.7 | 1,265.4 | 1,294.6 | 1,122.7 | 1,050.5 | 1,062.6 |

[a]Estimates include carbon stock changes in both Forest Land Remaining Forest Land and Land Converted to Forest Land.

[b]Estimates include carbon stock changes in mineral soils and organic soils on Cropland Remaining Cropland and liming emissions from all Cropland, Grassland, and Settlement categories.

[c]Estimates include carbon stock changes in both Settlements Remaining Settlements and Land Converted to Settlements.

Note: Totals may not equal sum of components due to independent rounding.

Source: U.S. Environmental Protection Agency, *Inventory of U.S. Greenhouse Gas Emissions and Sinks: 1990-2007*, EPA 430-R-09-004 (Washington, DC, April 2009), web site www.epa.gov/climatechange/emissions/usinventoryreport.html.

## Forest Lands and Harvested Wood Pools

### Summary

- Carbon sequestration attributed to forest land remaining forest land in 2007 totaled 910.1 MMTCO$_2$e (Figure 30 and Table 31). According to the EPA, that total has a 95-percent probability of being between 736 and 1,083 MMTCO$_2$e.
- Changes in underlying data from the USDA Forest Service Forestry Inventory Analysis Database (FIADB) have led to changes in estimates across the time series. Most States have added new data or modified existing data in the inventory.
- Version 3.0 of the FIADB includes more use of moving averages, which affect extrapolations of stocks and stock changes. The major changes in estimates for 1990-2007 relative to 1990-2006 is a spike in the 2002-2006 estimates for forest carbon.
- According to the USDA, average carbon densities have been updated from 331 million grams per hectare (Mg/ha) to 179 Mg/ha for Alaska and from 89 Mg/ha to 91 Mg/ha for the lower 48 States.
- The largest changes for the 1990-2007 period are in the estimates for soil organic carbon and dead wood, which increased by 153 percent and 139 percent, respectively.

**Carbon Sequestration in U.S. Forest Lands and Harvested Wood Pools, 1990, 2006, and 2007**

|  | 1990 | 2006 | 2007 |
|---|---|---|---|
| Estimated Sequestration (Million Metric Tons CO$_2$e) | 661.1 | 900.3 | 910.1 |
| Change from 1990 (Million Metric Tons CO$_2$e) |  | 239.2 | 249.0 |
| *(Percent)* |  | *36.2%* | *37.7%* |
| Average Annual Change from 1990 *(Percent)* |  | *2.0%* | *1.9%* |
| Change from 2006 (Million Metric Tons CO$_2$e) |  |  | 9.8 |
| *(Percent)* |  |  | *1.1%* |

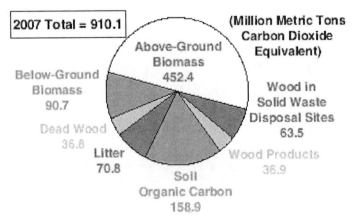

Source: U.S. EPA estimates.

Figure 30. Carbon Sequestration in U.S. Forest Lands and Harvested Wood Pools, 2007

# Emissions of Greenhouse Gases in the United States 2008

**Table 31. Net Carbon Dioxide Sequestration in U.S. Forests and Harvested Wood Pools, 1990-2007 (Million Metric Tons Carbon Dioxide Equivalent)**

| Carbon Pool | 1990 | 1995 | 2000 | 2002 | 2003 | 2004 | 2005 | 2006 | 2007 |
|---|---|---|---|---|---|---|---|---|---|
| **Forests** | **529.3** | **568.2** | **399.7** | **754.1** | **1,028.6** | **1,044.8** | **871.7** | **791.7** | **809.6** |
| Above-Ground Biomass | 321.5 | 390.9 | 352.1 | 438.9 | 529.2 | 531.6 | 469.4 | 442.7 | 452.4 |
| Below-Ground Biomass | 61.8 | 78.2 | 71.5 | 88.4 | 105.7 | 105.6 | 93.3 | 88.9 | 90.7 |
| Dead Wood | 15.4 | 27.3 | 18.2 | 31.2 | 42.6 | 44.5 | 39.4 | 35.6 | 36.8 |
| Litter | 67.8 | 37.2 | 14.8 | 68.5 | 92.2 | 93.0 | 79.6 | 68.7 | 70.8 |
| Soil Orgaic Carbon | 62.8 | 34.6 | -56.9 | 127.1 | 258.8 | 270.1 | 190.1 | 155.9 | 158.9 |
| **Harvested Wood** | **131.8** | **118.4** | **112.9** | **98.0** | **94.8** | **105.4** | **103.9** | **108.6** | **100.4** |
| Wood Products | 64.8 | 55.2 | 47.0 | 34.9 | 35.5 | 45.6 | 44.1 | 45.2 | 36.9 |
| Wood in Solid Waste Disposal Sites | 67.0 | 63.2 | 65.9 | 63.1 | 59.3 | 59.8 | 59.8 | 63.3 | 63.5 |
| **Total** | **661.1** | **686.6** | **512.6** | **852.1** | **1,123.4** | **1,150.2** | **975.7** | **900.3** | **910.1** |

Notes: The sums of the annual net stock changes in this table (shown in the "Total" row) represent estimates of the actual net flux between the total forest carbon pool and the atmosphere. Forest estimates are based on periodic measurements; harvested wood estimates are based on annual surveys and models. Totals may not equal sum of components due to independent rounding.

Source: U.S. Environmental Protection Agency, *Inventory of U.S. Greenhouse Gas Emissions and Sinks: 1990-2007*, EPA 430-R-09-004 (Washington, DC, April 2009), web site www.epa.gov/climatechange/emissions/usinventoryreport.html.

## Croplands and Grasslands

### Summary

- For 2007, aggregate carbon flux for the four agricultural categories shown in Table 32 was 45.1 MMTCO$_2$e.
- Land converted to grassland, cropland remaining cropland, and grassland remaining grassland sequestered 26.7, 19.7, and 4.7 MMTCO$_2$e, respectively, in 2007. Land converted to cropland emitted 5.9 MMTCO$_2$e (Figure 31 and Table 32).
- In the USDA National Resources Inventory (NRI), land is classified as cropland remaining cropland in a given year if it has been used as cropland for 20 years.
- In previous inventory reports, land use data for 1997 were used for all subsequent years; however, the availability of new NRI data has extended the information on

land use from 1997 to 2003. In Table 32, the data for 2004-2007 are extrapolations, except for the data on liming of soils, which extend through 2006.
- Annual area data (rather than 5-year increments) were used to estimate soil carbon stock changes for the current inventory, leading to more accurate estimates; and each NRI point was simulated separately, instead of simulating clusters of common cropping rotation histories and soil characteristics in a county.
- NRI area data were reconciled with the forest area estimates in the FIADB data and incorporated into the estimation of soil carbon stock changes, leading to adjustments in grassland areas in the NRI data, including land converted to grassland.

**Carbon Sequestration in U.S. Croplands and Grasslands, 1990, 2006, and 2007**

|  | 1990 | 2006 | 2007 |
|---|---|---|---|
| Estimated Sequestration (Million Metric Tons $CO_2e$) | 96.3 | 44.5 | 45.1 |
| Change from 1990 (Million Metric Tons $CO_2e$) |  | -51.8 | -51.2 |
| *(Percent)* |  | *-53.8%* | *-53.2%* |
| Average Annual Change from 1990 *(Percent)* |  | *-4.7%* | *-4.4%* |
| Change from 2006 (Million Metric Tons $CO_2e$) |  |  | 0.6 |
| *(Percent)* |  |  | *1.4%* |

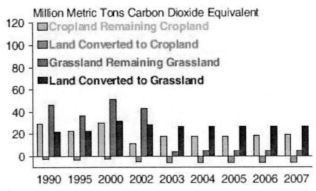

Source: U.S. EPA estimates.

Figure 31. Carbon Sequestration in U.S. Croplands and Grasslands, 1990-2007

**Table 32. Net Carbon Dioxide Sequestration in U.S. Croplands and Grasslands, 1990-2007 (Million Metric Tons Carbon Dioxide Equivalent)**

| Carbon Pool | 1990 | 1995 | 2000 | 2002 | 2003 | 2004 | 2005 | 2006 | 2007 |
|---|---|---|---|---|---|---|---|---|---|
| **Cropland Remaining Cropland** | **29.4** | **22.9** | **30.2** | **11.5** | **17.7** | **18.1** | **18.3** | **19.1** | **19.7** |
| Mineral Soils | 56.8 | 50.6 | 57.9 | 39.2 | 45.3 | 45.8 | 45.9 | 46.8 | 47.3 |
| Organic Soils | -27.4 | -27.7 | -27.7 | -27.7 | -27.7 | -27.7 | -27.7 | -27.7 | -27.7 |
| **Land Converted to Cropland** | **-2.2** | **-2.9** | **-2.4** | **-4.8** | **-5.9** | **-5.9** | **-5.9** | **-5.9** | **-5.9** |
| Mineral Soils | 0.3 | -0.3 | 0.3 | -2.2 | -3.3 | -3.3 | -3.3 | -3.3 | -3.3 |

# Emissions of Greenhouse Gases in the United States 2008

**Table 32. (Continued)**

| Carbon Pool | 1990 | 1995 | 2000 | 2002 | 2003 | 2004 | 2005 | 2006 | 2007 |
|---|---|---|---|---|---|---|---|---|---|
| Organic Soils | -2.4 | -2.6 | -2.6 | -2.6 | -2.6 | -2.6 | -2.6 | -2.6 | -2.6 |
| **Grassland Remaning Grassland** | **46.7** | **36.4** | **51.4** | **43.1** | **4.5** | **4.5** | **4.6** | **4.6** | **4.7** |
| Mineral Soils | 50.6 | 40.1 | 55.1 | 46.8 | 8.2 | 8.2 | 8.3 | 8.3 | 8.4 |
| Organic Soils | -3.9 | -3.7 | -3.7 | -3.7 | -3.7 | -3.7 | -3.7 | -3.7 | -3.7 |
| **Land Converted to Grassland** | **22.3** | **22.5** | **32.0** | **28.3** | **26.7** | **26.7** | **26.7** | **26.7** | **26.7** |
| Mineral Soils | 22.7 | 23.4 | 32.8 | 29.1 | 27.6 | 27.6 | 27.6 | 27.6 | 27.6 |
| Organic Soils | -0.5 | -0.9 | -0.9 | -0.9 | -0.9 | -0.9 | -0.9 | -0.9 | -0.9 |
| **Total Sequestration** | **96.3** | **78.9** | **111.2** | **78.1** | **42.9** | **43.4** | **43.6** | **44.5** | **45.1** |
| Liming of Soils | -4.7 | -4.4 | -4.3 | -5.0 | -4.6 | -3.9 | -4.3 | -4.2 | -4.1 |

Note: Negative values indicate emissions.

Source: U.S. Environmental Protection Agency, *Inventory of U.S. Greenhouse Gas Emissions and Sinks: 1990-2007*, EPA 430-R-09-004 (Washington, DC, April 2009), web site www.epa.gov/climatechange/emissions/usinventoryreport.html.

## Urban Trees, Yard Trimmings, and Food Scraps

### *Summary*

- Urban trees, yard trimmings, and food scraps sequestered 107.4 $MMTCO_2e$ in 2007 (Figure 32 and Table 33).
- Updated data from *Municipal Solid Waste Generation, Recycling, and Disposal in the United States: 2007 Facts and Figures*,[17] used in this chapter—including revisions to the amounts of food scraps generated in 2000 and 2004-2007—resulted in an average 1.0-percent decrease in carbon storage across the time series for yard trimmings and food scraps.
- Carbon sequestration in yard trimmings showed the largest change from 1990 to 2007, with a decrease of 70 percent attributable to an increase in municipal collection and composting of yard trimmings.

### Carbon Sequestration in U.S. Urban Trees, Yard Trimmings, and Food Scraps, 1990, 2006, and 2007

| | 1990 | 2006 | 2007 |
|---|---|---|---|
| Estimated Sequestration (Million Metric Tons $CO_2e$) | 84.1 | 105.8 | 107.4 |
| Change from 1990 (Million Metric Tons $CO_2e$) | | 21.7 | 23.3 |
| *(Percent)* | | *25.8%* | *27.7%* |
| Average Annual Change from 1990 *(Percent)* | | *1.4%* | *1.4%* |
| Change from 2006 (Million Metric Tons $CO_2e$) | | | 1.6 |
| *(Percent)* | | | *1.5%* |

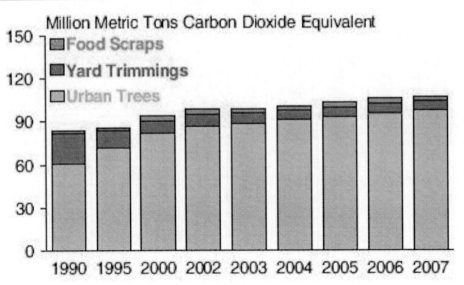

Source: U.S. EPA estimates.

Figure 32. Carbon Sequestration in U.S. Urban Trees, Yard Trimmings, and Food Scraps, 1990-2007

**Table 33. Net Carbon Dioxide Sequestration in U.S. Urban Trees, Yard Trimmings, and Food Scraps, 1990-2007 (Million Metric Tons Carbon Dioxide Equivalent)**

| Carbon Pool | 1990 | 1995 | 2000 | 2002 | 2003 | 2004 | 2005 | 2006 | 2007 |
|---|---|---|---|---|---|---|---|---|---|
| Urban Trees | 60.6 | 71.5 | 82.4 | 86.8 | 88.9 | 91.1 | 93.3 | 95.5 | 97.6 |
| Yard Trimmings | 21.2 | 12.5 | 8.2 | 8.7 | 7.1 | 6.4 | 6.6 | 6.8 | 6.3 |
| Grass | 1.9 | 0.8 | 0.4 | 0.6 | 0.4 | 0.3 | 0.4 | 0.5 | 0.4 |
| Leaves | 9.7 | 6.0 | 4.0 | 4.2 | 3.5 | 3.2 | 3.3 | 3.3 | 3.1 |
| Branches | 9.7 | 5.8 | 3.7 | 3.9 | 3.2 | 2.9 | 3.0 | 3.0 | 2.8 |
| Food Scraps | 2.2 | 1.4 | 3.1 | 3.1 | 3.0 | 3.5 | 3.5 | 3.6 | 3.5 |
| Total Net Flux | 84.1 | 85.5 | 93.7 | 98.5 | 99.0 | 101.0 | 103.5 | 105.8 | 107.4 |

Note: Totals may not equal sum of components due to independent rounding.

Source: U.S. Environmental Protection Agency, *Inventory of U.S. Greenhouse Gas Emissions and Sinks: 1990-2007*, EPA 430-R-09-004 (Washington, DC, April 2009), web site www.epa.gov/climatechange/emissions/usinventoryreport.html.

# GLOSSARY

**Acid stabilization:** A circumstance where the pH of the waste mixture in an animal manure management system is maintained near 7.0, optimal conditions for methane production.

**Aerobic bacteria:** Microorganisms living, active, or occurring only in the presence of oxygen.

**Aerobic decomposition:** The breakdown of a molecule into simpler molecules or atoms by microorganisms under favorable conditions of oxygenation.

**Aerosols:** Airborne particles.

**Afforestation:** Planting of new forests on lands that have not been recently forested.

**Agglomeration:** The clustering of disparate elements.

**Airshed:** An area or region defined by settlement patterns or geology that results in discrete atmospheric conditions.

**Albedo:** The fraction of incident light or electromagnetic radiation that is reflected by a surface or body. See *Planetary albedo*.

**Anaerobes:** Organisms that live and are active only in the absence of oxygen.

**Anaerobic bacteria:** Microorganisms living, active, or occurring only in the absence of oxygen.

**Anaerobic decomposition:** The breakdown of molecules into simpler molecules or atoms by microorganisms that can survive in the partial or complete absence of oxygen.

**Anaerobic lagoon:** A liquid-based manure management system, characterized by waste residing in water to a depth of at least six feet for a period ranging between 30 and 200 days.

**Anode:** A positive electrode, as in a battery, radio tube, etc.

**Anthracite:** The highest rank of coal; used primarily for residential and commercial space heating. It is a hard, brittle, and black lustrous coal, often referred to as hard coal, containing a high percentage of fixed carbon and a low percentage of volatile matter. The moisture content of fresh-mined anthracite generally is less than 15 percent. The heat content of anthracite ranges from 22 to 28 million Btu per ton on a moist, mineral-matter-free basis. The heat content of anthracite coal consumed in the United States averages 25 million Btu per ton, on the as-received basis (i.e., containing both inherent moisture and mineral matter). Note: Since the 1980's, anthracite refuse or mine waste has been used for steam electric power generation. This fuel typically has a heat content of 15 million Btu per ton or less.

**Anthropogenic:** Made or generated by a human or caused by human activity. The term is used in the context of global climate change to refer to gaseous emissions that are the result of human activities, as well as other potentially climate-altering activities, such as deforestation.

**API Gravity:** American Petroleum Institute measure of specific gravity of crude oil or condensate in degrees. An arbitrary scale expressing the gravity or density of liquid petroleum products. The measuring scale is calibrated in terms of degrees API; it is calculated as follows: Degrees API = (141.5/sp.gr.60 deg.F/60 deg.F) - 131.5.

**Asphalt:** A dark brown-to-black cement-like material obtained by petroleum processing and containing bitumens as the predominant component; used primarily for road construction. It includes crude asphalt as well as the following finished products: cements, fluxes, the asphalt content of emulsions (exclusive of water), and petroleum distillates blended with asphalt to make cutback asphalts. Note: The conversion factor for asphalt is 5.5 barrels per short ton.

**Associated natural gas:** See *Associated-dissolved natural gas* and *Natural gas*.

**Associated-dissolved natural gas:** Natural gas that occurs in crude oil reservoirs either as free gas (associated) or as gas in solution with crude oil (dissolved gas). See *Natural gas*.

**Aviation gasoline (finished):** A complex mixture of relatively volatile hydrocarbons with or without small quantities of additives, blended to form a fuel suitable for use in aviation reciprocating engines. Fuel specifications are provided in ASTM Specification D 910 and Military Specification MIL-G-5572. Note: Data on blending components are not counted in data on finished aviation gasoline.

**Balancing item:** Represents differences between the sum of the components of natural gas supply and the sum of the components of natural gas disposition. These differences may be due to quantities lost or to the effects of data reporting problems. Reporting problems include differences due to the net result of conversions of flow data metered at varying temperature and pressure bases and converted to a standard temperature and pressure base; the effect of variations in company accounting and billing practices; differences between billing cycle and calendar period time frames; and imbalances resulting from the merger of data reporting systems that vary in scope, format, definitions, and type of respondents.

**Biofuels:** Liquid fuels and blending components produced from biomass (plant) feedstocks, used primarily for transportation.

**Biogas:** The gas produced from the anaerobic decomposition of organic material in a landfill.

**Biogenic:** Produced by the actions of living organisms.

**Biomass:** Organic nonfossil material of biological origin constituting a renewable energy source.

**Biosphere:** The portion of the Earth and its atmosphere that can support life. The part of the global carbon cycle that includes living organisms and biogenic organic matter.

**Bituminous coal:** A dense coal, usually black, sometimes dark brown, often with well-defined bands of bright and dull material, used primarily as fuel in steam-electric power generation, with substantial quantities also used for heat and power applications in manu-facturing and to make coke. Bituminous coal is the most abundant coal in active U.S. mining regions. Its moisture content usually is less than 20 percent. The heat content of bituminous coal ranges from 21 to 30 million Btu per ton on a moist, mineral-matter-free basis. The heat content of bituminous coal consumed in the United States averages 24 million Btu per ton, on the as-received basis (i.e., containing both inherent moisture and mineral matter).

**$BOD_5$:** The biochemical oxygen demand of wastewater during decomposition occurring over a 5-day period. A measure of the organic content of wastewater.

**Bromofluorocarbons (halons):** Inert, nontoxic chemicals that have at least one bromine atom in their chemical makeup. They evaporate without leaving a residue and are used in fire extinguishing systems, especially for large computer installations.

**Bunker fuel:** Fuel supplied to ships and aircraft, both domestic and foreign, consisting primarily of residual and distillate fuel oil for ships and kerosene-based jet fuel for aircraft. The term "international bunker fuels" is used to denote the consumption of fuel for international transport activities. *Note*: For the purposes of greenhouse gas emissions inventories, data on emissions from combustion of international bunker fuels are subtracted from national emissions totals. Historically, bunker fuels have meant only ship fuel. See *Vessel bunkering*.

**Calcination:** A process in which a material is heated to a high temperature without fusing, so that hydrates, carbonates, or other compounds are decomposed and the volatile material is expelled.

**Calcium sulfate:** A white crystalline salt, insoluble in water. Used in Keene's cement, in pigments, as a paper filler, and as a drying agent.

**Calcium sulfite:** A white powder, soluble in dilute sulfuric acid. Used in the sulfite process for the manufacture of wood pulp.

**Capital stock:** Property, plant and equipment used in the production, processing and distribution of energy resources.

**Carbon black:** An amorphous form of carbon, produced commercially by thermal or oxidative decomposition of hydrocarbons and used principally in rubber goods, pigments, and printer's ink.

**Carbon budget:** Carbon budget: The balance of the exchanges (incomes and losses) of carbon between carbon sinks (e.g., atmosphere and biosphere) in the carbon cycle. See *Carbon cycle* and *Carbon sink*.

**Carbon cycle:** All carbon sinks and exchanges of carbon from one sink to another by various chemical, physical, geological, and biological processes. See *Carbon sink* and *Carbon budget*.

**Carbon dioxide ($CO_2$):** A colorless, odorless, nonpoisonous gas that is a normal part of Earth's atmosphere. Carbon dioxide is a product of fossil-fuel combustion as well as other processes. It is considered a greenhouse gas as it traps heat (infrared energy) radiated by the Earth into the atmosphere and thereby contributes to the potential for global warming. The global warming potential (GWP) of other greenhouse gases is measured in relation to that of carbon dioxide, which by international scientific convention is assigned a value of one (1). See *Global warming potential (GWP)* and *Greenhouse gases*.

**Carbon dioxide equivalent:** The amount of carbon dioxide by weight emitted into the atmosphere that would produce the same estimated radiative forcing as a given weight of another radiatively active gas. Carbon dioxide equivalents are computed by multiplying the weight of the gas being measured (for example, methane) by its estimated global warming potential (which is 21 for methane). "Carbon equivalent units" are defined as carbon dioxide equivalents multiplied by the carbon content of carbon dioxide (i.e., 12/44).

**Carbon flux:** See *Carbon budget*.

**Carbon intensity:** The amount of carbon by weight emitted per unit of energy consumed. A common measure of carbon intensity is weight of carbon per British thermal unit (Btu) of energy. When there is only one fossil fuel under consideration, the carbon intensity and the emissions coefficient are identical. When there are several fuels, carbon intensity is based on their combined emissions coefficients weighted by their energy consumption levels. See *Emissions coefficient* and *Carbon output rate*.

**Carbon output rate:** The amount of carbon by weight per kilowatthour of electricity produced.

**Carbon sequestration:** The fixation of atmospheric carbon dioxide in a carbon sink through biological or physical processes.

**Carbon sink:** A reservoir that absorbs or takes up released carbon from another part of the carbon cycle. The four sinks, which are regions of the Earth within which carbon behaves in a systematic manner, are the atmosphere, terrestrial biosphere (usually including freshwater systems), oceans, and sediments (including fossil fuels).

**Catalytic converter:** A device containing a catalyst for converting automobile exhaust into mostly harmless products.

**Catalytic hydrocracking:** A refining process that uses hydrogen and catalysts with relatively low temperatures and high pressures for converting middle boiling or residual material to high octane gasoline, reformer charge stock, jet fuel, and/or high grade fuel oil. The process uses one or more catalysts, depending on product output, and can handle high sulfur feedstocks without prior desulfurization.

**Cesspool:** An underground reservoir for liquid waste, typically household sewage.

**Chlorofluorocarbon (CFC):** Any of various compounds consisting of carbon, hydrogen, chlorine, and fluorine used as refrigerants. CFCs are now thought to be harmful to the earth's atmosphere.

**Clean Development Mechanism (CDM):** A Kyoto Protocol program that enables industrialized countries to finance emissions-avoiding projects in developing countries and receive credit for reductions achieved against their own emissions limitation targets. See *Kyoto Protocol*.

**Climate:** The average course or condition of the weather over a period of years as exhibited by temperature, humidity, wind velocity, and precipitation.

**Climate change:** A term that refers to a change in the state of the climate that can be identified (e.g., by using statistical tests) by changes in the mean and/or the variability of its properties, and that persists for an extended period, typically decades or longer. Climate change may be due to natural internal processes or external forcings, or to persistent anthropogenic changes in the composition of the atmosphere or in land use.

**Clinker:** Powdered cement, produced by heating a properly proportioned mixture of finely ground raw materials (calcium carbonate, silica, alumina, and iron oxide) in a kiln to a temperature of about 2,700°F.

**Cloud condensation nuclei:** Aerosol particles that provide a platform for the condensation of water vapor, resulting in clouds with higher droplet concentrations and increased albedo.

**Coal coke:** See *Coke (coal)*.

**Coalbed methane:** Methane is generated during coal formation and is contained in the coal microstructure. Typical recovery entails pumping water out of the coal to allow the gas to escape. Methane is the principal component of natural gas. Coalbed methane can be added to natural gas pipelines without any special treatment.

**Coke (coal):** A solid carbonaceous residue derived from low-ash, low-sulfur bituminous coal from which the volatile constituents are driven off by baking in an oven at temperatures as high as 2,000 degrees Fahrenheit so that the fixed carbon and residual ash are fused together. Coke is used as a fuel and as a reducing agent in smelting iron ore in a blast furnace. Coke from coal is grey, hard, and porous and has a heating value of 24.8 million Btu per ton.

**Coke (petroleum):** A residue high in carbon content and low in hydrogen that is the final product of thermal decomposition in the condensation process in cracking. This product is reported as marketable coke or catalyst coke. The conversion is 5 barrels (of 42 U.S. gallons each) per short ton. Coke from petroleum has a heating value of 6.024 million Btu per barrel.

**Combustion:** Chemical oxidation accompanied by the generation of light and heat.

**Combustion chamber:** An enclosed vessel in which chemical oxidation of fuel occurs.

**Conference of the Parties (COP):** The collection of nations that have ratified the Framework Convention on Climate Change (FCCC). The primary role of the COP is to keep implementation of the FCCC under review and make the decisions necessary for its effective implementation. See *Framework Convention on Climate Change (FCCC)*.

**Cracking:** The refining process of breaking down the larger, heavier, and more complex hydrocarbon molecules into simpler and lighter molecules.

**Criteria pollutant:** A pollutant determined to be hazardous to human health and regulated under EPA's National Ambient Air Quality Standards. The 1970 amendments to the Clean Air Act require EPA to describe the health and welfare impacts of a pollutant as the "criteria" for inclusion in the regulatory regime.

**Crop residue:** Organic residue remaining after the harvesting and processing of a crop.

**Cultivar:** A horticulturally or agriculturally derived variety of a plant.

**Deforestation:** The net removal of trees from forested land.

**Degasification system:** The methods employed for removing methane from a coal seam that could not otherwise be removed by standard ventilation fans and thus would pose a substantial hazard to coal miners. These systems may be used prior to mining or during mining activities.

**Degradable organic carbon:** The portion of organic carbon present in such solid waste as paper, food waste, and yard waste that is susceptible to biochemical decomposition.

**Desulfurization:** The removal of sulfur, as from molten metals, petroleum oil, or flue gases.

**Diffusive transport:** The process by which particles of liquids or gases move from an area of higher concentration to an area of lower concentration.

**Distillate fuel:** A general classification for one of the petroleum fractions produced in conventional distillation operations. It includes diesel fuels and fuel oils. Products known as No. 1, No.2, and No.4 diesel fuel are used in on-highway diesel engines, such as those in trucks and automobiles, as well as off-highway engines, such as those in railroad locomotives and agricultural machinery. Products known as No. 1, No. 2, and No. 4 fuel oils are used primarily for space heating and electric power generation.

**Efflux:** An outward flow.

**Electrical generating capacity:** The full-load continuous power rating of electrical generating facilities, generators, prime movers, or other electric equipment (individually or collectively).

**EMCON Methane Generation Model:** A model for estimating the production of methane from municipal solid waste landfills.

**Emissions:** Anthropogenic releases of gases to the atmosphere. In the context of global climate change, they consist of radiatively important greenhouse gases (e.g., the release of carbon dioxide during fuel combustion).

**Emissions coefficient:** A unique value for scaling emissions to activity data in terms of a standard rate of emissions per unit of activity (e.g., pounds of carbon dioxide emitted per Btu of fossil fuel consumed).

**Enteric fermentation:** A digestive process by which carbohydrates are broken down by microorganisms into simple molecules for absorption into the bloodstream of an animal.

**Eructation:** An act or instance of belching.

**ETBE (ethyl tertiary butyl ether):** $(CH_3)_3COC_2H$: An oxygenate blend stock formed by the catalytic etherification of isobutylene with ethanol.

**Ethylene:** An olefinic hydrocarbon recovered from refinery processes or petrochemical processes. Ethylene is used as a petrochemical feedstock for numerous chemical applications and the production of consumer goods.

**Ethylene dichloride:** A colorless, oily liquid used as a solvent and fumigant for organic synthesis, and for ore flotation.

**Facultative bacteria:** Bacteria that grow equally well under aerobic and anaerobic conditions.

**Flange:** A rib or a rim for strength, for guiding, or for attachment to another object (e.g., on a pipe).

**Flared:** Gas disposed of by burning in flares usually at the production sites or at gas processing plants.

**Flatus:** Gas generated in the intestines or the stomach of an animal.

**Flue gas desulfurization:** Equipment used to remove sulfur oxides from the combustion gases of a boiler plant before discharge to the atmosphere. Also referred to as scrubbers. Chemicals such as lime are used as scrubbing media.

**Fluidized-bed combustion:** A method of burning particulate fuel, such as coal, in which the amount of air required for combustion far exceeds that found in conventional burners. The fuel particles are continually fed into a bed of mineral ash in the proportions of 1 part fuel to 200 parts ash, while a flow of air passes up through the bed, causing it to act like a turbulent fluid.

**Flux material:** A substance used to promote fusion, e.g., of metals or minerals.

**Fodder:** Coarse food for domestic livestock.

**Forestomach:** See *Rumen*.

**Fossil fuel:** An energy source formed in the earths crust from decayed organic material. The common fossil fuels are petroleum, coal, and natural gas.

**Framework Convention on Climate Change (FCCC):** An agreement opened for signature at the "Earth Summit" in Rio de Janeiro, Brazil, on June 4, 1992, which has the goal of stabilizing greenhouse gas concentrations in the atmosphere at a level that would prevent significant anthropogenically forced climate change. See *Climate change*.

**Fuel cycle:** The entire set of sequential processes or stages involved in the utilization of fuel, including extraction, transformation, transportation, and combustion. Emissions generally occur at each stage of the fuel cycle.

**Fugitive emissions:** Unintended leaks of gas from the processing, transmission, and/or transportation of fossil fuels.

**Gasification:** A method for converting coal, petroleum, biomass, wastes, or other carbon-containing materials into a gas that can be burned to generate power or processed into chemicals and fuels.

**Gate station:** Location where the pressure of natural gas being transferred from the transmission system to the distribution system is lowered for transport through small diameter, low pressure pipelines.

**Geothermal:** Pertaining to heat within the Earth.

**Global climate change:** See *Climate change*.

**Global warming:** A gradual increase, observed or projected, in global surface temperature, as one of the consequences of radiative forcing caused by anthropogenic emissions. See *Climate change*.

**Global warming potential (GWP):** An index used to compare the relative radiative forcing of different gases without directly calculating the changes in atmospheric concentrations. GWPs are calculated as the ratio of the radiative forcing that would result from the emission of one kilogram of a greenhouse gas to that from the emission of one kilogram of carbon dioxide over a fixed period of time, such as 100 years.

**Greenhouse effect:** The result of water vapor, carbon dioxide, and other atmospheric gases trapping radiant (infrared) energy, thereby keeping the earth's surface warmer than it would otherwise be. Greenhouse gases within the lower levels of the atmosphere trap this radiation, which would otherwise escape into space, and subsequent re-radiation of some of this energy back to the Earth maintains higher surface temperatures than would occur if the gases were absent. See *Greenhouse gases*.

**Greenhouse gases:** Those gases, such as water vapor, carbon dioxide, nitrous oxide, methane, hydrofluorocarbons (HFCs), perfluorocarbons (PFCs) and sulfur hexafluoride, that are transparent to solar (short-wave) radiation but opaque to long-wave (infrared) radiation, thus preventing long-wave radiant energy from leaving the Earth's atmosphere. The net effect is a trapping of absorbed radiation and a tendency to warm the planet's surface.

**Gross gas withdrawal:** The full-volume of compounds extracted at the wellhead, including nonhydrocarbon gases and natural gas plant liquids.

**Gypsum:** Calcium sulfate dihydrate ($CaSO_4 \cdot 2H_2O$), a sludge constituent from the conventional lime scrubber process, obtained as a byproduct of the dewatering operation and sold for commercial use.

**Halogenated substances:** A volatile compound containing halogens, such as chlorine, fluorine or bromine.

**Halons:** See *Bromofluorocarbons*.

**Heating degree-days (HDD):** A measure of how cold a location is over a period of time relative to a base temperature, most commonly specified as 65 degrees Fahrenheit. The measure is computed for each day by subtracting the average of the day's high and low temperatures from the base temperature (65 degrees), with negative values set equal to zero. Each day's heating degree-days are summed to create a heating degree-day measure for a specified reference period. Heating degree-days are used in energy analysis as an indicator of space heating energy requirements or use.

**Herbivore:** A plant-eating animal.

**Hydrocarbon:** An organic chemical compound of hydrogen and carbon in either gaseous, liquid, or solid phase. The molecular structure of hydrocarbon compounds varies from the simple (e.g., methane, a constituent of natural gas) to the very heavy and very complex.

**Hydrochlorofluorocarbons (HCFCs):** Chemicals composed of one or more carbon atoms and varying numbers of hydrogen, chlorine, and fluorine atoms.

**Hydrofluorocarbons (HFCs):** A group of man-made chemicals composed of one or two carbon atoms and varying numbers of hydrogen and fluorine atoms. Most HFCs have 100-year Global Warming Potentials in the thousands.

**Hydroxyl radical (OH):** An important chemical scavenger of many trace gases in the atmosphere that are greenhouse gases. Atmospheric concentrations of OH affect the atmospheric lifetimes of greenhouse gases, their abundance, and, ultimately, the effect they have on climate.

**Intergovernmental Panel on Climate Change (IPCC):** A panel established jointly in 1988 by the World Meteorological Organization and the United Nations Environment Program to assess the scientific information relating to climate change and to formulate realistic response strategies.

**International bunker fuels:** See *Bunker fuels*.

**Jet fuel:** A refined petroleum product used in jet aircraft engines. It includes kerosene-type jet fuel and naphtha-type jet fuel.

**Joint Implementation (JI):** Agreements made between two or more nations under the auspices of the Framework Convention on Climate Change (FCCC) whereby a developed country can receive "emissions reduction units" when it helps to finance projects that reduce net emissions in another developed country (including countries with economies in transition).

**Kerosene:** A light petroleum distillate that is used in space heaters, cook stoves, and water heaters and is suitable for use as a light source when burned in wick-fed lamps. Kerosene has a maximum distillation temperature of 400 degrees Fahrenheit at the 10-percent recovery point, a final boiling point of 572 degrees Fahrenheit, and a minimum flash point of 100 degrees Fahrenheit. Included are No. 1-K and No. 2-K, the two grades recognized by ASTM Specification D 3699 as well as all other grades of kerosene called range or stove oil, which have properties similar to those of No. 1 fuel oil. See *Kerosene-type jet fuel*.

**Kerosene-type jet fuel:** A kerosene-based product having a maximum distillation temperature of 400 degrees Fahrenheit at the 10-percent recovery point and a final maximum boiling point of 572 degrees Fahrenheit and meeting ASTM Specification D 1655 and Military Specifications MIL-T-5624P and MIL-T-83133D (Grades JP-5 and JP-8). It is used for commercial and military turbojet and turboprop aircraft engines.

**Kyoto Protocol:** The result of negotiations at the third Conference of the Parties (COP-3) in Kyoto, Japan, in December of 1997. The Kyoto Protocol sets binding greenhouse gas emissions targets for countries that sign and ratify the agreement. The gases covered under the Protocol include carbon dioxide, methane, nitrous oxide, hydrofluorocarbons (HFCs), perfluorocarbons (PFCs) and sulfur hexafluoride.

**Ketone-alcohol (cyclohexanol):** An oily, colorless, hygroscopic liquid with a camphor-like odor. Used in soapmaking, dry cleaning, plasticizers, insecticides, and germicides.

**Leachate:** The liquid that has percolated through the soil or other medium.

**Lignite:** The lowest rank of coal, often referred to as brown coal, used almost exclusively as fuel for steam-electric power generation. It is brownish-black and has a high inherent moisture content, sometimes as high as 45 percent The heat content of lignite ranges from 9 to 17 million Btu per ton on a moist, mineral-matter-free basis. The heat content of lignite consumed in the United States averages 13 million Btu perton, on the as-received basis (i.e., containing both inherent moisture and mineral matter).

**Liquefied petroleum gases:** A group of hydrocarbon-based gases derived from crude oil refining or natural gas fractionation. They include ethane, ethylene, propane, propylene, normal butane, butylene, isobutane, and isobutylene. For convenience of transportation, these gases are liquefied through pressurization.

**Lubricants:** Substances used to reduce friction between bearing surfaces, or incorporated into other materials used as processing aids in the manufacture of other products, or used as carriers of other materials. Petroleum lubricants may be produced either from distillates or residues. Lubricants include all grades of lubricating oils, from spindle oil to cylinder oil to those used in greases.

**Methane:** A colorless, flammable, odorless hydrocarbon gas (CH4) which is the major component of natural gas. It is also an important source of hydrogen in various industrial processes. Methane is a greenhouse gas. See also *Greenhouse gases*.

**Methanogens:** Bacteria that synthesize methane, requiring completely anaerobic conditions for growth.

**Methanol:** A light alcohol that can be used for gasoline blending. See oxygenate.

**Methanotrophs:** Bacteria that use methane as food and oxidize it into carbon dioxide.

**Methyl chloroform (trichloroethane):** An industrial chemical ($CH_3CCl_3$) used as a solvent, aerosol propellant, and pesticide and for metal degreasing.

**Methyl tertiary butyl ether (MTBE):** A colorless, flammable, liquid oxygenated hydrocarbon containing 18.15 percent oxygen.

**Methylene chloride:** A colorless liquid, nonexplosive and practically nonflammable. Used as a refrigerant in centrifugal compressors, a solvent for organic materials, and a component in nonflammable paint removers.

**Mole:** The quantity of a compound or element that has a weight in grams numerically equal to its molecular weight. Also referred to as gram molecule or gram molecular weight.

224    United States Energy Information Administration

**Montreal Protocol:** The Montreal Protocol on Substances that Deplete the Ozone Layer (1987). An international agreement, signed by most of the industrialized nations, to substantially reduce the use of chlorofluorocarbons (CFCs). Signed in January 1989, the original document called for a 50-percent reduction in CFC use by 1992 relative to 1986 levels. The subsequent London Agreement called for a complete elimination of CFC use by 2000. The Copenhagen Agreement, which called for a complete phaseout by January 1, 1996, was implemented by the U.S. Environmental Protection Agency.

**Motor gasoline (finished):** A complex mixture of relatively volatile hydrocarbons with or without small quantities of additives, blended to form a fuel suitable for use in spark-ignition engines. Motor gasoline, as defined in ASTM Specification D 4814 or Federal Specification VV-G-1690C, is characterized as having a boiling range of 122 to 158 degrees Fahrenheit at the 10 percent recovery point to 365 to 374 degrees Fahrenheit at the 90 percent recovery point. "Motor Gasoline" includes conventional gasoline; all types of oxygenated gasoline, including gasohol; and reformulated gasoline, but excludes aviation gasoline. Note: Volumetric data on blending components, such as oxygenates, are not counted in data on finished motor gasoline until the blending components are blended into the gasoline.

**Multiple cropping:** A system of growing several crops on the same field in one year.

**Municipal solid waste:** Residential solid waste and some nonhazardous commercial, institutional, and industrial wastes.

**Naphtha less than 401 degrees Fahrenheit:** A naphtha with a boiling range of less than 401 degrees Fahrenheit that is intended for use as a petrochemical feedstock. Also see *Petrochemical feedstocks*.

**Naphtha-type jet fuel:** A fuel in the heavy naphtha boiling range having an average gravity of 52.8 degrees API, 20 to 90 percent distillation temperatures of 290 degrees to 470 degrees Fahrenheit, and meeting Military Specification MIL-T-5624L (Grade JP-4). It is used primarily for military turbojet and turboprop aircraft engines because it has a lower freeze point than other aviation fuels and meets engine requirements at high altitudes and speeds.

**Natural gas:** A mixture of hydrocarbons and small quantities of various nonhydrocarbons in the gaseous phase or in solution with crude oil in natural underground reservoirs.

**Natural gas liquids (NGLs):** Those hydrocarbons in natural gas that are separated as liquids from the gas. Includes natural gas plant liquids and lease condensate.

**Natural gas, pipeline quality:** A mixture of hydrocarbon compounds existing in the gaseous phase with sufficient energy content, generally above 900 Btu, and a small enough share of impurities for transport through commercial gas pipelines and sale to end-users.

**Nitrogen oxides (NO$_x$):** Compounds of nitrogen and oxygen produced by the burning of fossil fuels.

**Nitrous oxide ($N_2O$):** A colorless gas, naturally occurring in the atmosphere.

**Nonmethane volatile organic compounds (NMVOCs):** Organic compounds, other than methane, that participate in atmospheric photochemical reactions.

**Octane:** A flammable liquid hydrocarbon found in petroleum. Used as a standard to measure the antiknock properties of motor fuel.

**Oil reservoir:** An underground pool of liquid consisting of hydrocarbons, sulfur, oxygen, and nitrogen trapped within a geological formation and protected from evaporation by the overlying mineral strata.

**Organic content:** The share of a substance that is of animal or plant origin.

**Organic waste:** Waste material of animal or plant origin.

**Oxidize:** To chemically transform a substance by combining it with oxygen.

**Oxygenates:** Substances which, when added to gasoline, increase the amount of oxygen in that gasoline blend. Ethanol, Methyl Tertiary Butyl Ether (MTBE), Ethyl Tertiary Butyl Ether (ETBE), and methanol are common oxygenates.

**Ozone:** A molecule made up of three atoms of oxygen. Occurs naturally in the stratosphere and provides a protective layer shielding the Earth from harmful ultraviolet radiation. In the troposphere, it is a chemical oxidant, a greenhouse gas, and a major component of photochemical smog.

**Ozone precursors:** Chemical compounds, such as carbon monoxide, methane, nonmethane hydrocarbons, and nitrogen oxides, which in the presence of solar radiation react with other chemical compounds to form ozone.

**Paraffinic hydrocarbons:** Straight-chain hydrocarbon compounds with the general formula $C_nH2_{n+2}$.

**Perfluorocarbons (PFCs):** A group of man-made chemicals composed of one or two carbon atoms and four to six fluorine atoms, containing no chlorine. PFCs have no commercial uses and are emitted as a byproduct of aluminum smelting and semiconductor manufacturing. PFCs have very high 100-year Global Warming Potentials and are very long-lived in the atmosphere.

**Perfluoromethane:** A compound ($CF_4$) emitted as a byproduct of aluminum smelting.

**Petrochemical feedstocks:** Chemical feedstocks derived from petroleum principally for the manufacture of chemicals, synthetic rubber, and a variety of plastics.

226 United States Energy Information Administration

**Petroleum:** A broadly defined class of liquid hydrocarbon mixtures. Included are crude oil, lease condensate, unfinished oils, refined products obtained from the processing of crude oil, and natural gas plant liquids. Note: Volumes of finished petroleum products include nonhydrocarbon compounds, such as additives and detergents, after they have been blended into the products.

**Petroleum coke:** See *Coke (petroleum)*.

**Photosynthesis:** The manufacture by plants of carbohydrates and oxygen from carbon dioxide and water in the presence of chlorophyll, with sunlight as the energy source. Carbon is sequestered and oxygen and water vapor are released in the process.

**Pigiron:** Crude, high-carbon iron produced by reduction of iron ore in a blast furnace.

**Pipeline, distribution:** A pipeline that conveys gas from a transmission pipeline to its ultimate consumer.

**Pipeline, gathering:** A pipeline that conveys gas from a production well/field to a gas processing plant or transmission pipeline for eventual delivery to end-use consumers.

**Pipeline, transmission:** A pipeline that conveys gas from a region where it is produced to a region where it is to be distributed.

**Planetary albedo:** The fraction of incident solar radiation that is reflected by the Earth-atmosphere system and returned to space, mostly by backscatter from clouds in the atmosphere.

**Pneumatic device:** A device moved or worked by air pressure.

**Polystyrene:** A polymer of styrene that is a rigid, transparent thermoplastic with good physical and electrical insulating properties, used in molded products, foams, and sheet materials.

**Polyvinyl chloride (PVC):** A polymer of vinyl chloride. Tasteless. odorless, insoluble in most organic solvents. A member of the family vinyl resin, used in soft flexible films for food packaging and in molded rigid products, such as pipes, fibers, upholstery, and bristles.

**Post-mining emissions:** Emissions of methane from coal occurring after the coal has been mined, during transport or pulverization.

**Radiative forcing:** A change in average net radiation at the top of the troposphere (known as the tropopause) because of a change in either incoming solar or exiting infrared radiation. A positive radiative forcing tends on average to warm the earth's surface; a negative radiative forcing on average tends to cool the earth's surface. Greenhouse gases, when emitted into the atmosphere, trap infrared energy radiated from the earth's surface and therefore tend to produce positive radiative forcing. See *Greenhouse gases*.

**Radiatively active gases:** Gases that absorb incoming solar radiation or outgoing infrared radiation, affecting the vertical temperature profile of the atmosphere. See *Radiative forcing.*

**Ratoon crop:** A crop cultivated from the shoots of a perennial plant.

**Redox potential:** A measurement of the state of oxidation of a system.

**Reflectivity:** The ratio of the energy carried by a wave after reflection from a surface to its energy before reflection.

**Reforestation:** Replanting of forests on lands that have recently been harvested or otherwise cleared of trees.

**Reformulated gasoline:** Finished motor gasoline formulated for use in motor vehicles, the composition and properties of which meet the requirements of the reformulated gasoline regulations promulgated by the U.S. Environmental Protection Agency under Section 211(k) of the Clean Air Act. Note: This category includes oxygenated fuels program reformulated gasoline (OPRG) but excludes reformulated gasoline blendstock for oxygenate blending (RBOB).

**Renewable energy resources:** Energy resources that are naturally replenishing but flow-limited. They are virtually inexhaustible in duration but limited in the amount of energy that is available per unit of time. Renewable energy resources include: biomass, hydro, geothermal, solar, wind, ocean thermal, wave action, and tidal action.

**Residual fuel oil:** A general classification for the heavier oils, known as No.5 and No.6 fuel oils, that remain after the distillate fuel oils and lighter hydrocarbons are distilled away in refinery operations. It conforms to ASTM Specifications D 396 and D 975 and Federal Specification VV-F-815C. No. 5, a residual fuel oil of medium viscosity, is also known as Navy Special and is defined in Military Specification MIL-F-859E, including Amendment 2 (NATO Symbol F-770). It is used in steam-powered vessels in government service and inshore powerplants. No. 6 fuel oil includes Bunker C fuel oil and is used for the production of electric power, space heating, vessel bunkering, and various industrial purposes.

**Rumen:** The large first compartment of the stomach of certain animals in which cellulose is broken down by the action of bacteria.

**Sample:** A set of measurements or outcomes selected from a given population.

**Sequestration:** See *Carbon sequestration.*

**Septic tank:** A tank in which the solid matter of continuously flowing sewage is disintegrated by bacteria.

**Sinter:** A chemical sedimentary rock deposited by precipitation from mineral waters, especially siliceous sinter and calcareous sinter.

**Sodium silicate:** A grey-white powder soluble in alkali and water, insoluble in alcohol and acid. Used to fireproof textiles, in petroleum refining and corrugated paperboard manufacture, and as an egg preservative. Also referred to as liquid gas, silicate of soda, sodium metasilicate, soluble glass, and water glass.

**Sodium tripolyphosphate:** A white powder used for water softening and as a food additive and texturizer.

**Stabilization lagoon:** A shallow artificial pond used for the treatment of wastewater. Treatment includes removal of solid material through sedimentation, the decomposition of organic material by bacteria, and the removal of nutrients by algae.

**Still gas (refinery gas):** Any form or mixture of gases produced in refineries by distillation, cracking, reforming, and other processes. The principal constituents are methane, ethane, ethylene, normal butane, butylene, propane, propylene, etc. Still gas is used as a refinery fuel and a petrochemical feedstock. The conversion factor is 6 million Btu per fuel oil equivalent barrel.

**Stratosphere:** The region of the upper atmosphere extending from the tropopause (8 to 15 kilometers altitude) to about 50 kilometers. Its thermal structure, which is determined by its radiation balance, is generally very stable with low humidity.

**Stripper well:** An oil or gas well that produces at relatively low rates. For oil, stripper production is usually defined as production rates of between 5 and 15 barrels of oil per day. Stripper gas production would generally be anything less than 60 thousand cubic feet per day.

**Styrene:** A colorless, toxic liquid with a strong aromatic aroma. Insoluble in water, soluble in alcohol and ether; polymerizes rapidly; can become explosive. Used to make polymers and copolymers, polystyrene plastics, and rubber.

**Subbituminous coal:** A coal whose properties range from those of lignite to those of bituminous coal and used primarily as fuel for steam-electric power generation. It may be dull, dark brown to black, soft and crumbly, at the lower end of the range, to bright, jet black, hard, and relatively strong, at the upper end. Subbituminous coal contains 20 to 30 percent inherent moisture by weight. The heat content of subbituminous coal ranges from 17 to 24 million Btu per ton on a moist, mineral-matter-free basis. The heat content of subbituminous coal consumed in the United States averages 17 to 18 million Btu per ton, on the as-received basis (i.e., containing both inherent moisture and mineral matter).

**Sulfur dioxide ($SO_2$):** A toxic, irritating, colorless gas soluble in water, alcohol, and ether. Used as a chemical intermediate, in paper pulping and ore refining, and as a solvent.

**Sulfur hexafluoride (SF$_6$):** A colorless gas soluble in alcohol and ether, and slightly less soluble in water. It is used as a dielectric in electronics.

**Sulfur oxides (SO$_x$):** Compounds containing sulfur and oxygen, such as sulfur dioxide (SO$_2$) and sulfur trioxide (SO$_3$).

**Tertiary amyl methyl ether ((CH$_3$)$_2$(C$_2$H$_5$)COCH$_3$):** An oxygenate blend stock formed by the catalytic etherification of isoamylene with methanol.

**Troposphere:** The inner layer of the atmosphere below about 15 kilometers, within which there is normally a steady decrease of temperature with increasing altitude. Nearly all clouds form and weather conditions manifest themselves within this region. Its thermal structure is caused primarily by the heating of the earth's surface by solar radiation, followed by heat transfer through turbulent mixing and convection.

**Uncertainty:** A measure used to quantify the plausible maximum and minimum values for emissions from any source, given the biases inherent in the methods used to calculate a point estimate and known sources of error.

**Vapor displacement:** The release of vapors that had previously occupied space above liquid fuels stored in tanks. These releases occur when tanks are emptied and filled.

**Ventilation system:** A method for reducing methane concentrations in coal mines to non-explosive levels by blowing air across the mine face and using large exhaust fans to remove methane while mining operations proceed.

**Vessel bunkering:** Includes sales for the fueling of commercial or private boats, such as pleasure craft, fishing boats, tugboats, and ocean-going vessels, including vessels operated by oil companies. Excluded are volumes sold to the U.S. Armed Forces.

**Volatile organic compounds (VOCs):** Organic compounds that participate in atmospheric photochemical reactions.

**Volatile solids:** A solid material that is readily decomposable at relatively low temperatures.

**Waste flow:** Quantity of a waste stream generated by an activity.

**Wastewater:** Water that has been used and contains dissolved or suspended waste materials.

**Wastewater, domestic and commercial:** Wastewater (sewage) produced by domestic and commercial establishments.

**Wastewater, industrial:** Wastewater produced by industrial processes.

# 230 United States Energy Information Administration

**Water vapor:** Water in a vaporous form, especially when below boiling temperature and diffused (e.g., in the atmosphere).

**Wax:** A solid or semi-solid material derived from petroleum distillates or residues by such treatments as chilling, precipitating with a solvent, or de-oiling. It is a light-colored, more-or-less translucent crystalline mass, slightly greasy to the touch, consisting of a mixture of solid hydrocarbons in which the paraffin series predominates. Includes all marketable wax, whether crude scale or fully refined. The three grades included are microcrystalline, crystalline-fully refined, and crystalline-other. The conversion factor is 280 pounds per 42 U.S. gallons per barrel.

**Weanling system:** A cattle management system that places calves on feed starting at 165 days of age and continues until the animals have reached slaughter weight.

**Wellhead:** The point at which the crude (and/or natural gas) exits the ground. Following historical precedent, the volume and price for crude oil production are labeled as "wellhead," even though the cost and volume are now generally measured at the lease boundary. In the context of domestic crude price data, the term "wellhead" is the generic term used to reference the production site or lease property.

**Wetlands:** Areas regularly saturated by surface or groundwater and subsequently characterized by a prevalence of vegetation adapted for life in saturated-soil conditions.

**Wood energy:** Wood and wood products used as fuel, including roundwood (cordwood), limbwood, wood chips, bark, sawdust, forest residues, charcoal, pulp waste, and spent pulping liquor..

**Yearling system:** A cattle management system that includes a stocker period from 165 days of age to 425 days of age followed by a 140-day feedlot period.

## End Notes

[1] U.S. Environmental Protection Agency, "Transportation and Climate: Regulations and Standards: Vehicles/Engines," web site http://epa.gov/otaq/climate/regulations.htm. Standards expire after 5 years.

[2] Intergovernmental Panel on Climate Change, *Climate Change 2007: The Physical Science Basis: Errata* (Cambridge, UK: Cambridge University Press, 2008), web site http://ipcc-wg1.ucar.edu/wg1/Report/ AR4WG1_Errata_2008-12-01.pdf.

[3] U.S. Department of Energy, Office of Fossil Energy, National Energy Technology Laboratory, "Secretary Chu Announces First Awards from $1.4 Billion for Industrial Carbon Capture and Storage Projects" (News Release, October 2, 2009), web site www.netl.doe.gov/ publications/press/2009/09072-DOE_Announcesindustrial_CCS.html.

[4] U.S. Department of Energy, Office of Fossil Energy, National Energy Technology Laboratory, "Technologies: Carbon Sequestration," web site www.netl.doe.gov/technologies/carbon

[5] U.S. Department of Energy, Office of Fossil Energy, National Energy Technology Laboratory, "Carbon Sequestration: FAQ Information Portal," www.netl.doe.gov/technologies/carbon

[6] H.S. Eggleston, "Estimation of Emissions from $CO_2$ Capture and Storage: The 2006 IPCC Guidelines for National Greenhouse Gas Inventories," web site
http://unfccc.int/files/meetings/workshops/other_meetings/2006/application/pdf/ccs_20060723.pdf.

[7] U.S. Department of Energy, Office of Fossil Energy, National Energy Technology Laboratory, "Program Facts: Carbon Sequestration Through Enhanced Oil Recovery" (April 2008), web site www.netl.doe.gov/publications/factsheets/program.

[8] *Ibid.*

[9] U.S. Department of Energy, Office of Fossil Energy, National Energy Technology Laboratory, *2008 Carbon Sequestration Atlas of the United States and Canada*, Second Edition (November 2008), web site www.netl.doe.gov/technologies/carbon atlasII.pdf.

[10] International Energy Agency, *$CO_2$ Capture and Storage: A Key Abatement Option* (Paris, France, 2008), web site www.iea.org/ publications/free_new_Desc.asp?PUBS_ID=2145.

[11] *Ibid.*

[12] U.S. Department of Energy, Office of Fossil Energy, National Energy Technology Laboratory, *2008 Carbon Sequestration Atlas of the United States and Canada*, Second Edition (November 2008), web site www.netl.doe.gov/technologies/carbon atlasII.pdf.

[13] Massachusetts Institute of Technology, *The Future of Coal: Options for a Carbon-Constrained World* (Cambridge, MA, 2007), Appendix 4.A, "Unconventional $CO_2$ Storage Targets," pp. 159-160, web site http://web.mit.edu/coal

[14] Intergovernmental Panel on Climate Change, IPCC Special Report, *Carbon Dioxide Capture and Storage: Summary for Policymakers and Technical Summary* (Cambridge, UK: Cambridge University Press, 2005), "Technical Summary," web site www.ipcc.ch/publications_ and_data/publications_and_data_ reports_carbon_dioxide.htm.

[15] U.S. Department of Energy, Office of Fossil Energy, National Energy Technology Laboratory, *2008 Carbon Sequestration Atlas of the United States and Canada*, Second Edition (November 2008), web site www.netl.doe.gov/technologies/carbon atlasII.pdf.

[16] Energy Information Administration, *Annual Energy Outlook 2009*, DOE /EIA-0383(2009) (Washington, DC, March2009), Tables A4 and A5, web site www.eia.doe.gov/oiaf/aeo/pdf/appa.pdf.

[17] U.S. Environmental Protection Agency, *Municipal Solid Waste in the United States: 2007 Facts and Figures*, EPA 530-R-08-010 (Washington, DC, November 2008).

In: Sources and Reduction of Greenhouse Gas Emissions ISBN: 978-1-61668-856-1
Editor: Steffen D. Saldana © 2010 Nova Science Publishers, Inc.

*Chapter 8*

# THE USE OF AGRICULTURAL OFFSETS TO REDUCE GREENHOUSE GASES

## *Joseph Kile*

Chairman Holden, Congressman Goodlatte, and Members of the Subcommittee, thank you for the invitation to testify on the use of agricultural offsets as part of a capand-trade program for reducing greenhouse gases.

H.R. 2454, the American Clean Energy and Security Act of 2009, which was passed by the House of Representatives, would set an annual limit, or cap, on greenhouse-gas emissions for each year between 2012 and 2050 and would distribute "allowances," or rights to produce those emissions. After the allowances were distributed, regulated entities—those that generate electricity or refine petroleum products, for example— would be free to trade them, so entities that could reduce their emissions at lower costs would sell allowances to others facing higher costs.

The provisions of H.R. 2454 reflect the fact that a variety of other actions—including changing agricultural practices and reducing deforestation—can also reduce the concentration of greenhouse gases in the atmosphere. Those actions have the potential to "offset" the extent to which more costly actions, such as reducing the use of fossil fuels, would have to be undertaken to meet a chosen target for total greenhouse-gas emissions. Under the bill, regulated entities would be allowed to use offsets—meaning reductions in greenhouse gases from activities *not* subject to limits on emissions—in lieu of reducing their emissions or purchasing allowances. Yet the difficulty of verifying offsets raises concerns about whether the specified overall limit would actually be met. Such concerns may be especially acute when, as under H.R. 2454, allowable offsets include actions taken outside the United States.

My testimony makes the following key points:

- Researchers have concluded that a cap-and-trade program that allowed for offsets— such as those that might be generated by changes in agricultural practices and forestry—could reduce greenhouse gases more cheaply than a cap-and-trade program that did not include offsets, but instead relied entirely on reducing the consumption of fossil fuels.

- Because of concerns that the use of offsets could undermine the environmental goals of a cap-and-trade program, four challenges would have to be addressed if offsets are to play a meaningful role in reducing the concentration of greenhouse gases in the atmosphere. In particular, offsets would have to bring about reductions in greenhouse gases that (1) would not have otherwise occurred; (2) could be quantified; (3) were permanent rather than merely a delay in the release of greenhouse gases into the atmosphere; and (4) accounted for "leakage," that is, higher emissions elsewhere or in different sectors of the economy as a result of the activities producing the offsets.
- On the basis of data from the Environmental Protection Agency (EPA), the Congressional Budget Office (CBO) expects that, under the provisions of H.R. 2454, most offsets would be generated by changes in forestry and agricultural practices. Of the offsets from those sectors, fewer than half would be produced domestically in most years, and only about 10 percent of the domestically produced offsets would be from changes in agricultural practices. The remaining offsets from those sectors would come from international sources and would be more evenly split between agriculture and forestry.
- CBO estimates that the savings generated by offsets under H.R. 2454 would be substantial—reducing the price of allowances and the net cost of the program to the economy by about 70 percent. By CBO's estimates, regulated entities would use offsets for about 45 percent of the total emission reductions that they would be required to make over the 20 12–2050 period covered by the policy.
- Any assessment of the use of offsets is subject to many uncertainties, which are inherent in the models used, about such things as the types of activities that would be eligible to generate offsets and the amount supplied by those activities, the prospects for concluding agreements with other nations to allow the use of international offsets, and the cost of ensuring that activities generating offsets actually reduce greenhouse gases.

## POTENTIAL BENEFITS OF OFFSETS IN REDUCING THE COST OF MEETING A TARGET FOR EMISSIONS

Offsets used as a part of a cap-and-trade program for greenhouse-gas emissions have the potential to reduce the cost of meeting the cap by substituting cheaper reductions in greenhouse gases for more expensive ones. The effect of greenhouse gases on the climate does not depend on where and how those gases are produced, but rather on the concentration of those gases in the atmosphere. Consequently, the cheapest way to reduce greenhouse gases by a chosen amount is to create a system that encourages reductions wherever and however they are least costly to make.

In principle, a comprehensive cap could apply to all sources of greenhouse gases. In practice, however, policies currently in effect in parts of the United States and in other countries, as well as those being considered by the Congress, cap only emissions from significant sources of greenhouse gases that can be easily and reliably measured.

The electric power industry, for instance, which produces over one-third of all greenhouse gases in the United States, can use systems that continuously monitor emissions (such as methods currently required under the Acid Rain program) to accurately measure the release of carbon dioxide. In contrast, entities whose emissions are much less significant or more difficult to monitor systematically are generally excluded from existing and proposed caps. Nonetheless, some of those entities may be able to reduce greenhouse gases more cheaply than the electric power industry or other industries subject to a cap. Owners of livestock are one example. When livestock waste decomposes, methane (which is more damaging to the climate on a per-ton basis than carbon dioxide) is produced, but manure can be collected and processed with special bacteria in airtight holding tanks or covered lagoons that allow operators to trap and recover methane. If capturing methane was cheaper than reducing carbon dioxide emissions from other sources by an amount that would have an equivalent impact on the climate, then taking steps to capture methane would reduce the cost of meeting a specified cap on greenhouse gases. As another example, greenhouse gases might be reduced at relatively low cost in developing countries through practices that would preserve existing forests and encourage reforestation.

The potential for reducing costs in a cap-and-trade program through the use of offsets would depend on the stringency of the cap over time and on the scope and amount of allowed offsets. The more stringent the cap, the greater the opportunity to reduce costs by using offsets. The sooner that significant emission reductions were required under the cap, the more expensive compliance would be (because there would be less time to develop and adopt new lower-emission technologies)—and the greater the opportunity to reduce costs by using offsets. Similarly, that opportunity grows with increases in the types of allowable offsets, the number of potential providers, and the proportion of compliance for which offsets could be used.

There are many potential types of offsets. Within the United States, offsets can be generated by changing forest management practices and planting trees to increase carbon storage or changing livestock management and crop production, among other methods. For example, farmers can alter various crop management practices to reduce the amount of nitrous oxide produced and released by soils through decreasing the use of fertilizers or adopting practices involving little or no tilling. Outside of the United States, in developing countries, important potential sources of offsets include reducing deforestation and changing forest management practices, planting trees, and reducing methane and nitrous oxide emissions from livestock, cropland, and rice paddies.

To illustrate the potential savings from reducing greenhouse gases partly through using offsets rather than exclusively through reducing emissions from carbon- intensive fuels, one can compare the estimated cost of emission reductions for capand-trade proposals that would allow the use of offsets and proposals that would not. Different researchers, using a number of different modeling approaches, have analyzed a variety of proposals and developed a range of estimated costs (see Figure 1). The pattern of the estimates is clear: When offsets are allowed, the costs of achieving a given reduction in greenhouse gases are lower—substantially so for large reductions.

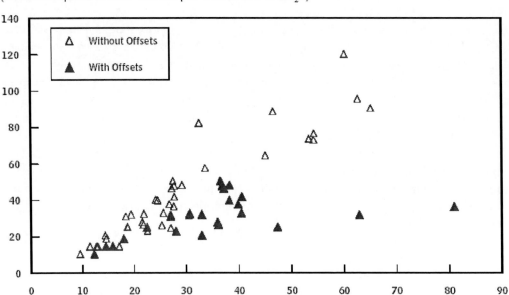

Source: Congressional Budget Office based on estimates from the National Commission on Energy Policy, the Environmental Protection Agency, the Energy Information Administration, the Nicholas Institute for Environmental Policy Solutions, and the Massachusetts Institute of Technology.

Notes: The figure shows, for 2030, the allowance prices and emission reductions under various cap-and-trade proposals, including variations on S. 280, the Climate Stewardship and Innovation Act of 2007, and S. 2191, America's Climate Security Act of 2007. Costs are reported in terms of the price per metric ton of carbon dioxide equivalent ($CO_2e$) emissions associated with achieving a given reduction in greenhouse gases. A metric ton of $CO_2e$ is the amount of a given greenhouse gas (for example, methane or nitrous oxide) that makes the same contribution to global warming as a metric ton of carbon dioxide.

The estimates do not account for the costs of measures to address concerns about the credibility of offsets.

Figure 1. Various Estimates of the Costs of Reducing Greenhouse-Gas Emissions Under Cap-and-Trade Programs With and Without Offsets

## POTENTIAL LIMITATIONS OF OFFSETS

Despite the large cost savings that may be realized from including offsets in a cap-and-trade program, some observers are concerned that the use of offsets can undermine the program's environmental goals. Those concerns arise because the reductions in greenhouse gases from offsets are generally more difficult to verify than the reductions from sources whose emissions are subject to the cap. Moreover, some types of offsets are more difficult to verify than others. For example, although it is relatively easy to measure the amount of methane captured in the United States from using special processes to treat animal waste, it is quite difficult to measure the amount of carbon removed from the atmosphere because of efforts to plant trees or avoid deforestation in developing countries.

Offsets are used by a number of existing climate programs, which employ a variety of strategies, varying in rigor and cost, for verifying the reductions in greenhouse gases claimed by an entity offering an offset.[1]

The Clean Development Mechanism was created in December 1997 under the United Nations Framework Convention on Climate Change, to assist countries in meeting the goal for reducing emissions under the Kyoto Protocol. Industrialized countries can purchase offsets from developing countries and use them to meet a portion of their commitment to reduce greenhouse gases.

The Regional Greenhouse Gas Initiative in the United States, established in 2005, requires power plants that rely on fossil fuels and are located in 10 Northeastern member states to reduce emissions. Members can purchase offsets generated in participating states and, under certain circumstances, elsewhere in the United States and internationally to meet a portion of their compliance obligation.

The Chicago Climate Exchange was established in 2003. Members have made voluntary, but legally binding, commitments to reduce their greenhouse gases. Members can use domestic and international offsets to help meet those commitments.

The Voluntary Carbon Standard was developed in 2007 to establish uniform and transparent standards for a worldwide voluntary market made up of a number of mechanisms through which buyers from the public and private spheres can achieve self-defined objectives by funding activities that reduce greenhouse gases. Projects that do so can have their offsets certified by adhering to the standards.

Verifying that offsets actually reduce greenhouse-gas emissions generally involves addressing four issues:

- Offsets would need to bring about *additional* reductions in greenhouse gases. That is, they would need to result in reductions that would not have occurred in the absence of the program that grants credit for offsets.
- Offsets would need to be *quantifiable* so that any reductions in greenhouse gases could be reliably measured.
- Offsets would need to be *permanent* rather than simply delay the release of greenhouse gases into the atmosphere.
- Offsets would need to be credited in a way that accounted for *leakage* in the form of higher emissions elsewhere or in different sectors of the economy as a result of the offset activity.

## Identifying Additional Reductions Attributable to the Policy

Different climate programs use a variety to strategies to ensure that offsets credited in a cap-and-trade program satisfy "additionality"—that is, that they effect reductions in greenhouse gases that would not have occurred otherwise. Simple strategies for identifying reductions attributable to offset policies include accepting only activities that are not mandated by other laws, activities that reduce greenhouse gases after a specified date, and activities that are not common practice. Other possible strategies involve performance standards or the use of specific technologies. Still more complex assurances can be sought through demonstrations that the production of offsets—by planting trees, for example— would constrain an alternative use of resources that (apart from the value of the offsets) would be more profitable—such as using that land as pasture for livestock.

The United Nations Clean Development Mechanism, for example, employs all three of the simple checks. In addition, it requires that providers of offsets either document that their projects could not be implemented without the offset program's support or demonstrate that the projects are not prompted by intrinsic financial gains. To document the need for the program's support, offset providers must offer evidence of barriers to implementation. Those barriers may relate to investment (such as limited access to capital markets), technology (such as a lack of skilled labor or of access to materials and equipment), institutions (such as uncertain land ownership and tenure), or other factors. As evidence, the Clean Development Mechanism accepts market and statistical data, sector studies, legislative and regulatory information, and assessments by independent experts. Alternatively, offset providers can show that the financial benefits of producing the offsets (aside from selling them to entities subject to the cap) are less than the benefits available through alternative uses of the resources. Evidence must be based on standard market measures that are not linked to subjective expectations of profitability, and they must be bolstered by an analysis showing how the conclusions would vary with reasonable changes to key assumptions.

## Quantifying Reductions

Processes employed by different climate programs for quantifying reductions vary in their level of detail, degree of transparency, and procedures for external verification. Depending on the activity, offsets may be estimated on the basis of general relationships (such as estimates of the amount of carbon storage expected when minimizing the extent to which soil is disturbed by agriculture in different geographic regions) or measured directly (for example, the amount of methane captured from the decomposition of animal waste in holding tanks). Direct measurement may provide greater certainty but often comes at greater cost. Quantification processes that are more transparent promote oversight by interested parties, and many programs require that third parties verify the reductions of greenhouse gases reported by offset providers.

The Regional Greenhouse Gas Initiative, for instance, requires that offset providers use preapproved, publicly available methodologies for calculating offsets, have quality control programs, and hire accredited third parties to validate the calculations. The initiative then follows those steps with a separate determination to award credit for offsets.

## Ensuring That Reductions Are Permanent

Concerns about the permanence of reductions in greenhouse gases brought about by offsets are heightened if no one is liable for unintended or unforeseen releases. Ascertaining permanence is a particular challenge for carbon offsets generated from land use, because carbon stored in plants and soils can be released to the atmosphere by environmental changes such as forest fires and pest infestations as well as by human activities such as logging and plowing.

Climate programs address concerns about permanence in various ways. Some programs require legal assurances that carbon will remain stored. Others assign expiration dates to offsets, and once those dates have passed, entities subject to the cap can no longer use those offsets to meet compliance obligations and must replace them.[2] Some programs hold in reserve a portion of the credits earned by each offset activity and use that pooled reserve to compensate for any reversals of carbon storage.[3] For example, the Voluntary Carbon Standard calls for holding in reserve between 10 percent and 60 percent of the offsets produced by an agriculture or forestry project, depending on the project's risk of reversal. That risk is regularly reevaluated and the reserve amount adjusted as needed to account for changes in the project's financial, technical, and management situation; the economic risk of changing land values; the risk posed by regulatory and social instability; and the risk of natural disturbances.

## Accounting for "Leakage," or Related Increases in Emissions

Leakage—increases in emissions elsewhere that stem from the activities producing offsets—diminishes the net effect of offsets in reducing greenhouse gases, but it can be hard to identify and quantify, which makes it extremely difficult to address. The smaller the scope of leakage—within the holdings of the offset provider, for example—the easier it is to account for, but when leakage occurs on a national or international level or in economic sectors other than the one generating the offset, accounting for it is a bigger challenge. For instance, offsets produced by capturing methane emissions from livestock waste may not result in increased emissions elsewhere; however, preserving trees in one location would reduce the supply of timber on the world market, thereby raising its price and encouraging increased production elsewhere, which would be difficult to prevent or measure.

Programs try to deal with leakage in two ways: by requiring certain design features that minimize it and by applying discounts when issuing offsets to account for leakage that cannot be avoided. The Chicago Climate Exchange, for example, requires offset providers to manage their forestry holdings in a sustainable way. The program also requires projects to quantify leakage, but only within a developer's own land holdings. That approach ignores changes in land use that are less proximate to the offset but nonetheless attributable to the offset project.

## THE EFFECT OF OFFSETS ON THE COST OF H.R. 2454

In analyzing the cap-and-trade program in H.R. 2454, the American Clean Energy and Security Act of 2009, which was passed by the House of Representatives, CBO estimates that the availability and use of offsets would reduce the net cost of complying with the cap by about 70 percent between 2012 and 2050. The net cost includes the gross cost of complying with the cap minus the sum of the allowance value that would be returned to U.S. households and the net revenues resulting from the domestic production of offsets.[4]

H.R. 2454 would allow regulated entities to substitute offsets in lieu of up to 2 billion greenhouse-gas allowances each year. By comparison, total greenhouse-gas emissions in the United States were about 7 billion tons in 2007. Under the bill, domestic offsets could be used in place of up to 1 billion allowances per year and international offsets, in place of an additional 1 billion allowances. In recognition of the greater challenge of verifying international offsets, after 2017 the legislation would require 1.25 tons of reductions from international offsets to substitute for an allowance representing 1 ton of emissions—thus discounting international offsets by 20 percent. If fewer than 900 million domestic offsets were used, the use of international offsets could be increased to make up the shortfall but could never substitute for more than 1.5 billion allowances in a given year. In no case could domestic and international offsets together substitute for more than 2 billion allowances.

CBO expects that regulated entities would take advantage of those provisions when the costs were less than those for other methods of compliance—such as reducing their own emissions or purchasing allowances. CBO estimates that regulated entities would use domestic offsets in place of about 230 million allowances in 2012 and about 300 million allowances in 2020. Annual use of domestic offsets would probably not reach the limit of 1 billion tons until after 2040. Regulated entities would use international offsets in place of about 190 million allowances in 2012 and about 340 million allowances in 2020. The constraint of 2 billion metric tons on the overall use of offsets would become restrictive for the first time shortly after 2030. Over the 2012–2050 period, by CBO's estimates, offsets would account for about 45 percent of the total emission reductions resulting from the cap, including reductions made by regulated entities as well as those made through offsets. A little fewer than half of those offsets would be produced domestically (see Figure 2).

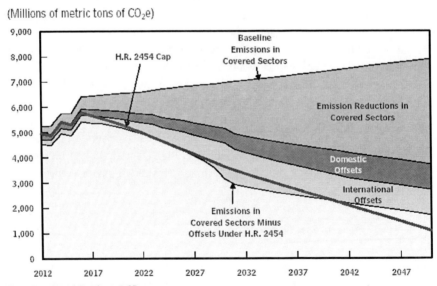

Source: Congressional Budget Office.
Notes: $CO_2e$ = carbon dioxide equivalent.
　　　The figure includes both cap-and-trade programs specified under H.R. 2454: the one for hydrofluorocarbons and the one for all other greenhouse gases.

Figure 2. Estimated U.S. Emissions Under H.R. 2454, the American Clean Energy and Security Act of 2009

By reducing the cost of complying with the cap, the use of offsets would have a significant effect on allowance prices. Together, the provisions allowing the use of domestic and international offsets would decrease the price of greenhouse-gas allowances by about 70 percent over the 20 12–2050 period because they would provide a cheaper alternative for reducing greenhouse gases than relying exclusively on reductions from regulated entities.[5]

Domestic offsets would probably come predominantly from the forestry sector, where producers would find it profitable to make changes in forest management and increase the planting of trees to increase carbon storage. Only about 10 percent of the offsets generated in the United States would come from agriculture. In the supply of international offsets, ones deriving from agriculture would probably be roughly equal in importance to ones from forestry. Those agricultural offsets would be generated primarily through the reduction of methane and nitrous oxide emissions from livestock, cropland, and rice paddies.

## Estimating the Supply of Offsets

CB O's approach to estimating the supply of offsets incorporates three factors: the direct costs of an activity that produces an offset, such as the cost of planting trees; the forgone value of other uses of the land; and the costs associated with verifying and bringing offsets to the market.

CBO's analysis drew on data from the Environmental Protection Agency, which are the most comprehensive available.[6] The data incorporate direct costs and the forgone value of other uses of the land. EPA's estimates of the costs of offsets supplied by the agriculture and forestry sectors in the United States and by the forestry sector outside the United States were generated by models that simulate profit-maximizing decisions by landowners and acknowledge, to different degrees, the choices that they face among different land uses (including different strategies for generating offsets) and the market responses associated with those choices. For example, a landowner takes into account information on how the value of the current use of the land compares with that of, say, growing crops for biofuels or growing trees to store carbon if a climate program is in place. EPA's estimates of the number of offsets supplied by the agriculture sector outside the United States came from engineering studies that focus on direct costs—for which the quality of data varies by region and by practice—and are less effective at accounting for alternative uses of resources that may be more profitable to landowners.[7]

CBO adjusted EPA's data for the costs of verifying and bringing offsets to the market, in two ways. First, for both international and domestic offsets, CBO added an estimated verification cost of $5 per metric ton of carbon dioxide equivalent ($CO_2e$).[8] (By way of comparison, that $5 verification cost is less than 10 percent of CBO's estimate of what the allowance price would be in 2012 *without* offsets.) CB O's estimate reflects information from the few available studies that use data from pilot projects involving offsets and projects in the agriculture, forestry, waste, and energy sectors, but there is no consensus on how to define, quantify, and predict such costs.[9] The studies define costs differently and may include expenses for feasibility studies, technical assistance, verification, administration, regulatory approval, and efforts to locate offset buyers and sellers and negotiate transactions.[10] Those costs, which vary by type of project and region, are lower in more mature markets—

indicating a potential benefit in adopting verification procedures with which there is some familiarity gained through existing offset markets. Some researchers have found, however, that the apparent influence of a mature market on the costs is actually attributable to economies of scale and that projects generating greater numbers of offsets are simply the ones that have lower per-ton verification costs.

Second, CBO adjusted EPA's projected supplies of international offsets to account for the challenges of bringing offsets to the cap-and-trade market. Under H.R. 2454, developing countries generating international offsets for the market would have to be party to an agreement with the United States. CBO expects that such agreements would address developing countries' institutional and technical capacity to verify offsets, that negotiating the agreements would take a significant amount of time, and that it would not be possible to reach agreements to produce carbon offsets from the energy sectors of developing countries. CBO concluded that the number of agreements and the scope of their coverage would increase over the 2012–2050 period covered by the legislation but that throughout the period the supply of offsets would be lower than that estimated by EPA.[11] CBO's assessment, which is subject to significant uncertainty, is based on indicators of regulatory bodies' capacity to verify offsets and on information from the Department of State, EPA, and outside experts on negotiating agreements.[12]

## The Projected Use of Offsets

To illustrate the role of offsets under H.R. 2454, CBO has estimated their impact in 2030 after making adjustments for the costs of verifying and bringing offsets to the market and taking into account the fact that other developed countries would also wish to purchase international offsets (see Table 1). The legislation would establish a cap on greenhouse-gas emissions in 2030 of 3,427 million metric tons of $CO_2e$, so the government would distribute 3,427 million allowances in that year. Without offsets, 3,555 million metric tons of emissions would occur in 2030, CBO estimates, which would be equal to the number of allowances distributed that year plus 128 million allowances that entities would have banked in previous years and chose to use in 2030.

With offsets, as allowed for in the bill, sources with compliance obligations would emit 5,031 million metric tons and purchase offsets for 1,790 million metric tons— about one-third supplied domestically and about two-thirds supplied internationally, CBO estimates. About 60 percent of those domestic offsets would come from forestry and agriculture—the vast majority (roughly 90 percent) from forestry. About 80 percent of those international offsets would come from agriculture and forestry—the majority (roughly 60 percent) from agriculture.

If the offsets represented true incremental reductions, then net emissions would be 3,241 million metric tons (5,031 minus 1,790). The sources subject to the cap would use 3,241 million allowances to cover their net emissions and would bank 186 million allowances (3,427 distributed minus 3,241 used) to cover future emissions.

# The Use of Agricultural Offsets to Reduce Greenhouse Gases

**Table 1. Effects of H.R. 2454, the American Clean Energy Security Act, With and Without Offsets, 2030**

| | With Offsets | Without Offsets |
|---|---|---|
| | **Billions of 2007 Dollars** | |
| Net Cost[a] | 101 | 248 |
| | **Million Metric Tons $CO_2e$** | |
| Net Cap on Greenhouse Gases | 3,427 | 3,427 |
| Emissions from Sources Subject to Limits | 5,031 | 3,555 |
| Allowances Banked[b] | 186 | -128 [c] |
| Emissions Covered by Offsets | 1,790 | 0 |
| | **Dollars/Metric Ton $CO_2e$** | |
| Allowance Price | 40 | 138 |

Source: Congressional Budget Office.

Notes: Emissions are represented in terms of carbon dioxide equivalent ($CO_2e$). A metric ton of $CO_2e$ is the amount of a given greenhouse gas (for example, methane or nitrous oxide) that makes the same contribution to global warming as a metric ton of carbon dioxide.

Whereas the dollar figures in this table (as well as the text) are reported in constant 2007 dollars, those in CBO's cost estimates, including the one for H.R. 2454, the American Clean Energy and Security Act of 2009, as reported by the House Committee on Energy and Commerce on May 21, 2009 (June 5, 2009), are in nominal dollars.

a. As measured here, the United States' net cost includes the gross cost of complying with the cap minus the sum of the allowance value that would be returned to U.S. households under H.R. 2454 and the net revenues resulting from the domestic production of offsets. The net cost also represents the loss in purchasing power that households would experience as a result of the policy. As measured here, the net cost does not include the costs that some current investors and workers in sectors of the economy that produce energy and energy-intensive goods and services would incur as the economy moved away from the use of fossil fuels or the full range of effects on the economy, nor does it include the benefits of the reduction in greenhouse gases and the associated slowing of climate change. For more information, see Congressional Budget Office, *The Economic Effects of Legislation to Reduce Greenhouse-Gas Emissions* (September 2009).

b. Under H.R. 2454, allowances could be banked and used to cover future emissions. (Borrowing future allowances for current use could also occur for up to five years, with certain restrictions.)

c. The negative amount indicates that entities would be using allowances that they banked in previous years.

## The Impact of Offsets on Net Costs and the Price of Allowances

The substantial use of offsets would significantly reduce the net cost of the cap-and-trade program that H.R. 2454 would establish. Without offsets, net costs would be an estimated $248 billion in 2030 (expressed in 2007 dollars), or about 1 percent of gross domestic product in that year. By CBO's estimate, the availability of offsets would reduce those costs by about 60 percent during that year—to an estimated $101 billion. On average during the overall period that the legislation would be in effect, offsets would reduce net costs by about 70 percent.

With offsets, more emissions would be allowed from sources subject to the cap, thus making allowances less valuable. Without offsets, the price of an allowance in 2030 would be $138 per metric ton (in 2007 dollars), CBO estimates; with offsets, the allowance price would be only $40 per metric ton.

Finally, if international offsets were not available to regulated entities, the use of domestic offsets would expand. Entities subject to the cap would use an estimated 891 million domestic offsets in 2030 (more than the use of domestic offsets projected under H.R. 2454 but not as much as the use of international offsets under the legislation), and the allowance price and the net cost of the policy would be greater than that under the legislation. This alternative would benefit offset producers in the domestic agriculture and forestry sectors, but the program would be less effective at lowering net costs to the economy as a whole.

## SOURCES OF UNCERTAINTY

The potential for offsets to reduce net costs depends critically on the types and sources of offsets allowed and on the costs of producing and verifying offsets. H.R. 2454 provides neither detailed specifications for the types and sources of offsets to be included in the cap-and-trade program nor the methodologies necessary to verify those offsets; it assigns primary responsibilities for those determinations to two federal agencies. For domestic offsets from changes in agriculture and forestry, that responsibility would fall to the Department of Agriculture, which would take into account the recommendations of its Greenhouse Gas Emission Reduction and Sequestration Advisory Committee, established under the legislation. For all other offsets, that responsibility would fall to EPA. That agency would consult with appropriate federal agencies; take into account the recommendations of the Offsets Integrity Advisory Board, also established by the legislation; and accept international offsets only if the country providing them had negotiated an agreement or arrangement with the United States.

CBO's estimates of the costs to produce offsets are based on data from EPA that take into account a wide range of types and sources of activities that could generate offsets. CBO adjusted those data to reflect its best judgment of how regulators might identify classes of offsets and how methodologies required for verification might affect costs. Actual developments might turn out quite differently.

There are uncertainties inherent in the modeling used to generate initial estimates of the supply of offsets—such as the extent to which they are able to account for competition among different land uses and other market responses.[13] Moreover, the data used in modeling are themselves uncertain. For example, recently revised estimates of past deforestation rates imply lower potential for generating offsets through avoided deforestation. Also, the types and sources of offsets that would ultimately be allowed under a cap-and-trade program in the United States could be different from those envisioned in EPA's data and CBO's estimates. Verification costs, too, are uncertain because of a lack of relevant experience. All of those factors have implications for the ultimate impact of offsets on the net cost of the policy to reduce the concentration of greenhouse gases that would be established by H.R. 2454.

## End Notes

[1] See Congressional Budget Office, *The Use of Offsets to Reduce Greenhouse Gases,* Issue Brief (August 3, 2009).

[2] In addition to providing for the use of standard offsets, H.R. 2454 also provides for the use of expiring offsets generated by agricultural practices that sequester greenhouse gases.

[3] H.R. 2454 lists that approach as one mechanism that regulators should consider using to address concerns about the permanence of reductions.

[4] The net cost represents the loss in purchasing power that households would experience as a result of the policy. See Congressional Budget Office, *The Economic Effects of Legislation to Reduce Greenhouse-Gas Emissions* (September 2009) for a discussion of how the loss in purchasing power resulting from H.R. 2454 would be distributed among households in different income brackets.

[5] Under H.R. 2454, regulated entities would be allowed to hold for later use as many allowances as they chose. Thus, their profit-maximizing behavior would cause the price of an allowance to increase at the same rate as the return they expected to receive on comparable alternative investments. As a result, even though the composition of reductions in greenhouse gases (that is, from regulated entities, from domestic offsets, and from international offsets) would change over time, the use of offsets would lower the price of allowances in any given year by the same amount.

[6] See the data annex for EPA's analysis of H.R. 2454 in the 111th Congress, the American Clean Energy and Security Act of 2009, available at www.epa.gov/climatechange/economics/. The data sources are described in three publications: Environmental Protection Agency, *Greenhouse Gas Mitigation Potential in U.S. Forestry and Agriculture*, EPA 430-R-05-006 (November 2005); Environmental Protection Agency, *Global Mitigation of Non-$CO_2$ Greenhouse Gases*, EPA 430-R-06-005 (June 2006); and Brent Sohngen and Robert Mendelsohn, "A Sensitivity Analysis of Carbon Sequestration," in *Human-Induced Climate Change: An Interdisciplinary Assessment*, edited by Michael E. Schlesinger and others (Cambridge: Cambridge University Press, 2007).

[7] Estimates of the supply of offsets from outside the agriculture and forestry sectors, both within and outside of the United States, have also been derived from those engineering models.

[8] A metric ton of carbon dioxide equivalent is the amount of a given greenhouse gas, such as methane or nitrous oxide, that makes the same contribution to global warming as a metric ton of carbon dioxide.

[9] See Oscar Cacho and others, *Economic Potential of Land-Use Change and Forestry for Carbon Sequestration and Poverty Reduction,* Technical Report 68 (Australian Center for International Agricultural Research, 2008); Camille Antinori and Jayant Sathaye, *Assessing Transaction Costs of Project-Based Greenhouse Gas Emissions Trading* (Lawrence Berkeley National Laboratory, Environmental Energy Technologies Division, January 25, 2007); Neeff Till and others, *Update on Markets for Forestry Offsets* (Tropical Agricultural Research and Higher Education Center, September 2007); and Axel Michaelowa and Frank Jotzo, "Transaction Costs, Institutional Rigidities, and the Size of the Clean Development Mechansim," *Energy Policy,* vol. 33, no. 4 (March 2005), pp. 511–523.

[10] Verification costs estimated by the four studies range from $0.10 to $4.30 per metric ton of carbon dioxide equivalent.

[11] CBO's adjustment also takes into account provisions for allocations of allowances to support emission reductions from reduced deforestation. Under H.R. 2454, entities receiving such support would be prohibited from generating offsets for direct sale into the U.S. market.

[12] CBO also modified EPA's projected supply of offsets to reflect the judgment that activities producing offsets could not be undertaken at negative cost—that is, there are no extensive opportunities for suppliers to adopt practices that would reduce greenhouse gases while also yielding a profit. In EPA's data, the projected availability of offsets at negative cost, which probably derives from not accounting for some barriers to adoption or from omitting some costs, is particularly significant for the practice of controlling methane and nitrous oxide emissions from livestock and cropland in developing countries.

[13] One consideration is the potential for concentration of market power in the hands of a limited number of offset providers—if, for instance, a few parties control significant expanses of forests or if requirements for verification significantly limit entry into the offset market.

# CHAPTER SOURCES

The following chapters have been previously published:

Chapter 1 – This is an edited, excerpted and augmented edition of a United States Congressional Research Service publication, Report Order Code R40813, dated September 17, 2009.

Chapter 2 – This is an edited, excerpted and augmented edition of a United States Congressional Research Service publication, Report Order Code R40874, dated October 26, 2009.

Chapter 3 – This is an edited, excerpted and augmented edition of a United States Congressional Research Service publication, Report Order Code R40936, dated November 30, 2009.

Chapter 4 – This is an edited, excerpted and augmented edition of a United States Congressional Research Service publication, Report Order Code R40667, dated January 4, 2010.

Chapter 5 – This is an edited, excerpted and augmented edition of a United States Department of Agriculture, Office of the Chief Economist, Office of Energy Policy and New Uses, Agriculture Economic Report Number 843, dated February 2008.

Chapter 6 – This is an edited, excerpted and augmented edition of a United States Congressional Budget Office Issue Brief, dated November 23, 2009.

Chapter 7 – This is an edited, excerpted and augmented edition of a United States Energy Information Administration, Office of Integrated Analysis and Forecasting, U.S. Department of Energy publication, DOE/EIA-0573(2008), dated December 2009.

Chapter 8 – These remarks were delivered as Statement of Joseph Kile, Assistant Director for Microeconomic Studies, before the Subcommittee on Conservation, Credit, Energy, and Research, Committee on Agriculture, U.S. House of Representatives, dated December 3, 2009.

# INDEX

## A

abatement, 4, 8, 13, 26, 27, 36, 54, 80, 128
accounting, vii, 1, 8, 54, 140, 153, 183, 214, 239, 241, 245
accreditation, 57
acid, 23, 26, 29, 34, 35, 197, 228
acidity, 199
acquisitions, 33, 52
adaptation, 6, 27, 61, 155
additives, 214, 224, 226
adjustment, 39, 52, 179, 245
aerosols, 142, 201
Africa, 151
agricultural sector, 80
agriculture, vii, viii, 1, 2, 3, 4, 9, 11, 14, 15, 18, 22, 27, 29, 34, 35, 36, 47, 48, 49, 63, 64, 80, 86, 92, 120, 142, 144, 145, 158, 183, 187, 191, 234, 238, 239, 241, 242, 244, 245
air emissions, 48, 96
air pollutants, 5, 67
air quality, 90, 91, 99
airports, 38
Alaska, 10, 16, 208
alcohol, 197, 223, 228, 229
alternative energy, 52, 68, 188
alternatives, viii, 2, 27, 127, 128
aluminum, 34, 35, 52, 55, 72, 142, 181, 200, 203, 225
American Recovery and Reinvestment Act, 8, 66, 87, 152, 188
ammonia, 24, 29, 35, 81, 95, 96, 112, 197
ammonium, 26
anaerobic bacteria, 77
anaerobic digesters, 76, 90, 91, 102
annual rate, 72, 168, 206
ash, 35, 142, 181, 217, 220
Asia, 9, 59, 150

assessment, ix, 32, 44, 96, 97, 101, 106, 234, 242
assets, 109
assumptions, 87, 105, 106, 108, 111, 130, 131, 132, 133, 238
Attorney General, 101
attribution, 144
Australia, 43, 44, 71, 148, 150, 155
Austria, 117
authority, viii, 5, 14, 15, 16, 21, 28, 30, 41, 54, 57, 63, 152
automobiles, 4, 41, 218
availability, 13, 26, 61, 77, 92, 100, 112, 122, 124, 130, 132, 133, 209, 239, 243, 245

## B

background, 102
bacteria, 8, 102, 105, 212, 213, 219, 227, 228, 235
banking, 36, 69, 123, 129, 130, 131, 132
barriers, x, 11, 12, 75, 83, 238, 245
base year, 48
basic research, 95
behavior, 121, 126, 128, 135, 245
bicarbonate, 35
binding, ix, 16, 31, 36, 41, 53, 54, 57, 61, 65, 68, 73, 222, 237
biodiesel, 90
biofuel, 127, 152
biological processes, 7, 8, 215
biomass, 8, 15, 16, 20, 29, 46, 68, 83, 85, 93, 96, 152, 214, 220, 227
biosphere, 215, 216
bloodstream, 219
bonds, 15, 20, 57
boreholes, 13
borrowing, 123, 128, 129
Brazil, 9, 10, 45, 46, 47, 48, 71, 72, 151, 155, 220
breakdown, 99, 213
Britain, 71

bromine, 28, 215, 221
building code, 53
Bulgaria, 36
burn, 11, 20, 103, 188
burning, 93, 120, 121, 124, 135, 187, 195, 219, 220, 224
Bush, President George W., 74
butyl ether, 219, 223

# C

calcium, 181, 217
calcium carbonate, 181, 217
Canada, 48, 49, 72, 94, 106, 113, 150, 155, 164, 165, 166, 231
capacity building, 53
capital goods, 126
capital markets, 238
carbohydrate, 118
carbohydrates, 219, 226
carbon atoms, 221, 222, 225
carbon monoxide, 81, 225
catalyst, 29, 197, 216, 217
cattle, 11, 46, 80, 81, 186, 193, 230
cell, 19
cellulose, 227
Census, 73
chain of production, 125
chemical industry, 190
childrearing, 121
China, 9, 10, 50, 51, 52, 53, 72, 149, 150, 155
chlorine, 28, 217, 221, 225
chloroform, 223
chlorophyll, 226
classes, 62, 66, 128, 244
classification, 218, 227
Clean Air Act, viii, 4, 5, 7, 8, 11, 12, 14, 18, 21, 22, 27, 28, 29, 67, 68, 86, 152, 202, 218, 227
clean energy, ix, x, 6, 7, 66, 75, 76, 85, 155
cleaning, 142, 203, 223
climate change, vii, viii, x, 1, 2, 6, 8, 11, 21, 22, 23, 26, 27, 28, 31, 37, 39, 40, 48, 49, 50, 53, 60, 63, 65, 66, 67, 68, 72, 74, 80, 81, 85, 86, 89, 90, 91, 130, 133, 134, 135, 152, 155, 220, 222, 243
closure, 51
clustering, 213
clusters, 210
coal, vii, 1, 2, 3, 4, 5, 6, 13, 14, 15, 19, 38, 45, 46, 49, 51, 63, 64, 73, 96, 125, 131, 135, 137, 140, 142, 144, 145, 162, 166, 167, 171, 172, 183, 184, 195, 213, 214, 217, 218, 220, 223, 226, 228, 229, 231
codes, 51, 68
coke, 35, 158, 170, 172, 190, 191, 214, 217, 226

collaboration, 20, 84
colonization, 47
combustion, 4, 17, 18, 19, 20, 23, 27, 29, 35, 86, 87, 103, 110, 112, 118, 128, 138, 140, 142, 144, 145, 146, 154, 158, 159, 162, 172, 181, 183, 184, 188, 195, 198, 215, 219, 220
communication, 96, 114
community, 11, 70, 85, 86, 156
competition, 37, 108, 244
competitiveness, 14, 37, 40, 154
competitor, 108
competitors, 33, 37
complexity, 83, 130, 158
compliance, 4, 34, 36, 41, 55, 66, 70, 73, 117, 129, 235, 237, 239, 240, 242
complications, 103
components, 45, 56, 84, 112, 139, 146, 161, 163, 169, 170, 173, 175, 176, 178, 179, 180, 182, 184, 185, 188, 189, 191, 193, 195, 197, 198, 199, 201, 203, 204, 205, 207, 209, 212, 214, 224
composition, 11, 52, 126, 133, 217, 227, 245
composting, 103, 211
compounds, 28, 91, 201, 215, 217, 221, 224, 225, 226, 229
compression, 103
concentration, 13, 52, 81, 82, 102, 105, 199, 218, 233, 234, 244, 245
concrete, 39, 54, 79
condensation, 217
conditioning, 144, 155
conductor, 60
conductors, 64
confidence, 84
conformity, 70
confusion, 93
Congress, 1, 2, 5, 6, 7, 14, 18, 19, 21, 27, 28, 30, 31, 65, 67, 68, 75, 76, 81, 83, 85, 86, 101, 113, 129, 130, 135, 234, 245
Congressional Budget Office, 129, 132, 134, 135, 136, 234, 236, 240, 243, 245, 247
connectivity, 54
consensus, 131, 241
conservation, 27, 29, 50, 57, 64, 84, 87, 95, 131
construction, 16, 39, 57, 83, 84, 85, 94, 97, 99, 124, 214
consultants, 94
consumer goods, 219
consumers, 3, 53, 66, 93, 97, 101, 120, 121, 125, 126, 130, 226
consumption, vii, x, 8, 18, 34, 40, 48, 57, 64, 92, 119, 120, 121, 122, 125, 126, 127, 128, 131, 133, 137, 140, 142, 145, 155, 158, 166, 168, 171, 172, 176, 179, 181, 184, 215, 233

# Index 251

contamination, 95
control, vii, viii, ix, 1, 3, 4, 5, 27, 30, 31, 32, 33, 49, 51, 55, 67, 71, 81, 86, 93, 98, 102, 106, 107, 112, 114, 120, 124, 125, 162, 197, 245
conversion, 6, 24, 26, 29, 48, 81, 112, 140, 142, 158, 214, 217, 228, 230
cooling, 36, 120, 160, 166, 167
corn, 25, 26, 29, 76, 93, 113, 191
corporate average fuel economy, 152
cost saving, 66, 106, 236
costs, x, 3, 5, 7, 11, 12, 13, 15, 22, 29, 33, 37, 44, 71, 75, 76, 79, 84, 94, 95, 96, 99, 101, 104, 106, 107, 108, 111, 113, 119, 120, 121, 122, 123, 124, 125, 126, 127, 128, 130, 131, 133, 134, 135, 203, 233, 235, 236, 240, 241, 242, 243, 244, 245
cotton, 25, 29
covering, 3, 4, 56, 107
credibility, 125, 236
credit, 4, 15, 20, 36, 83, 85, 87, 108, 114, 135, 217, 237, 238
crop production, 22, 26, 235
crops, 23, 25, 29, 93, 99, 103, 224, 241
crude oil, 93, 213, 214, 223, 224, 226, 230
crystalline, 215, 230
cultivation, 18, 19, 23
currency, 71, 72, 73
customers, 93, 100
cyclohexanol, 223
Cyprus, 36
Czech Republic, 36

## D

dairies, 102, 109, 111
dairy industry, 99
data collection, 18
decisions, 39, 48, 124, 125, 126, 218, 241
decomposition, 11, 188, 213, 214, 215, 218, 228, 238
deduction, 36
deficit, 7, 71
definition, 18, 33, 45, 96, 154, 179
deforestation, vii, x, 45, 46, 47, 61, 71, 119, 213, 233, 235, 237, 244, 245
degradation, 106
delivery, 226
denitrification, 22, 24, 26, 193
Denmark, 34, 38, 91, 92, 93, 98, 99, 103, 113, 116, 155
density, 124, 213
Department of Agriculture, 3, 14, 29, 76, 86, 87, 91, 94, 97, 116, 244, 247
Department of Commerce, 97, 116, 143
Department of Energy, 14, 15, 19, 42, 71, 87, 94, 115, 130, 152, 162, 230, 231, 247

Department of the Interior, 6, 7
deposits, 16
depreciation, 16
destruction, x, 6, 28, 47, 89, 92, 104
detection, 205
detergents, 226
developed countries, 34, 63, 155, 242
developed nations, 155
developing countries, 36, 48, 155, 217, 235, 237, 242, 245
diamonds, 55
diesel engines, 218
diesel fuel, 140, 172, 218
digestion, vii, ix, x, 1, 11, 19, 20, 75, 76, 83, 84, 85, 86, 87, 91, 92, 95, 104, 107, 186
direct cost, 121, 128, 241
direct costs, 128, 241
directives, 113
discounting, 240
disorder, 12
displacement, 229
disposition, 214
dissociation, 16
distillation, 218, 222, 224, 228
distortions, 135
distribution, vii, 1, 42, 46, 47, 93, 99, 121, 125, 126, 142, 144, 153, 183, 204, 215, 220, 226
district heating, 92
District of Columbia, 151, 179
diversification, 46
draft, 36, 152, 153, 154
drainage, 13, 23
duration, 66, 227
Duration, 164, 165

## E

earnings, 133, 134
earth, 217, 221, 226, 229
eating, 221
economic activity, 63, 121, 123, 126, 128, 130
economic crisis, 65, 155
economic downturn, 47
economic evaluation, 95
economic growth, 36, 126, 130, 132, 141, 148, 159, 160, 171, 172
economic growth rate, 148
economic incentives, 11, 63
economic integration, 49, 74
economics, vii, x, 1, 2, 6, 26, 89, 90, 130, 134, 245
economies of scale, 242
ecosystem, 29
Education, 118, 245
effluent, 77, 95, 104, 112

252 Index

Egypt, 155
election, 49, 56
electric arc furnaces, 160
electromagnetic, 213
emitters, 11, 81, 151
employment, 42, 121, 126, 133
emulsions, 214
encouragement, 47, 89
end-users, 224
energy consumption, 34, 38, 50, 51, 54, 140, 144,
    146, 161, 178, 179, 216
energy efficiency, 6, 7, 27, 39, 40, 42, 43, 46, 53, 54,
    55, 57, 61, 63, 64, 66, 67, 124, 152, 153
Energy Independence and Security Act, 15, 152
Energy Policy Act of 2005, 16, 29, 96, 101, 113
Energy Policy and Conservation Act, 68, 74
energy recovery, 67
energy supply, 64, 141, 159
Enhanced Oil Recovery, 231
environment, x, 7, 29, 60, 70, 71, 72, 76, 92, 119
environmental change, vii, x, 119, 239
environmental issues, 86
environmental policy, 94
Environmental Protection Act, 49
Environmental Protection Agency, viii, 2, 14, 17, 18,
    19, 20, 22, 23, 24, 28, 29, 49, 66, 74, 86, 87, 94,
    95, 96, 115, 117, 128, 130, 132, 135, 151, 184,
    193, 201, 203, 204, 205, 207, 209, 211, 212, 224,
    227, 230, 231, 234, 236, 241, 245
environmental standards, 37
environmental technology, 60
EPA, viii, 2, 3, 5, 9, 12, 14, 18, 19, 20, 21, 22, 24,
    25, 26, 27, 28, 29, 30, 55, 57, 66, 67, 68, 69, 74,
    77, 78, 80, 87, 94, 102, 130, 151, 152, 158, 184,
    188, 193, 200, 202, 204, 205, 207, 208, 209, 210,
    211, 212, 218, 231, 234, 241, 242, 244, 245
equilibrium, 128
erosion, 26
estimating, 121, 135, 219, 241
Estonia, 36
etching, 142
ethanol, 26, 48, 49, 76, 90, 112, 113, 158, 172, 188,
    191, 219
etherification, 219, 229
ethylene, 158, 190, 223, 228
EU, ix, 31, 32, 33, 34, 35, 36, 37, 38, 39, 40, 41, 42,
    61, 63, 65, 69, 70
Eurasia, 150
Euro, 71
Europe, x, 38, 40, 51, 53, 63, 70, 89, 90, 92, 104,
    148, 150
European Commission, 33, 34, 36, 37, 39, 40, 41
European Community, 34

European Court of Justice, 70
European Parliament, 34, 70, 71
European policy, 39
European Union, viii, ix, 9, 10, 31, 32, 33, 34, 35,
    37, 38, 40, 42, 57, 74, 155
evaporation, 225
exchange rate, 33, 71, 72, 73
exercise, 125
expenditures, 121, 131
expertise, 85, 105
exports, 50, 52, 55, 127
extraction, 16, 19, 63, 220

# F

family, 64, 226
family income, 64
farmers, 93, 235
farms, 91, 92, 93, 94, 95, 101, 104, 105, 107, 110,
    111, 152
fermentation, 2, 3, 6, 9, 11, 17, 18, 80, 86, 186, 219
fertilization, 23, 26, 158, 191, 193
fertilizers, viii, 21, 22, 26, 29, 55, 158, 191, 197, 235
fibers, 37, 197, 226
filament, 37
films, 226
finance, 15, 51, 70, 125, 217, 222
financial resources, 27
financial support, 61, 85, 95
financing, x, 4, 43, 45, 46, 47, 57, 75, 76, 87, 96,
    106, 107, 111, 153
Finland, 34, 38
fires, 9, 239
firms, 45, 52, 120, 122, 123, 125, 126, 128, 130, 132,
    134, 135
fishing, 229
fixation, 29, 216
fixed rate, 42
flatulence, 17, 86
flexibility, 36, 120, 123, 135
flotation, 219
fluctuations, 166, 206
flue gas, 162, 218
fluid, 220
fluorine, 201, 217, 221, 222, 225
fluorine atoms, 221, 222, 225
foams, 226
focusing, 121
food, 35, 37, 42, 91, 92, 95, 103, 104, 107, 186, 211,
    218, 220, 223, 226, 228
food industry, 91
forest management, 47, 235, 241
forest resources, 47

Index 253

forests, 48, 53, 55, 62, 72, 120, 131, 158, 206, 213, 227, 235, 245
fossil, vii, x, 4, 18, 35, 46, 49, 50, 51, 54, 55, 67, 71, 91, 107, 108, 119, 120, 121, 122, 123, 124, 125, 126, 127, 128, 135, 140, 142, 145, 148, 153, 159, 162, 173, 174, 175, 177, 181, 215, 216, 219, 220, 224, 233, 237, 243
France, ix, 32, 34, 37, 38, 155, 231
freshwater, 216
friction, 223
fuel efficiency, 49, 51, 53, 55
full capacity, 107
funding, 14, 19, 42, 43, 49, 54, 65, 66, 67, 83, 95, 101, 112, 153, 162, 237
funds, 7, 14, 54, 72, 95

**G**

gases, vii, viii, ix, x, 1, 8, 11, 18, 19, 21, 22, 28, 31, 32, 35, 37, 49, 51, 56, 62, 65, 76, 80, 81, 86, 119, 120, 125, 132, 135, 137, 138, 142, 144, 146, 155, 156, 200, 216, 218, 219, 220, 221, 222, 223, 226, 227, 228, 233, 234, 235, 237
gasification, 162
gasoline, 29, 38, 55, 56, 140, 172, 195, 214, 216, 223, 224, 225, 227
GDP, 36, 50, 60, 62, 64, 72, 74, 126, 133, 141, 142, 143
GDP per capita, 36
General Accounting Office, 29
generation, 8, 12, 35, 36, 38, 39, 42, 44, 46, 48, 49, 51, 55, 56, 57, 60, 61, 62, 64, 65, 82, 84, 85, 94, 97, 99, 100, 106, 107, 117, 120, 131, 142, 144, 145, 154, 160, 161, 174, 213, 214, 217, 218, 223, 228
Germany, ix, 32, 39, 40, 91, 92, 93, 98, 99, 103, 155
global climate change, 213, 219
global economy, 128
goals, 50, 61, 68, 123, 125, 234, 236
goods and services, 120, 121, 125, 126, 133, 243
government, 2, 5, 14, 26, 39, 41, 43, 44, 45, 47, 48, 49, 50, 51, 52, 53, 54, 55, 56, 57, 60, 61, 62, 63, 64, 65, 68, 70, 72, 91, 93, 94, 98, 121, 124, 125, 126, 129, 154, 227, 242
government intervention, 26
grades, 222, 223, 230
grants, 6, 44, 68, 83, 85, 94, 95, 96, 99, 106, 107, 108, 237
graph, 108
grasslands, 206
gravity, 213, 224
grazing, 86
Greece, 36

greenhouse gases, vii, viii, x, xi, 1, 6, 8, 17, 18, 21, 22, 28, 34, 40, 74, 80, 86, 90, 118, 119, 120, 122, 124, 129, 130, 131, 135, 137, 145, 146, 151, 156, 201, 216, 219, 222, 233, 234, 235, 236, 237, 238, 239, 240, 241, 243, 244, 245
grids, 40, 60
gross domestic product, 50, 121, 126, 131, 141, 243
groundwater, 26, 230
groups, 3, 129, 132, 134
growth, 9, 41, 50, 51, 60, 64, 68, 72, 97, 112, 119, 122, 133, 137, 141, 148, 160, 162, 167, 168, 172, 188, 202, 223
growth pressure, 68
guidance, 9
guidelines, 35, 39, 53, 153, 158, 184, 193, 199
Gulf of Mexico, 16, 20

**H**

half-life, 80
halogens, 221
Hawaii, 10
health, 47, 94, 105, 218
heat, x, 8, 10, 22, 23, 39, 42, 49, 61, 75, 76, 80, 81, 82, 86, 89, 91, 92, 103, 104, 108, 112, 140, 213, 214, 216, 217, 220, 223, 228, 229
heat transfer, 229
heating, 36, 46, 82, 84, 87, 91, 93, 103, 106, 120, 138, 140, 162, 166, 167, 213, 217, 218, 221, 227, 229
high density polyethylene, 78
higher quality, 4
highways, 38, 124
House, 8, 18, 19, 69, 95, 116, 129, 135, 153, 233, 239, 243, 247
household sector, 41
households, 34, 38, 64, 120, 121, 122, 123, 125, 126, 128, 130, 131, 132, 134, 239, 243, 245
housing, 35, 36, 64, 167
human activity, 213
humidity, 217, 228
Hungary, 36
hybrid, 60
hydrocarbons, 214, 215, 224, 225, 227, 230
hydrocracking, 216
hydroelectric power, 52
hydrogen, 35, 81, 82, 91, 201, 216, 217, 221, 222, 223

**I**

implementation, vii, x, 1, 46, 51, 57, 70, 75, 76, 85, 97, 132, 154, 218, 238
imports, 37, 39, 52, 55, 92, 154, 172

254 Index

impurities, 103, 224
in transition, 222
incentives, vii, 1, 2, 8, 15, 28, 37, 50, 52, 54, 56, 57, 60, 65, 66, 83, 92, 94, 96, 97, 101, 124
inclusion, 28, 218
income, 38, 121, 122, 125, 126, 134, 166, 245
income tax, 38
India, 9, 10, 53, 54, 55, 72, 149, 150, 155
indication, 107, 134
indicators, 65, 242
industrial emissions, 183, 190
industrial processing, 49
industrial sectors, 48, 145, 160, 181
industrial wastes, 224
industrialized countries, 53, 217
industry, vii, x, 1, 14, 15, 16, 37, 38, 45, 46, 52, 56, 57, 58, 60, 72, 85, 89, 91, 93, 94, 96, 112, 113, 140, 152, 200, 234
inefficiencies, 66, 203
infestations, 239
inflation, 20, 42, 110, 113, 126, 131, 135
information exchange, 11
infrastructure, 6, 33, 43, 63
innovation, 7, 40, 43
institutions, 14, 129, 238
instruments, 50, 53, 61
intellectual property, 43
interdependence, ix, 75
internal processes, 217
interview, 72
investment, 15, 36, 42, 43, 53, 63, 66, 76, 84, 99, 104, 106, 107, 108, 109, 121, 124, 126, 128, 238
investors, 54, 243
ions, 7
Ireland, 34, 70
iron, 35, 38, 42, 52, 55, 158, 190, 191, 217, 226
isobutane, 223
isobutylene, 219, 223
Italy, 155

## J

Japan, 56, 57, 58, 73, 148, 150, 155, 222
joint ventures, 155
judgment, 244, 245
jurisdiction, 6
justification, x, 89, 90

## K

kerosene, 215, 222
Korea, 60

## L

labeling, 40, 55, 57
labor, 33, 126, 133, 238
labor force, 133
land, x, 7, 9, 77, 90, 91, 92, 103, 104, 105, 119, 120, 124, 153, 158, 206, 208, 209, 210, 217, 218, 238, 239, 241, 244
land use, x, 9, 90, 91, 119, 153, 158, 206, 209, 217, 238, 239, 241, 244
landfills, vii, 1, 2, 3, 4, 5, 9, 11, 12, 14, 15, 18, 67, 96, 144, 145, 183, 188, 219
land-use, 120, 124, 158, 206
language, 70, 97
lasers, 205
Latvia, 36
laughing, 23
law enforcement, 47
laws, 46, 51, 54, 67, 68, 237
leaching, 26
leadership, 50, 63
leakage, 36, 37, 71, 127, 234, 237, 239
learning, 112
legal issues, vii, 1, 13
legislation, vii, viii, 1, 2, 5, 8, 21, 22, 27, 31, 34, 35, 47, 57, 65, 67, 68, 76, 83, 85, 92, 131, 153, 240, 242, 243, 244
legislative proposals, 2, 28, 129
leisure, 121
lifetime, 8, 13, 19, 23, 42, 80, 82, 133
light trucks, 195
light-emitting diodes, 60
limestone, 142, 159
limitation, ix, 31, 217
liquid fuels, 229
liquids, 87, 103, 218, 221, 224, 226
Lithuania, 36
litigation, 68
livestock, ix, 3, 7, 8, 15, 18, 75, 76, 77, 80, 81, 83, 84, 85, 86, 87, 90, 91, 92, 94, 95, 96, 97, 98, 99, 101, 102, 103, 104, 105, 112, 158, 183, 186, 220, 235, 238, 239, 241, 245
loans, 83, 85, 92, 95, 96, 97
local authorities, 68
local government, 6, 12, 50, 56, 61, 95
logging, 46, 47, 239
Louisiana, 187
lubricating oil, 223

## M

machinery, 57, 122, 124, 218
maintenance, 12, 85, 94, 106, 107

management, vii, viii, x, 1, 2, 3, 6, 11, 14, 15, 18, 21, 23, 26, 27, 29, 39, 53, 55, 76, 80, 84, 86, 89, 90, 94, 105, 183, 186, 191, 212, 213, 230, 235, 239

mandates, 26, 36, 97, 155

manufacturing, 46, 49, 51, 57, 60, 61, 63, 120, 127, 133, 135, 142, 171, 203, 214, 225

manure, vii, ix, 1, 2, 3, 11, 18, 23, 29, 75, 76, 77, 78, 79, 80, 81, 86, 91, 92, 94, 95, 97, 99, 101, 102, 103, 104, 105, 106, 107, 111, 112, 158, 193, 212, 213, 235

market, vii, x, 1, 2, 3, 4, 18, 33, 36, 40, 43, 44, 48, 52, 56, 60, 63, 66, 71, 72, 73, 89, 90, 96, 97, 105, 106, 108, 112, 120, 122, 123, 124, 125, 126, 154, 202, 237, 238, 239, 241, 242, 244, 245

market share, 33, 60

marketing, 103, 105, 106

markets, 13, 40, 48, 96, 121, 125, 133, 242

mature economies, 148

measurement, 4, 5, 8, 18, 27, 28, 80, 156, 227, 238

measures, ix, x, 27, 31, 32, 33, 34, 36, 37, 39, 40, 42, 46, 47, 54, 55, 56, 57, 60, 62, 63, 64, 66, 68, 70, 73, 74, 102, 108, 119, 121, 236, 238

media, 73, 219

metals, 62, 218, 220

methanol, 158, 190, 225, 229

Mexico, 61, 62, 150, 154, 155, 162, 164

microstructure, 217

Middle East, 151

military, 179, 222, 224

mineral water, 228

mining, vii, 1, 6, 13, 15, 19, 57, 137, 183, 184, 214, 218, 226, 229

model, 27, 29, 64, 66, 67, 117, 128, 130, 133, 152, 219

modeling, 128, 132, 235, 244

models, 128, 129, 130, 131, 133, 135, 209, 234, 241, 245

modernization, 40, 64

moisture, 11, 82, 86, 106, 213, 214, 223, 228

moisture content, 213, 214, 223

molecular structure, 221

molecular weight, 156, 223

molecules, 28, 213, 218, 219

money, 26, 71, 84, 107

motivation, 94, 101, 112

## N

nation, 11, 42, 74

National Ambient Air Quality Standards, 218

NATO, 227

natural gas, 2, 3, 4, 8, 9, 10, 11, 12, 13, 15, 16, 18, 19, 20, 29, 38, 40, 42, 45, 62, 63, 64, 82, 86, 92, 93, 101, 103, 112, 118, 131, 137, 140, 142, 144, 145, 153, 162, 166, 167, 172, 181, 183, 184, 214, 217, 220, 221, 223, 224, 226, 230

natural resources, 48

negotiating, 97, 155, 242

negotiation, 61

Netherlands, 114

New Zealand, 148, 150

Niels Bohr, 116

nitrates, 24

nitrification, 22, 24, 26

nitrogen, 7, 22, 23, 24, 26, 29, 81, 92, 104, 158, 191, 193, 199, 224, 225

nitrogen fixation, 23, 24

nitrogen gas, 24, 29, 81

nitrogen oxides, 225

nitrous oxide, viii, ix, xi, 7, 21, 22, 27, 29, 31, 32, 86, 120, 135, 137, 138, 142, 144, 145, 146, 158, 191, 192, 193, 195, 197, 198, 199, 221, 222, 235, 236, 241, 243, 245

North America, 61, 73, 90, 92, 106, 116, 130, 148, 150

Norway, 38, 47

nutrient imbalance, 104

nutrients, 83, 85, 104, 105, 112, 228

## O

Obama Administration, 16, 65, 68

Obama, President, 65, 67, 152, 155

objectives, 8, 237

obligation, 33, 41, 62, 66, 97, 113, 237

observations, ix, 31, 32, 111

oceans, vii, x, 119, 216

octane, 216

ODS, 22, 27, 28, 30

OECD, 73, 148, 150, 151

Office of Management and Budget, 152

oil, vii, 1, 2, 7, 10, 11, 12, 13, 14, 15, 18, 19, 20, 38, 44, 49, 60, 62, 63, 64, 92, 93, 96, 108, 127, 153, 162, 165, 184, 214, 215, 216, 218, 222, 223, 226, 227, 228, 229

oil production, 12, 15, 184

oil sands, 49

oils, 218, 226, 227

Omnibus Appropriations Act,, 16

operator, 11, 99, 100, 105

order, ix, 32, 34, 35, 36, 40, 42, 49, 52, 56, 63, 87, 104, 124, 154

organic chemicals, 35

organic compounds, 5, 92, 96, 104, 225, 229

organic matter, 8, 86, 91, 214

organic solvents, 226

Organization for Economic Cooperation and Development, 148

oxidation, 24, 29, 197, 217, 218, 227
oxides, 219, 224, 229
oxygen, 8, 24, 76, 91, 95, 162, 199, 212, 213, 215, 223, 224, 225, 226, 229
ozone, viii, 21, 22, 23, 27, 28, 30, 155, 200, 201, 202, 225

## P

Pacific, 180
particles, 213, 217, 218, 220
partnership, 14, 43, 57, 76, 96, 154
pasteurization, 104
pasture, 29, 86, 238
pathogens, 91, 104
payback period, 85
PCA, 74
peat, 105, 106
peer review, 106
penalties, 36, 38, 41, 51
performance benchmarking, 60
permit, 7, 8, 37, 54, 103, 123
personal communication, 114
pesticide, 38, 223
petroleum distillates, 214, 230
pH, 212
phosphorus, 104
planning, 16, 55, 94, 96
plants, 24, 35, 40, 46, 50, 51, 53, 55, 60, 65, 73, 92, 93, 101, 102, 107, 112, 144, 162, 171, 172, 197, 219, 226, 239
plastics, 145, 177, 197, 225, 228
Poland, 36, 155
policy choice, 122, 134
pollutants, 38, 49, 55, 68, 152
pollution, 5, 6, 36, 49, 50, 51, 55, 73, 77, 86, 91, 95
polymer, 226
polymers, 228
polystyrene, 228
poor performance, 84
population, 16, 53, 64, 81, 122, 166, 186, 199, 227
population growth, 166
population size, 199
portfolio, 48, 68, 97, 99, 108
Portugal, 36
poultry, 77, 91, 102, 106
power plants, 50, 92, 93, 103, 125, 126, 135, 152, 162, 195, 237
precedent, 230
precipitation, 9, 217, 228
preference, 5
present value, 108
preservative, 228
pressure, 32, 48, 166, 214, 220, 226

prevention, 61
price changes, 126, 127
price floor, 123
price signals, 126
prices, x, 13, 29, 40, 42, 43, 47, 51, 52, 89, 90, 92, 93, 97, 98, 105, 106, 108, 111, 112, 121, 123, 126, 127, 130, 131, 132, 133, 135, 137, 148, 159, 160, 172, 195, 236, 241
private firms, 91
private investment, 61, 95
producers, 11, 26, 27, 35, 37, 52, 61, 66, 68, 71, 76, 82, 83, 84, 85, 94, 95, 96, 120, 121, 122, 125, 126, 130, 241, 244
production costs, 71, 87, 95, 113
productive capacity, 126
productivity, 11, 33
profit, 76, 84, 104, 107, 108, 113, 241, 245
profitability, 84, 107, 108, 238
profits, 35, 121, 134
program, ix, 2, 3, 4, 5, 7, 8, 15, 16, 19, 20, 26, 28, 31, 32, 39, 41, 42, 43, 46, 50, 52, 53, 54, 57, 61, 62, 64, 67, 81, 84, 86, 87, 94, 98, 99, 101, 120, 121, 123, 124, 125, 126, 129, 131, 133, 134, 153, 154, 217, 227, 231, 233, 234, 235, 236, 237, 238, 239, 241, 243, 244
proliferation, 68
propane, 223, 228
property rights, 124
propylene, 223, 228
protocol, 95, 102, 105, 115
PSD, 27, 29
public education, 67
public health, 152
public sector, 41, 42, 64, 91
public support, 93
Puerto Rico, 180
pulp, 35, 91, 215, 230
purchasing power, 126, 134, 243, 245
PVC, 226

## Q

quality control, 238
quotas, 40, 42, 54, 60

## R

radiation, 18, 213, 221, 225, 226, 227, 228, 229
Radiation, 200, 201, 203, 204, 205
radio, 213
range, ix, 8, 32, 39, 79, 84, 92, 99, 102, 112, 113, 127, 128, 130, 131, 132, 135, 148, 222, 224, 228, 235, 243, 244, 245
rash, 108

# Index 257

rate of return, 106, 132
raw materials, 52, 217
real income, 168
reason, 48, 105, 135
recovery, viii, 2, 6, 7, 13, 15, 21, 22, 27, 28, 40, 65, 87, 155, 162, 166, 188, 217, 222, 224
recycling, 188
refining, vii, 1, 12, 60, 62, 158, 216, 218, 223, 228
reflection, 227
reforms, 51, 61
regeneration, 55
region, 99, 112, 213, 226, 228, 229, 241
Registry, 87
regression, 110
regression equation, 110
regulation, vii, viii, 1, 5, 7, 11, 21, 28, 30, 67, 73, 122, 152
regulations, 12, 13, 22, 29, 30, 45, 46, 49, 50, 51, 66, 67, 68, 69, 122, 124, 227, 230
regulators, 244, 245
regulatory bodies, 242
regulatory framework, 28, 68, 124
relationship, 27, 105, 110, 111
reliability, x, 11, 75, 76, 82, 84, 97
renewable energy, 6, 34, 36, 40, 42, 46, 48, 54, 56, 60, 61, 62, 64, 65, 66, 76, 81, 84, 85, 86, 87, 90, 91, 92, 93, 98, 108, 112, 127, 131, 152, 153, 188, 214
Requirements, 113, 154
reserves, 13, 46
residues, 91, 223, 230
resources, 6, 7, 15, 19, 29, 44, 46, 51, 52, 53, 63, 65, 77, 87, 96, 97, 99, 116, 117, 121, 125, 133, 155, 174, 215, 227, 238, 241
restaurants, 152, 167
restructuring, 52
retail, 84, 98, 99, 100, 107, 112
retention, 81, 86
returns, 6, 102
revenue, x, 12, 36, 38, 41, 43, 44, 63, 85, 89, 107, 108, 122, 125, 126
rice, 18, 19, 187, 235, 241
risk, 36, 37, 108, 239
Romania, 36
rubber, 215, 225, 228
runoff, 26, 158
rural areas, 54
Russia, 62, 63, 64, 65, 73, 150, 155

## S

safety, 3, 82, 84, 97, 123
sales, 53, 58, 97, 100, 107, 108, 114, 134, 144, 146, 166, 169, 170, 173, 175, 176, 229

salt, 10, 215
Samoa, 180
Sarkozy, Nicolas, 37, 38, 39
savings, 40, 42, 46, 52, 55, 128, 234, 235
school, 152, 167
search, 73, 117
Secretary of the Treasury, 16
security, 7, 73, 90, 130
sedimentation, 228
semiconductor, 203, 225
Senate, 7, 19, 43, 71, 153
sensitivity, 33, 70, 133
sensors, 22
separation, 105
sewage, 92, 102, 199, 216, 227, 229
shareholders, 125
shares, 34, 54, 138, 144
sharing, viii, 31, 34, 90
sheep, 3, 11, 102
silica, 181, 217
sludge, 91, 92, 107, 221
smog, 225
social development, 63
social infrastructure, 40
sodium, 35, 228
software, 94
soil, viii, 21, 23, 24, 26, 27, 29, 66, 77, 104, 105, 112, 208, 210, 223, 230, 238
solid phase, 221
solid waste, 3, 5, 6, 12, 15, 20, 91, 140, 142, 145, 151, 161, 188, 193, 198, 218, 219, 224
solvents, 48, 142, 201
somatic cell, 105
South Africa, 155
South Dakota, 154
South Korea, 60, 150, 155
soybean, 23, 25, 46
space, 81, 82, 103, 105, 213, 218, 221, 222, 226, 227, 229
Spain, 34, 70
specific gravity, 213
stabilization, 212
stakeholders, 66, 152
standards, vii, ix, 1, 2, 5, 8, 29, 32, 34, 36, 37, 43, 49, 51, 53, 54, 55, 56, 58, 61, 62, 65, 66, 67, 68, 69, 70, 82, 84, 87, 95, 97, 122, 124, 152, 153, 237, 238
statistics, 52
steel, 35, 38, 40, 42, 52, 55, 57, 60, 79, 158, 190, 191
steel industry, 52
stimulus, 149, 151
stock, 51, 207, 208, 209, 210, 215, 216, 219, 229
stomach, 219, 227

storage, 10, 11, 19, 29, 43, 49, 53, 76, 77, 81, 92, 95, 105, 127, 135, 152, 162, 165, 166, 211, 235, 238, 239, 241
stoves, 222
strategies, x, 2, 11, 12, 46, 69, 119, 222, 237, 241
strength, 5, 219
subsidy, 53, 108, 188
substance use, 220
substitutes, 125, 201
substitution, 46, 160
success rate, 90, 93
sulfur, ix, 7, 22, 31, 32, 49, 139, 142, 145, 200, 201, 216, 217, 218, 219, 221, 222, 225, 229
sulfur dioxide, 49, 229
sulfuric acid, 215
summer, 16, 137, 159
suppliers, 33, 41, 95, 97, 245
supply, 7, 46, 71, 82, 90, 92, 93, 104, 106, 112, 125, 126, 137, 214, 239, 241, 242, 244, 245
supply disruption, 104
Supreme Court, 5, 18, 68, 152
surplus, 52, 62, 63, 100
sustainable development, 46, 47
swamps, 102
Sweden, 34, 38, 91, 92, 93, 103
switching, 126
Switzerland, 38

## T

tanks, 104, 229, 235, 238
tar, 157
targets, ix, 31, 32, 34, 36, 41, 42, 46, 48, 50, 51, 53, 54, 55, 56, 57, 63, 64, 65, 66, 68, 73, 97, 123, 125, 128, 133, 154, 217, 222
tariff, 52
tax credit, 15, 16, 20, 43, 83, 85, 108, 125, 188
tax incentive, vii, 1, 12, 15, 16, 60, 66, 67
tax rates, 125
tax system, 135
taxation, 34, 47
technical assistance, 15, 19, 27, 29, 32, 66, 94, 95, 112, 241
technology transfer, 11, 26
television, 67
temperature, 11, 26, 34, 78, 81, 102, 123, 199, 214, 215, 217, 220, 221, 222, 227, 229, 230
textiles, 55, 62, 228
thermal decomposition, 217
threat, 28, 152
threshold, 123, 152
thresholds, 67, 153
timber, 47, 239
time frame, 23, 214

time periods, 128
time series, 208, 211
timing, 123
Title I, 74
Title II, 74
Title V, 6, 7, 27, 29, 152
tolls, 56
total costs, 108, 125
total energy, 140, 162
total utility, 101
trade, viii, ix, 2, 3, 4, 8, 21, 28, 31, 32, 33, 36, 37, 43, 44, 45, 48, 51, 52, 53, 56, 62, 63, 66, 68, 69, 70, 71, 73, 120, 121, 122, 123, 124, 125, 126, 127, 129, 132, 134, 135, 153, 154, 233, 234, 235, 236, 237, 239, 240, 242, 243, 244
trade policy, 121
trading, 35, 36, 37, 38, 41, 54, 56, 60, 61, 68, 69, 108, 154
traffic, 38, 68
training, 53
transaction costs, 5
transactions, 241
transformation, 220
transition, 64, 154
transmission, 64, 99, 144, 204, 220, 226
transparency, 6, 64, 238
transport, 10, 34, 35, 36, 39, 42, 44, 46, 53, 55, 64, 77, 144, 153, 162, 215, 218, 220, 224, 226
transportation, 12, 15, 27, 36, 48, 49, 60, 68, 77, 90, 103, 107, 120, 124, 140, 145, 146, 152, 153, 154, 160, 166, 172, 175, 214, 220, 223
trees, 124, 126, 206, 211, 218, 227, 235, 237, 238, 239, 241
trimmings, 211
turbulent mixing, 229

## U

U.S. economy, 4, 11, 18, 128, 133, 141, 142, 171
UK, 41, 42, 43, 71, 139, 157, 230, 231
UN, 72
uncertainty, xi, 13, 18, 119, 123, 130, 135, 195, 206, 242
unemployment rate, 70
uniform, 237
United Kingdom, ix, 32, 41, 155
United Nations, 39, 49, 53, 64, 73, 142, 155, 222, 237, 238
USDA, viii, 3, 22, 25, 27, 29, 76, 81, 84, 86, 87, 89, 94, 95, 98, 103, 113, 117, 118, 208, 209
utility costs, 99

# Index

## V

variability, 47, 217
vegetation, 23, 62, 230
vehicles, 27, 35, 37, 38, 42, 51, 53, 55, 56, 57, 58, 62, 66, 67, 68, 69, 83, 93, 126, 152, 195, 202, 227
velocity, 217
ventilation, 3, 13, 184, 218
venture capital, 54
vessels, 144, 227, 229
vinyl chloride, 226
viscosity, 227
vision, 42, 71
volatilization, 26
vulnerability, 61

## W

wages, 126, 133
Wake Island, 180
waste disposal, 3, 34
waste management, 81, 96, 99, 101, 142, 145, 183, 188, 191, 198
waste treatment, 48, 92

wastewater, 18, 91, 144, 145, 158, 188, 198, 199, 215, 228
water heater, 67, 222
water quality, 90, 94, 101, 112
water vapor, 7, 162, 217, 221, 226
web, 139, 143, 149, 151, 157, 184, 193, 201, 203, 204, 205, 207, 209, 211, 212, 230, 231
welfare, 121, 131, 152, 218
wells, 13, 18, 19
white blood cells, 105
White House, 65, 66, 74, 152
wholesale, 84, 96, 100
wind, 40, 46, 65, 82, 97, 99, 107, 108, 109, 174, 217, 227
wind farm, 40, 97
wind speeds, 82
withdrawal, 221
wood, 47, 48, 138, 184, 208, 209, 215, 230
wood products, 230
workers, 125, 133, 243
World Bank, 47
World Trade Organization, 33
World War I, 92
WTO, 33, 37, 70